Algebra II Through Competitions

http://www.mymathcounts.com/index.php

Copyright © 2013 by mymahcounts@gmail.com

Contributors

Yongcheng Chen, Author.
Sam Chen, Author.
Guiling Chen, Owner, mymathcounts.com, Typesetter, Editor

All rights reserved. Printed in the United States of America

Reproduction of any portion of this book without the written permission of the authors is strictly prohibited, except as may be expressly permitted by the U.S. Copyright Act.

ISBN-13: 978-1489512628
ISBN-10: 1489512624

Please contact mymahcounts@gmail.com for suggestions, corrections, or clarifications.

TABLE OF CONTENTS

Chapter 1 Word Problems	1
Chapter 2 Simplifications and Calculations	17
Chapter 3 Lines & Functions	33
Chapter 4 Absolute Values	46
Chapter 5 Quadratic Equations/Functions	60
Chapter 6 Systems of Equations	73
Chapter 7 Inequalities	86
Chapter 8 Function Composition and Operation	99
Chapter 9 Polynomials	114
Chapter 10 Radicals	136
Chapter 11 Exponential Equations	154
Chapter 12 Logarithmic Functions	167
Chapter 13 Rational Functions	182
Chapter 14 Binomial Theorem	197
Chapter 15 Complex Numbers	207
Chapter 16 Arithmetic and Geometric Sequences	225
Chapter 17 Sequences and Series	241
Chapter 18 Combinations and Permutations	259
Chapter 19 Probability	273
Chapter 20 Trigonometry	292
Chapter 21 Analytic Geometry Circles	326
Chapter 22 Ellipses and Hyperbolas	342
Chapter 23 Number Theory	361
Chapter 24 Geometry	376
Index	391

This page is intentionally left blank.

Algebra II Through Competitions   **Chapter 1 Word Problems**

1. DISTANCE AND SPEED

1.1. Basic Formulas related to distance, rate, and time

$d = rt$ (1.1)

d stands for distance (given in units such as miles, feet, kilometers, meters, etc.), t stands for time (hours, minutes, seconds, etc.), and r stands for the (**uniform, fixed, steady, constant, or average**) rate of speed.

$r = \dfrac{d}{t}$ (1.2)

$t = \dfrac{d}{r}$ (1.3)

In the following figure, we add d_1 and d_2 to get the total distance d: $d = d_1 + d_2$, and we add t_1 and t_2 to get the total time t: $t = t_1 + t_2$.
However, we do not add r_1 and r_2: $r \ne r_1 + r_2$.

$A-B$	r_1	×	t_1	=	d_1
				+	+
$B-C$	r_2	×	t_2	=	d_2
				=	=
$A-C$	r	×	t	=	d

$r = \dfrac{d}{t} = \dfrac{d_1 + d_2}{t_1 + t_2} = \dfrac{d_1 + d_2}{\dfrac{d_1}{r_1} + \dfrac{d_2}{r_2}} = \dfrac{r_1 r_2 (d_1 + d_2)}{d_1 r_2 + d_2 r_1}$ (1.2.1)

If $d_1 = d_2$, $r = \dfrac{r_1 r_2 (d_1 + d_2)}{d_1 r_2 + d_2 r_1} = \dfrac{2 d r_1 r_2}{d(r_2 + r_1)} = \dfrac{2 r_1 r_2}{r_1 + r_2}$ (1.2.2)

Example 1. The distance from Charlotte to Asheville is about 120 miles. If R.C. has two hours to make the trip, but drives 50 mph for the first 60 miles, how fast must R.C. go on the second half of the trip to reach Asheville in time?
A. 70 mph   B. 75 mph   C. 77.2 mph   D. 80 mph   E. none of these

Solution: B.

$C-X$	50	×	t_1	=	60
			+		+
$X-A$	r_2	×	t_2	=	$120-60$
			=		=
$C-A$	r	×	2	=	120

Method 1: $t = t_1 + t_2 \quad \Rightarrow \quad \dfrac{60}{50} + \dfrac{120-60}{r_2} = 2 \quad \Rightarrow \quad r_2 = 75$.

Method 2: In the first 60 miles, by (1.3), R.C. has driven 60/50 = 6/5 hours. In the next 60 miles, R.C. has only 4/5 hours to drive, by (1.2), he must go 60/ (4/5) = 75 mph.

Example 2. The average speed of a biker over the whole 55 mile trail was 15 mph. On the trail's 30 mile uphill section the biker was averaging 10 mph. What was the average speed on the downhill section? (The trail was either uphill or downhill, there were no level sections.)
A. 20 mph   B. 21 mph   C. 25 mph   D. 36 mph   E. None of these

Solution: E.

uphill	10	×	t_1	=	30
			+		+
downhill	r_2	×	t_2	=	$55-30$
			=		=
Whole trip	15	×	t	=	55

Method 1: Method 1: $t = t_1 + t_2 \quad \Rightarrow \quad \dfrac{30}{10} + \dfrac{55-30}{r_2} = \dfrac{55}{15} \quad \Rightarrow \quad r_2 = \dfrac{75}{2}$.

Method 2: The 30 mile uphill section took 3 hours while the downhill 25 miles took h hours, while the average speed was $\dfrac{55}{3+h} = 15 \Rightarrow 55 = 45 + 15h \Rightarrow h = 2/3$ hours. The average speed downhill then was $\dfrac{25}{2/3} = \dfrac{75}{2} = 37.5$.

Example 3: A car travels 120 miles from A to B at 30 miles per hour but returns the same distance at 40 miles per hour. The average speed for the round trip is closest to:
(A) 33 mph   (B) 34 mph   (C) 35 mph   (D) 36 mph   (E) 37 mph

Solution: (B).
If a car travels a distance d at rate r_1 and returns the same distance at rate r_2, then

Algebra II Through Competitions   **Chapter 1 Word Problems**

$$\text{average speed} = \frac{\text{total distance}}{\text{total time}} = \frac{2d}{d/r_1 + d/r_2} = \frac{2r_1 r_2}{r_1 + r_2}; \quad \therefore x = \frac{2 \cdot 30 \cdot 40}{70} = \frac{240}{7} \sim 34\text{mph}$$

1.2. Two people walking in the same direction:

Denote Alex's speed as r_A and Bob's speed as r_B with $r_A > r_B$. Let the distance between them be d.

In order to simplify solving this problem, one may think that Bob is not moving, and that Alex walks in a relative speed of $(r_A - r_B)$ to cover a distance of d in time t.

$$d = (r_A - r_B)t \tag{1.4}$$

Therefore, the time needed for Alex to catch Bob is: $t = \dfrac{d}{r_A - r_B}$ (1.5)

Example 4. A car leaves a town traveling at 40 mph. Two hours later, a second car leaves the same town, on the same road, traveling 60 mph. The second car drives how many hours to overtake the first car?
A. 2 hours   B. 1 hour   C. 1.5 hours   D. 4 hours   E. 3 hours

Solution: D.
Within two hours, the first car travles a distance of 2 × 40 miles.
By (1.4), $d = (r_A - r_B)t$ $\Rightarrow$ $80 = (60 - 40)t$ $\Rightarrow$ $t = 4$ hours.

1.3. Two people walking in the opposite directions (towards each other)

Alex's speed is r_A and Bob's speed is r_B, and distance between them is d.
This time, one may think that Bob is not moving and Alex walks in a relative speed of $(r_A + r_B)$ to cover a distance of d with time t.

$$d = (r_A + r_B)t \tag{1.6}$$

The time needed for Alex to meet Bob is: $t = \dfrac{d}{r_A + r_B}$ (1.7)

Example 5: Alex and Bob run at a circular path of perimeter 500 meters. Their speeds are different. If they start at the same time and same place but run in opposite directions, they will meet for the first time in 75 seconds. Given that Bob's speed is 180 meters/minute, find Alex's speed.

Solution: 220 meters/min.
Converting 75 seconds into minutes yields 1.25 minutes. When they run toward each other, the relative speed of one person while the other is stationary is $r_A + r_B$. By (1.6), we have $d = (r_A + r_B)t \Rightarrow 500 = (r_A + 180) \times 1.25 \Rightarrow r_A = \dfrac{500}{1.25} - 180 = 220$.

1.4. Boat going with the current of a river

The engine of the boat produces a speed r_E for the boat in still water. The current's speed of the water is r_C, and the distance travelled in time t is d.
In order to solve these kinds of problems, one may think that the boat is moving at a relative speed of $(r_E + r_C)$ to cover a distance of d with time t.

$$d = (r_E + r_C)t \tag{1.8}$$

The time needed for the boat to travel a distance d is: $t = \dfrac{d}{r_E + r_C}$ (1.9)

1.5. Boat going against the current

The engine of the boat produces a speed r_E for the boat in still water. The speed of the river current is r_C, and the distance travelled in time t is d.
One may think that the boat is moving in a relative speed of $(r_E - r_C)$ to cover a distance of d.

$$d = (r_E - r_C)t \tag{1.10}$$

The time needed for the boat to travel a distance d is: $t = \dfrac{d}{r_E - r_C}$ (1.11)

Example 6. An airplane, flying with a tail wind, travels 1200 miles in 5 hours; the return trip, against the wind, takes 6 hours. Find the cruising speed of the plane and the speed of the wind (assume that both are constant).
a) 220 mph, 20 mph   b) 220 mph, 40 mph   c) 230 mph, 20 mph
d) 230 mph, 40 mph   e) 240 mph, 20 mph

Solution: A.
Let r_E be the cruising speed and r_C be the wind speed.

By (1.9), we have $t = \dfrac{d}{r_E + r_C} \Rightarrow 5 = \dfrac{1200}{r_E + r_C} \Rightarrow r_E + r_C = \dfrac{1200}{5} = 240$ (1)

By (1.11), we have $t = \dfrac{d}{r_E - r_C} \Rightarrow 6 = \dfrac{1200}{r_E - r_C} \Rightarrow r_E - r_C = \dfrac{1200}{6} = 200$ (2)

From here, it's a system of equations. $r_E = 220$, $r_C = 20$.

2. DILUTLNG SOLUTIONS

The "C-V-S" Method: C is the concentration or the strength of the solution. V is the volume of the solution. S is the substance of the solution.

$C \times V = S$, $V_A + V_B = V$, and $S_A + S_B = S$. There is no relationship for C_A, C_B, and C.

Name	C	$\times$	V	$=$	S
A	C_A	$\times$	V_A	$=$	$C_A \times V_A = S_A$
			+		+
B	C_B	$\times$	V_B	$=$	$C_B \times V_B = S_B$
			‖		‖
Mixture	x	$\times$	$(V_A + V_B)$	$=$	$S_A + S_B$

Example 7. Thirty ounces of vinegar with a strength of 30% was mixed with 50 ounces of a 20% vinegar solution. What was the percentage of the resulting solution?

Solution: 23.75%.

Name	C	$\times$	V	$=$	S
A	0.3	$\times$	30	$=$	$0.3 \times 30 = 9$
			+		+

B	0.2	×	50	=	0.2 × 50 = 10
			‖		‖
Mixture	x	×	80	=	80x

$$9 + 10 = 80x \quad \Rightarrow x = 0.2375 = 23.75\%.$$

Example 8. How many liters of a 20% alcohol solution must be added to 90 liters of a 50% alcohol solution to form a 45% solution?

Solution: 18 liters.

Name	C	V	S
A	0.2	x	0.2 × x
B	0.5	90	45
Mixture	0.45	90 + x	0.45(90 + x)

$$0.45(90 + x) = 0.2x + 45 \quad \Rightarrow \quad x = 18 \text{ liters}.$$

Example 9. A chemist needs a 15% alcohol solution and has only a 60% solution. How much water should be added to obtain 10 liters of the weaker solution?
Solution: 7.5 liters.

Name	C	V	S
A	0	x	0
B	0.6	y	0.6y
Mixture	0.15	10	1.5

$0.6y = 1.5 \quad \Rightarrow \quad y = 2.5.$
$x + y = 10. \quad \Rightarrow \quad x = 7.5.$

3. WORK

1. If one person or a machine can complete a job in t unit of time, then the rate of work is
$$r = \frac{1}{t} \quad \text{or} \quad r \times t = 1 \tag{3.1}$$
where t is the total time needed to complete the job.

2. If the person or the machine spends a T unit of time doing a job, then the portion of the job done is W, and
$$T \times r = T \times \frac{1}{t} = W \tag{3.2}$$

Algebra II Through Competitions   **Chapter 1 Word Problems**

3. If person A of the rate r_1 and person B of the rate r_2 work on a job together for a T unit of time, then $T \times (r_1 + r_2) = T \times (\frac{1}{t_1} + \frac{1}{t_2}) = W$ (3.3)

4. If three people of different rates work together for T hours, and the amount of job done is W, then $T \times (r_1 + r_2 + r_3) = T \times (\frac{1}{t_1} + \frac{1}{t_2} + \frac{1}{t_3}) = W$ (3.4)

5. If three people of the same rate work together for T hours and the amount of job done is W, then $T \times r = T \times (\frac{1}{t} + \frac{1}{t} + \frac{1}{t}) = T \times \frac{3}{t} = W$ (3.5)

Example 10. Working at a constant rate, Kay can do a job in 8 days. Working at a constant rate, Tom can do the same job in 24 days. If there is no loss of efficiency in working together, how many days will it take these two to do that job when working together?

Solution: 6.

By (3.1), $r = \frac{1}{t}$. Kay's rate of work is $r_k = \frac{1}{8}$. Tom's rate of work is $r_T = \frac{1}{24}$.

By (3.3), $T \times (\frac{1}{8} + \frac{1}{24}) = 1$ $\Rightarrow$ $T \times \frac{4}{24} = 1$ $\Rightarrow$ $T = 6$.

Example 11. It takes Amy and Bill 15 hours to paint a house, it takes Bill and Chandra 20 hours, and it takes Chandra and Amy 30 hours. How long will it take if all three work together?
(A) 9 hours and 40 minutes   (B) 10 hours   (C) 12 hours
(D) 13 hours and 20 min   (E) 14 hours

Solution: D.
If Amy, Bill and Chandra can paint a whole house at a rate of $1/a$; $1/b$, and $1/c$ of the house per hour, respectively, then $1/a + 1/b = 1/15$; $1/b + 1/c = 1/20$, and $1/c + 1/a = 1/30$. When all three work together, they can paint at a rate of $(1/a + 1/b + 1/c) = (1/2)(1/15 + 1/20 + 1/30) = 3/40$. So it will take $40/3$ hours, that is, 13 hours 20 min.

4. AGES

The key point to solve the age problems is that everyone is growing with the time at the same rate.

	x years ago	Now	y years later
Alex's age:	$a - x$	a	$a + y$

Betsy's age: $\quad b-x \qquad\qquad b \qquad\qquad b+y$

Example 12. The combined age of a man and his sister is 55 years. He is 3 times as old as she was when he was as old as she is now. What is the product of their ages?
A. 567 B. 726 C. 484 D. 750 E. 574.

Solution: B.

	x years ago	Now
His age:	$b-x$	b
His sister's age:	$s-x$	s

He is 3 times as old as she was: $\quad b = 3(s-x)$ (1)
When he was as old as she is now: $\quad b - x = s$ (2)

We also know that $b + s = 55$ (3)

From (1) and (2), we get: $2b = 3s \quad\Rightarrow\quad b = 3s/2$ (4)
Substituting (4) into (3): $s = 22$ and $b = 33$. The product is 726.

Example 13. One year from today, Amy will be two-thirds as old as Ann. Five years from today, Amy will be four-fifths as old as Ann. How old is Amy today?
A. 1 B. 2 C. 3 D. 4 E. 5

Solution: C.

	Now	1 year later	5 years later
Amy's age:	a	$a+1$	$a+5$
Ann's age:	b	$b+1$	$b+5$

One year from today, Amy will be two-thirds as old as Ann: $a + 1 = \dfrac{2}{3}(b+1)$ (1)

Five years from today, Amy will be four-fifths as old as Ann: $a + 5 = \dfrac{4}{5}(b+5)$ (2)

$(2) - (1)$: $4 = \dfrac{4}{5}(b+5) - \dfrac{2}{3}(b+1) = \dfrac{4}{5}b + 4 - \dfrac{2}{3}b - \dfrac{2}{3} = \dfrac{12-10}{15}b + \dfrac{12}{3} - \dfrac{2}{3} = \dfrac{2}{15}b + \dfrac{10}{3}$, or

$4 - \dfrac{10}{3} = \dfrac{2}{15}b \quad\Rightarrow\quad \dfrac{2}{3} = \dfrac{2}{15}b \quad\Rightarrow\quad b = 5$. So $a = 3$.

Algebra II Through Competitions — Chapter 1 Word Problems

PROBLEMS

Problem 1. Josh's average speed while driving from Charlotte to Asheville was 60 mph. The two cities are roughly 120 miles apart. If he drove 45 mph for 36 miles, how fast did he drive the remaining distance?
A. 66 3/7 mph   B. 70 mph   C. 72 3/7 mph   D. 75 mph   E. 82 mph

Problem 2. The distance from Asheville to Raleigh is approximately 250 miles. Acme Bus Lines advertises that it can make the trip in 4.5 hours. But the driver drives 50 mph for the first 120 miles. Approximately how fast must he drive for the remainder of the trip to reach Raleigh in time?
A. 62.3 mph   B. 55 mph   C. 60 mph   D. 61.1 mph   E. 61.9 mph

Problem 3. Emma starts her trip 30 minutes before Paul. If she drives 60 mph and Paul drives 70 mph, how far will each have traveled when Paul catches up with Emma?
A. 3 miles   B. 130 miles   C. 175 miles   D. 180 miles   E. 210 miles

Problem 4. How much of a 40% acid solution should be mixed with pure water to obtain 80 ml of a 30% acid solution?
A. 24 ml   B. 28 ml   C. 32 ml   D. 60 ml   E. 80 ml

Problem 5. The capacity of a car's radiator is nine liters. The mixture of antifreeze and water is 30% antifreeze. The temperature is predicted to drop rapidly requiring the mixture to be 65% antifreeze. How much of the mixture in the radiator must be drawn off and replaced with pure antifreeze?
A. 3.5 liters   B. 4.5 liters   C. 5.0 liters   D. 6.0 liters   E. none of these

Problem 6. A car radiator has a capacity of 17 quarts of fluid. It is full and contains a 41% solution of coolant. Approximately how much of the solution must be drained and replaced by 100% coolant to bring the mixture to a 50% solution?
A. 2.6 qt.   B. 2.125 gal.   C. 0.8 qt.   D. 3.2 qt.   E. 4 qt.

Problem 7. Mary has d liters of punch that is d% grape juice. How many liters of grape juice must she add to make the punch $3d$% grape juice?

A. $\dfrac{d^2}{100-3d}$ B. $\dfrac{2d^2}{100-3d}$ C. $\dfrac{d^2}{100+3d}$ D. $\dfrac{3d^2}{100+d}$ E. none of the above

Problem 8. A chemist has at hand two acid solutions. Solution A is 30% acid, and solution B is 65% acid. How much of each solution does she need to make 40 ml of 50% acid solution?
A. 20 ml of A, 20 ml of B B. 12 ml of A, 26 ml of B
C. 15.5 ml of A, 24.5 ml of B D. 17.1 ml of A, 22.9 ml of B

Problem 9. Fresh grapes contains 80% water by weight, whereas dried grapes contains 15% water by weight. How many pounds of dried grapes can be obtained from 34 pounds of fresh grapes ?
A. 8 B. 9 C. 10 D. 11 E. 12

Problem 10. One bottle contains 1000 grams of 15% alcohol solution. Alcohol solution A of 100 grams and alcohol solution B of 400 grams were added to the bottle to form a 14% alcohol solution. The percentage of alcohol solution A is twice as much as the percentage of alcohol solution B. What was the percentage of the solution A?

Problem 11. 100 kg of a fruit contained 90% water one week ago. How many kg of the fruit containing 80% water are there now?

Problem 12. Lynn purchased two candles of equal length. One of the candles will burn up completely in 5 hours, while the other candle requires 7 hours to burn up completely. If the candles are lit at the same time, approximately how long will they burn before one of the candles is twice the length of the other?
A. 3.9 hr B. 3.2 hr C. 2.8 hr D. 3hr E. none of the above

Problem 13. Sara and Kaleigh are painting a house together. If Sara works 8 days, Kaleigh will need 4 days to complete the job. If Sara works 4 days, Kaleigh will need 6 days to finish. How many days will Sara work if she paints the entire house by herself?
A. 8 days B. 10 days C. 12 days D. 14 days E. 16 days

Problem 14. If Jack can do a job in 104 hours and it takes Jack and Jill working together 40 hours to do the same job, how many hours will Jill take to do the job alone?
A. 64 B. 80 C. 28 D. 72 E. 65.

Problem 15. Abigale is 10 years older than Bertha who is twice as old as Charlotte who is as old Dorothy and Eva combined. If Abigale is 5 times as old as Eva, and Bertha is 23 years older than Dorothy, how old is Charlotte?
A. 7   B. 15   C. 21   D. 30   E. 40

Problem 16. Jake's age is X years, which is also the sum of the ages of his two children. His age Y years ago was twice the sum of their ages then. What is X/Y?
A. 2   B. 3   C. 3/2   D. 4/3   E. 4.

Algebra II Through Competitions **Chapter 1 Word Problems**

SOLUTIONS:

Problem 1. Solution: B.
Let X be the place 36 miles away from Charlotte.

$C - X$	45	×	t_1	=	36
			+		+
$X - A$	r_2	×	t_2	=	$120 - 36$
			=		=
$C - A$	60	×	t	=	120

Method 1: $t = t_1 + t_2$ $\Rightarrow$ $\dfrac{36}{45} + \dfrac{120-36}{r_2} = \dfrac{120}{60}$ $\Rightarrow$ $r_2 = 70$.

Method 2: By (1.2), $r = \dfrac{d}{t}$ $\Rightarrow$ $60 = \dfrac{120}{\dfrac{36}{45} + \dfrac{120-36}{r_2}}$ $\Rightarrow$ $\dfrac{36}{45} + \dfrac{120-36}{r_2} = 2$

$\Rightarrow$ $\dfrac{84}{r_2} = 2 - \dfrac{36}{45}$ $\Rightarrow$ $r_2 = 70$.

Problem 2. Solution: E.
Let X be the place 36 miles away from Asheville.

$A - X$	50	×	t_1	=	120
			+		+
$X - R$	r_2	×	t_2	=	$250 - 120$
			=		=
$A - R$	r	×	4.5	=	250

Method 1: $t = t_1 + t_2$ $\Rightarrow$ $\dfrac{120}{50} + \dfrac{250-120}{r_2} = 4.5$ $\Rightarrow$ $r_2 = 61.9$.

Method 2: In the first stretch of the trip, the driver has driven 120 miles and it has taken him 120/50 = 2.4 hours to do it. Thus, he has 250–120 = 130 miles left to go and 4.5 – 2.4 = 2.1 hours to with which to finish the trip. The average speed for the second stretch of the trip must then be 130/2.1 = 61.9 mph.

Problem 3. Solution: E.
Within 30 minutes or 0.5 hours, Emma travles a distance of $0.5 \times 60 = 30$ miles.
By (1.4), $d = (r_A - r_B)t$ $\Rightarrow$ $30 = (70 - 60)t$ $\Rightarrow$ $t = 3$ hours.
When Paul catches up with Emma, Paul has traveled $3 \times 70 = 210$ miles.

Problem 4. Solution: D.

Name	C	V	S
A	0.4	x	$0.4 \times x$
B	0.0	y	0
Mixture	0.3	80	0.3(80)

$0.3(80) = 0.4x + 0 \Rightarrow x = 60$ ml.

Problem 5. Solution: B.
Let x be the amount of antifreeze to be drained off.

Name	C	V	S
A_1	0.3	9	2.7
	0.3	x	$0.3x$
A_2	0.3	$9-x$	$0.3(9-x)$
B	1.0	x	$1.0x$
Mixture	0.65	9	0.65(9)

$0.3(9-x) + 1.0x = 0.65(9) \Rightarrow x = 4.5$ liters.

Problem 6. Solution: A.
Let x be the amount of antifreeze to be drained off.

Name	C	V	S
A_1	0.41	17	6.97
	0.41	x	$0.41x$
A_2	0.41	$17-x$	$0.41(17-x)$
B	1.0	x	$1.0x$
Mixture	0.5	17	0.5(17)

$0.41(17-x) + 1.0x = 0.5(17) \Rightarrow x \approx 2.593 \approx 2.6$ qt..

Problem 7. Solution: B.

Name	C	V	S
A	d	d	d^2
B	100	x	$100x$
Mixture	$3d$	$d+x$	$3d(d+x)$

$d^2 + 100x = 3d(d+x) \Rightarrow d^2 + 100x = 3d^2 + 3dx \Rightarrow 100x - 3dx = 2d^2$

$\Rightarrow x(100 - 3d) = 2d^2 \Rightarrow x = \dfrac{2d^2}{100 - 3d}$.

Problem 8. Solution: D.

Name	C	V	S
A	0.3	x	$0.3 \times x$
B	0.65	y	$0.65 \times y$
Mixture	0.5	40	0.5×40

$0.3 \times x + 0.65 \times y = 0.5 \times 40$ (1)
$x + y = 40$ (2)
Solving the system of equations (1) and (2): $x = 17.14$ ml and $y = 22.86$ ml.

Problem 9. Solution: A.

Name	C	V	S
Grape	0.2	34	0.2×34
Water	0	y	0
Final	0.85	$34 - y$	$0.85(34 - y)$

$0.2 \times 34 = 0.85(34 - y) \quad \Rightarrow \quad 34 - y = 8.$

Problem 10. Solution: 20%.

Name	C	V	S
A	2x	100	200 x
B	x	400	400 x
D	0.15	1000	150
Mixture	0.14	1500	210

$200x + 400x + 150 = 210 \quad \Rightarrow \quad x = 0.1 = 10\%. \quad \Rightarrow \quad 2x = 20\%.$

Problem 11. Solution: 50.

Name	C	V	S
A	0.1	100	10
B	0.2	x	0.2 x

The substance does not change before or after the evaporating: $10 = 0.2x \quad \Rightarrow \quad x = 50.$

Problem 12. Solution: A.
Following formula (3.1), we have the burning rates of the two candles as:
$r_1 = \dfrac{1}{t_1} = \dfrac{1}{7}$, and $r_2 = \dfrac{1}{t_2} = \dfrac{1}{5}$.

According to formula (3.2), in T hours after being lighted, the heights of burning candles are $H_1 = T \times \dfrac{1}{t_1} = \dfrac{T}{7}$, and $H_2 = T \times \dfrac{1}{t_2} = \dfrac{T}{5}$.

We are given that at this moment, the height of the first candle is twice the height of the second candle, so we have: $1 - H_1 = 2(1 - H_2) \Rightarrow 1 - \dfrac{T}{7} = 2(1 - \dfrac{T}{5})$.

$T = \dfrac{35}{9} = 3.\overline{8} \approx 3.9$ hours.

Problem 13. Solution: E.
Let s and k be the time Sara/ Kaleigh needs to finish the job alone, respectively.

If Sara works 8 days, Kaleigh will need 4 days to complete the job: $\dfrac{1}{s} \times 8 + \dfrac{1}{k} \times 4 = 1$ \hfill (1)

If Sara works 4 days, Kaleigh will need 6 days to finish: $\dfrac{1}{s} \times 4 + \dfrac{1}{k} \times 6 = 1$ \hfill (2)

(1) × 3: $\dfrac{1}{s} \times 24 + \dfrac{1}{k} \times 12 = 3$ \hfill (3)

(2) × 2: $\dfrac{1}{s} \times 8 + \dfrac{1}{k} \times 12 = 2$ \hfill (4)

(3) − (4): $\dfrac{1}{s} \times 16 = 1 \Rightarrow s = 16$.

Problem 14. Solution: E.
Let J and j be the time Jack/ Jill needs to finish the job alone, respectively.

Jack can do a job in 104 hours: $\dfrac{1}{J} \times 104 = 1 \Rightarrow \dfrac{1}{J} = \dfrac{1}{104}$ \hfill (1)

Jack and Jill working together 40 hours to do the same job: $(\dfrac{1}{J} + \dfrac{1}{j}) \times 40 = 1$ \hfill (2)

Substituting (1) into (2): $(\dfrac{1}{104} + \dfrac{1}{j}) \times 40 = 1 \Rightarrow \dfrac{40}{j} = 1 - \dfrac{40}{104} = 1 - \dfrac{5}{13} = \dfrac{8}{13} \Rightarrow$

$\dfrac{5}{j} = \dfrac{1}{13} \Rightarrow j = 5 \times 13 = 65$.

Problem 15. Solution: B.
Let a, b, c, d and e be the ages of Abigale, Bertha, Charlotte, Dorothy, and Eve respectively. Then $a = b + 10$, $b = 2c$, $c = d + e$, $a = 5e$ and $b = d + 23$, i.e. $d = b - 23$ and $e = 0.2(b+10)$. Thus $c = b - 23 + 0.2b + 2 \Rightarrow c = 1.2b - 21 \Rightarrow c = 1.2(2c) - 21 \Rightarrow c = 15$.

Problem 16.
Solution: B.

	x years ago	Now
Jake's age:	$X - Y$	X
His Children's age:	$C - 2Y$	C

Jake's age is X years, which is also the sum of the ages of his two children: $X = C$ (1)
His age Y years ago was twice the sum of their ages then: $X - Y = 2(C - 2Y)$ (2)
Substituting (1) into (2): $X - Y = 2(X - 2Y)$ $\Rightarrow$ $X - Y = 2X - 4Y$ $\Rightarrow$ $4Y - Y = 2X - X$ $\Rightarrow$ $3Y = X$ $\Rightarrow$ $X/Y = 3$.

1. SUM OF FRACTIONS

Useful formulas:

$$\frac{1}{n(n+1)} = \frac{1}{n} - \frac{1}{n+1} \quad \Rightarrow \quad \frac{1}{3\times 4} = \frac{1}{3} - \frac{1}{4}$$

$$\frac{1}{n} = \frac{1}{2n} + \frac{1}{2n} \quad \Rightarrow \quad \frac{1}{3} = \frac{1}{2\times 3} + \frac{1}{2\times 3} = \frac{1}{6} + \frac{1}{6}$$

$$\frac{1}{n(n+k)} = \frac{1}{k}\left(\frac{1}{n} - \frac{1}{n+k}\right) \quad \Rightarrow \quad \frac{1}{3\times 5} = \frac{1}{2}\left(\frac{1}{3} - \frac{1}{5}\right)$$

Example 1. Find the sum: $\frac{1}{1\times 2} + \frac{1}{2\times 3} + \cdots + \frac{1}{49\times 50}$.

Solution:

$$\frac{1}{1\times 2} + \frac{1}{2\times 3} + \cdots + \frac{1}{49\times 50} = \frac{1}{1} - \frac{1}{2} + \frac{1}{2} - \frac{1}{3} + \frac{1}{3} - \frac{1}{4} + \cdots - \frac{1}{50} = 1 - \frac{1}{50} = \frac{49}{50}$$

Example 2. Find the sum: $\frac{1}{1\times 3} + \frac{1}{3\times 5} + \cdots + \frac{1}{11\times 13}$.

Solution:

$$\frac{1}{1\times 3} + \frac{1}{3\times 5} + \cdots + \frac{1}{11\times 13} = \frac{1}{2}\left(\frac{1}{1} - \frac{1}{3} + \frac{1}{3} - \frac{1}{5} + \frac{1}{5} - \cdots + \frac{1}{11} - \frac{1}{13}\right) = \frac{1}{2}\times\left(1 - \frac{1}{13}\right) = \frac{6}{13}.$$

Example 3. Calculate: $\frac{1}{2\times 4} + \frac{1}{4\times 6} + \cdots + \frac{1}{98\times 100}$.

Solution:

$$\frac{1}{2\times 4} + \frac{1}{4\times 6} + \cdots + \frac{1}{98\times 100}$$
$$= \frac{1}{2}\left(\frac{1}{2} - \frac{1}{4} + \frac{1}{4} - \frac{1}{6} + \frac{1}{6} - \cdots + \frac{1}{98} - \frac{1}{100}\right) = \frac{1}{2}\times\left(\frac{1}{2} - \frac{1}{100}\right) = \frac{1}{2}\left(\frac{50}{100} - \frac{1}{100}\right) = \frac{49}{200}.$$

Example 4. Calculate: $\frac{3}{1\times 4} + \frac{3}{4\times 7} + \frac{3}{7\times 10} + \cdots + \frac{3}{19\times 22}$.

Solution:

$$\frac{3}{1\times 4} + \frac{3}{4\times 7} + \frac{3}{7\times 10} + \cdots + \frac{3}{19\times 22} = \frac{1}{1} - \frac{1}{4} + \frac{1}{4} - \frac{1}{7} + \frac{1}{7} \cdots - \frac{1}{19} + \frac{1}{19} - \frac{1}{22} = 1 - \frac{1}{22} = \frac{21}{22}$$

2. REPEATING DECIMALS

A rational number a/b in lowest terms results in a repeating decimal if a prime other than 2 or 5 is a factor of the denominator.

$$\frac{1}{3} = 0.33333\ldots = 0.\overline{3}.$$

$$\frac{1}{7} = 0.142857142857\ldots = 0.\overline{142857}.$$

Repeating block

If the denominator has only factors other than 2 and 5, then this fraction can become repeating decimal. The length of the repeating block (period) is the smallest number of nines needed for the number containing 9's to be divisible by the denominator.

For $\frac{1}{3}$, the repeating block is 1 since $\frac{9}{3} = 3$. $\qquad (\frac{1}{3} = 0.\overline{3})$

For $\frac{1}{7}$, the repeating block is 6 since $\frac{999999}{7} = 142857$. $\qquad (\frac{1}{7} = 0.\overline{142857})$.

Converting Repeating Decimals to Fractions

The fraction of any single digit repeating decimal is the digit over 9. $\Rightarrow 0.\overline{5} = \frac{5}{9}$

The fraction of any 2-digit repeating decimal is the digits over 99. $\Rightarrow 0.\overline{55} = \frac{55}{99}$

$$0.\overline{3123} = \frac{3123}{9999} = \frac{347}{1111}$$

$$0.17\overline{857142} = 0.\underbrace{17}_{2\text{ digits}}\underbrace{\overline{857142}}_{6\text{ digits}} = \frac{17857142 - 17}{\underbrace{999999}_{6\text{ 9's}}\underbrace{00}_{2\text{ 0's}}}$$

Example 5. Convert the repeating decimal $0.\overline{1}$ to a fraction.

Solution:
Method 1: The fraction of the single digit repeating decimal is the digit over 9:
$0.\overline{1} = \frac{1}{9}$.

Method 2:
Let $x = 0.\overline{1} = 0.111111...$ (1)
Multiply both sides by 10: $10x = 1.11111...$ (2)
(2) – (1): $9x = 1$ $\Rightarrow$ $x = 1/9$

Example 6. If you write the repeating decimal $0.0\overline{216}$ as a fully reduced common fraction, the difference between the denominator and the numerator is:
A) 9774   B) 1213   C) 783   D) 181   E) 29

Solution: D.
We set $x = 0.0\overline{216} = 0.0216216216....$ and multiply both sides by 10 and 10000:
$10x = 0.216216216...$ (1)

$10000x = 216.216216...$ (2)
Next we subtract:
$10000x = 216.216216...$
$-\ \ 10x = \ \ \ 0.216216...$

$9990x = 216$
Next we divide both sides by 9990 to get: $x = 216/9990 = 4/185$.

The difference is $185 - 4 = 181$.

Example 7. Express $0.72\overline{45}$ as a common fraction.

Method 1: $0.72\overline{45} = \dfrac{7245 - 72}{9900} = \dfrac{7173}{9900} = \dfrac{797}{1100}$

Method 2: Let $x = 0.72\overline{45}$ (1)
$100x = 72.\overline{45}$ (2)
$10000x = 7245.\overline{45}$ (3)
(3) – (2): $9900x = 7173$ (4)
$x = \dfrac{7173}{9900} = \dfrac{797}{1100}$

Example 8. Write the repeating decimal $0.3727272...$ as a fraction.
A. 372/999   B. 38/101   C. 136/495   D. 41/110   E. None of these

Solution: D.

$0.3\overline{72} = \dfrac{372-3}{990} = \dfrac{369}{990} = \dfrac{41}{110}$.

Example 9. Calculate: $2.\overline{48} + 2.\overline{83}$

Solution:

Since $\begin{array}{r} 2.\overline{48} \\ +2.\overline{83} \\ \hline 5.31 \end{array}$, and the sum of the first repeating digits (4 in $2.\overline{48}$ and 8 in $2.\overline{83}$) in the addends carries 1, so the last digit of the resulting number needs to be increased by 1: $2.\overline{48} + 2.\overline{83} = 5.\overline{32}$

Example 10. Calculate: $3.\overline{215} - 1.\overline{307}$
Solution:

Since $\begin{array}{r} 3.\overline{215} \\ -1.\overline{307} \\ \hline 1.908 \end{array}$, and the first repeating digit (2 in $3.\overline{215}$) in the top number is smaller than the corresponding digit (3 in $1.\overline{307}$) in the bottom number, so the last digit of the resulting number needs to be decreased by 1: $3.\overline{215} - 1.\overline{307} = 1.\overline{907}$.

Example 11. Calculate: $1.\overline{36} + 2.\overline{375}$

Solution:
Since the first addend has 2 repeating digits and the second addend has 3 repeating digits, the sum should have $2 \times 3 = 6$ repeating digits (the least common multiple of 2 and 3).

$\begin{array}{r} 1.\overline{363636} \\ +2.\overline{375375} \\ \hline 3.\overline{739011} \end{array}$

So $1.\overline{36} + 2.\overline{375} = 1.\overline{363636} + 2.\overline{375375} = 3.\overline{739011}$

Example 12. Calculate: $0.3\overline{42} + 2.\overline{35}$.

Solution:

Algebra II Through Competitions   Chapter 2 Simplifications And Calculations

$2.\overline{35} = 2.3535353..... = 2.3\overline{53}$ and $\dfrac{0.342 + 2.3\overline{53}}{2.6\overline{95}}$, so $0.3\overline{42} + 2.\overline{35} = 0.3\overline{42} + 2.3\overline{53} = 2.6\overline{95}$

Example 13. Calculate: $2.2\overline{5} - 1.3\overline{6}$

Solution:

Method 1: $2.2\overline{5} - 1.3\overline{6} = 2.25\overline{0} - 1.3\overline{6} = 2.25\overline{0} - 1.36\overline{6} == 0.88\overline{3}$

Method 2: $2.2\overline{5} - 1.3\overline{6} = \dfrac{225}{100} - \dfrac{136-13}{90} = \dfrac{2025 - 1230}{900} = \dfrac{795}{900}$.

We see that both expressions are the same: $0.88\overline{3} = \dfrac{883 - 88}{900} = \dfrac{795}{900}$.

Example 14. Calculate: $0.\overline{27} \div 3.\overline{2}$

Solution:

$0.\overline{27} \div 3.\overline{2} = \dfrac{27}{99} \div 3\dfrac{2}{9} = \dfrac{27}{99} \div \dfrac{29}{9} = \dfrac{27}{99} \times \dfrac{9}{29} = \dfrac{27}{319}$.

3. CONTINUED FRACTIONS

The simple continued fraction representation of a number is given by:

$$a_0 + \cfrac{1}{a_1 + \cfrac{1}{a_2 + \cfrac{1}{a_3 + \cfrac{1}{a_4 + ...}}}}$$

where a_0 is an integer, any other a_i members are positive integers.

Example 15. The solution set for the equation $\dfrac{1}{\dfrac{4}{x-2} + 2} = \dfrac{1}{2}$ is:

A. {2}   B. {4}   C. the empty set   D. {0}   E. {0.25}

Solution: C

$$\frac{1}{\frac{4}{x-2}+2}=\frac{1}{2} \quad\Rightarrow\quad \frac{4}{x-2}+2=2 \quad\Rightarrow\quad \frac{4}{x-2}=0.$$

This equation has no solution.

Example 16. For how many values of b does the equation
$$\frac{2}{2-\frac{2}{2-\frac{2}{2+x}}}=b \text{ have no solution?}$$

A. 0 B. 1 C. 2 D. 3 E. 4

Solution: D.
For the following four cases, the equation will have no solution:
Case 1: $x = \infty$.
Case 2: $2 + x = 0 \quad\Rightarrow\quad x = -2$.
Case 3: $2 - \frac{2}{2+x} = 0 \quad\Rightarrow\quad x = -1$.
Case 4: $2 - \frac{2}{2-\frac{2}{2+x}} = 0 \quad\Rightarrow\quad 1 = \frac{1}{2-\frac{2}{2+x}} \quad\Rightarrow\quad 2 - \frac{2}{2+x} = 1 \quad\Rightarrow\quad \frac{2}{2+x} = 1 \quad\Rightarrow\quad$
$2 + x = 2 \quad\Rightarrow\quad x = 0$.

When we simplify $\dfrac{2}{2-\dfrac{2}{2-\dfrac{2}{2+x}}} = b$, we get: $\dfrac{2}{2-\dfrac{2}{\frac{4+2x-2}{2+x}}} = b \quad\Rightarrow\quad \dfrac{2}{2-\dfrac{4+2x}{2+2x}} = b$

$\Rightarrow \dfrac{2}{2-\dfrac{2+x}{1+x}} = b \quad\Rightarrow\quad \dfrac{2}{\frac{2+2x-(2+x)}{1+x}} = b \quad\Rightarrow\quad \dfrac{2+2x}{x} = b \quad\Rightarrow\quad x = \dfrac{2}{b-2}$.

Thus, we see that when $b = 2$, $x = \infty$; $b = 1$, $x = -2$; $b = 0$, $x = -1$.
However, no value of b can lead to $x = 0$. So the answer is D.

Example 17. Determine the value of the continued fraction: $1 + \dfrac{2}{2+\dfrac{2}{2+\dfrac{2}{2+\dfrac{2}{\dots}}}}$.

A. $\sqrt{5}$ B. $\dfrac{1+\sqrt{5}}{2}$ C. $\sqrt{3}$ D. $\dfrac{1+\sqrt{3}}{2}$ D. $\dfrac{1-\sqrt{5}}{2}$.

Solution: C.

Let $\dfrac{2}{2+\dfrac{2}{2+\dfrac{2}{2+\dfrac{2}{\ldots}}}} = x$ $\Rightarrow$ $\dfrac{2}{2+x} = x$ $\Rightarrow$ $x^2 + 2x - 2 = 0$ $\Rightarrow$

$x^2 + 2x + 1 - 3 = 0$ $\Rightarrow$ $(x+1)^2 = 3$ $\Rightarrow$ $x = \sqrt{3} - 1$

So $1 + \dfrac{2}{2+\dfrac{2}{2+\dfrac{2}{2+\dfrac{2}{\ldots}}}} = 1 + \sqrt{3} - 1 = \sqrt{3}$.

4. SIMPLIFICATION

Example 18. Simplify $\dfrac{\dfrac{-2}{x+1}}{\dfrac{5}{x}+4}$.

A. $\dfrac{-2x}{4x^2+9x+5}$ B. $\dfrac{-2x-40}{5x+5}$ C. $\dfrac{-8x-10}{20x-40}$ D. $\dfrac{-2x}{9x+9}$ E. none of these

Solution: A.

$\dfrac{\dfrac{-2}{x+1}}{\dfrac{5}{x}+4} = \dfrac{-2}{x+1} \cdot \dfrac{x}{5+4x} = \dfrac{-2x}{4x^2+9x+5}$.

Example 19. Simplify: $\sqrt{\dfrac{2^{x+4} - 2(2^{x+1})}{2(2^{x+3})}}$.

A. $\dfrac{3}{8}$ B. $\dfrac{\sqrt{3}}{4}$ C. 2^x D. $\dfrac{x\sqrt{3}}{4}$ E. $\dfrac{\sqrt{3}}{2}$

Solution: E.

$$\sqrt{\frac{2^{x+4}-2(2^{x+1})}{2(2^{x+3})}} = \sqrt{\frac{2^{x+4}-2^{x+2}}{2^{x+4}}} = \sqrt{\frac{2^{x+4}(1-2^{-2})}{2^{x+4}}} = \sqrt{1-\frac{1}{4}} = \sqrt{\frac{3}{4}} = \frac{\sqrt{3}}{2}$$

5. SPECIAL OPERATIONS

Example 20. If Δ is defined by the equation $x \Delta y = x + xy + y$ for all real numbers x and y, what is the value of z if $8 \Delta z = 3$?
A. -5 B. −5/9 C. 3/8 D. 5/9 E. 5

Solution: B.
$x \Delta y = x + xy + y \Rightarrow 8 \Delta z = 8 + 8z + z = 8 + 9z = 3$. So $z = -5/9$.

Example 21. Define an operation for positive real numbers as $a * b = \dfrac{a+b}{ab}$. Then $1 * (2 * 4)$ is equal to:
A. 7/8 B. 11/12 C. 7/3 D. 8 E. none of these

Solution: C.

$$2 * 4 = \frac{2+4}{2 \cdot 4} = \frac{3}{4}.$$

$$1 * \frac{3}{4} = \frac{1+\frac{3}{4}}{1 \cdot \frac{3}{4}} = \frac{\frac{7}{4}}{\frac{3}{4}} = \frac{7}{3}.$$

Algebra II Through Competitions   **Chapter 2 Simplifications And Calculations**

PROBLEMS

Problem 1. Find the sum of the first 2006 terms of the following infinite series:
$$\frac{1}{1\cdot 3}+\frac{1}{3\cdot 5}+\frac{1}{5\cdot 7}+\cdots+\frac{1}{(2n-1)(2n+1)}+\cdots$$
A. $\dfrac{1}{32192286}$ B. $\dfrac{1}{16096143}$ C. $\dfrac{2006}{16096143}$ D. $\dfrac{2006}{4013}$ E. $\dfrac{2006}{4011}$

Problem 2. Evaluate $\dfrac{1}{1\cdot 4}+\dfrac{1}{4\cdot 7}+\cdots+\dfrac{1}{(3n-2)\cdot(3n+1)}$.

A. $\dfrac{n}{(3n+1)}$ B. $\dfrac{n}{(3n-1)}$ C. $\dfrac{2n}{(3n+1)}$ D. $\dfrac{n}{(n+1)}$ E. $\dfrac{n}{(n-1)}$

Problem 3. Let $f(x)=\dfrac{1}{x(x+1)}$. Find $f(1)+f(2)+\ldots f(2011)$.

A. 1005/2011   B. 2010/2011   C. 2011/2012   D. 4047/2012   E. 6066/2011

Problem 4. Find the product $\left(1-\dfrac{1}{2}\right)\left(1-\dfrac{1}{3}\right)\left(1-\dfrac{1}{4}\right)\cdots\left(1-\dfrac{1}{2008}\right)$.

A. $\dfrac{1}{2008!}$ B. $\dfrac{1}{2007}$ C. $\dfrac{2007!}{2^{2008}}$ D. $\dfrac{1}{2008}$ E. none of these

Problem 5. Evaluate
$$(\dfrac{1}{2}+\dfrac{1}{3}+\ldots+\dfrac{1}{2007})(1+\dfrac{1}{2}+\dfrac{1}{3}+\ldots+\dfrac{1}{2006})-(1+\dfrac{1}{2}+\dfrac{1}{3}+\ldots+\dfrac{1}{2007})(\dfrac{1}{2}+\dfrac{1}{3}+\ldots+\dfrac{1}{2006}).$$

A. $-\dfrac{1}{2007}$ B. $\dfrac{1}{2007}$ C. $\dfrac{2}{2007}$ D. $\dfrac{3}{2007}$ E. none of these

Problem 6. Find $\dfrac{1}{\sqrt{2}-\sqrt{1}}-\dfrac{1}{\sqrt{3}-\sqrt{2}}+\dfrac{1}{\sqrt{4}-\sqrt{3}}-\cdots+\dfrac{1}{\sqrt{100}-\sqrt{99}}$.

A. 10   B. 11   C. $3\sqrt{11}$ D. $1+3\sqrt{11}$ E. 9

Problem 7. Write the repeating decimal 0.35353535 . . . as a fraction.
A. 353/909   B. 7/18   C. 353/990   D. 353/999   E. 35/99

Algebra II Through Competitions **Chapter 2 Simplifications And Calculations**

Problem 8. If you write the repeating decimal $.\overline{518}$ as a fully reduced common fraction, the difference between the denominator and the numerator is
A. 5   B. 13   C. 14   D. 481   E. none of these

Problem 9. The number 2.53535.... can be written as a fraction. When this fraction is reduced to lowest terms, the sum of the numerator and denominator is:
A. 7   B. 350   C. 141   D. 257   E. 349

Problem 10. The sum of the repeated decimal fractions, $3.1\overline{45}$ and $4.\overline{154}$, is:
A. 7.3   B. $7.\overline{299}$ C. $7.2\overline{991}$ D. $7.\overline{2996086}$ E. $7.\overline{2996087}$

Problem 11. Write the repeating decimal 1.363636... as a fraction.
A. 135/999   B. 136/99   C. 47/11   D. 15/11   E. none of these

Problem 12. Evaluate $\dfrac{1}{2 - \dfrac{1}{2 - \dfrac{1}{2 - \dfrac{1}{2 - \dfrac{1}{2 - \cdots}}}}}$

A. –1   B. –1/2   C. 0   D. 1/2   E. 1

Problem 13. If $x = 1 + \dfrac{3}{1 + \dfrac{3}{1 + \cdots}}$, then $x = \dfrac{k + \sqrt{w}}{p}$, where k, w, and p are positive integers. Find the value of $(k + w + p)$.

Problem 14. Simply: $\sqrt[x]{\sqrt{\sqrt[3]{25}}}$ where x is a natural number greater than 1.
A. 25^{3x} B. $25^{\frac{1}{3x}}$ C. $5^{\frac{1}{3x}}$ D. 5^{3x} E. none of these

Problem 15. Simplify the following rational function. $\dfrac{x^3 - y^3}{x^4 + x^2 y^2 + y^4}$.

A. $\dfrac{x - y}{x^2 + xy + y^2}$ B. $\dfrac{x - y}{x^2 - xy + y^2}$ C. $\dfrac{1}{x^2 - xy + y^2}$ D. $\dfrac{1}{x^2 + y^2}$ E. none of these

Problem 16. Simplify $\dfrac{\dfrac{x^2}{y}+\dfrac{y^2}{x}}{y^2-xy+x^2}$.

A. x^2-y^2 B. $\dfrac{xy}{x-y}$ C. $\dfrac{x+y}{xy}$ D. $x-y$ E. none of these

Problem 17. Simplify $\dfrac{1-\dfrac{1}{m+1}}{1+\dfrac{1}{m-1}}$ ($m \neq 0, 1, -1$)

A. -1 B. 1 C. $\dfrac{m-1}{m+1}$ D. $\dfrac{1-m}{1+m}$ E. 2

Problem 18. The rational expression $\dfrac{\dfrac{x}{x-2}+1}{\dfrac{3}{x^2-4}+1}$ simplifies to

A. $\dfrac{x^2+2x+3}{3}$ B. $\dfrac{2x^2(4-x)}{12-x^2}$ C. $\dfrac{2(x-1)^2(x+1)}{(x-2)^2(x+2)}$ D. $\dfrac{2(x+2)}{x+1}$ E. $\dfrac{2(x-1)}{3(x-2)}$.

Problem 19. Simplify the following: $\dfrac{\dfrac{1}{x+2}+\dfrac{1}{x-5}}{\dfrac{2x^2-x-3}{x^2-3x-10}}$.

A. $\dfrac{2x-3}{(2x+3)(x-1)}$ B. $\dfrac{1}{x+1}$ C. $\dfrac{2x-3}{x-1}$ D. $\dfrac{1}{x-1}$

Problem 20. Let $n\downarrow$ denote the largest prime number less than n and $n\uparrow$ denote the smallest prime number greater than n. Find the value of $((10 + 20\uparrow)\downarrow + 30)\uparrow$.
A. 57   B. 59   C. 60   D. 61   E. 67.

Problem 21. Let A and B be positive real numbers, and define $A \oplus B = \dfrac{AB}{A+B}$. Solve $X^2 \oplus (2X \oplus 1) = \dfrac{1}{4}$
A. $-1/2$ B. $1/2$ C. $3/2$ D. $2/3$ E. None of these

27

Problem 22. Let $\oplus$ be defined as $A \oplus B = \dfrac{AB}{A+B}$. Find the general expression for $A \oplus A \oplus A \oplus \ldots \oplus A$, Where n A's are being combined using $\oplus$.

A. nA B. $\dfrac{A^{n-1}}{n}$ C. $\dfrac{A}{n}$ D. $\dfrac{A^n}{1+A+A^2+\cdots+A^{n-1}}$ E. None of these

SOLUTIONS

Problem 1. Solution: D.

By $\frac{1}{n(n+k)} = \frac{1}{k}(\frac{1}{n} - \frac{1}{n+k})$, $\frac{1}{1\cdot 3} + \frac{1}{3\cdot 5} + \frac{1}{5\cdot 7} + \cdots + \frac{1}{(2n-1)(2n+1)} + \cdots$

$= \frac{1}{2}(\frac{1}{1} - \frac{1}{3} + \frac{1}{3} - \frac{1}{5} \cdots + \frac{1}{4011} - \frac{1}{4013}) = \frac{1}{2}(\frac{1}{1} - \frac{1}{4013}) = \frac{1}{2}(\frac{4012}{4013}) = \frac{2006}{4013}$.

Problem 2. Solution: A.

By $\frac{1}{n(n+k)} = \frac{1}{k}(\frac{1}{n} - \frac{1}{n+k})$, $\frac{1}{1\cdot 4} + \frac{1}{4\cdot 7} + \cdots + \frac{1}{(3n-2)\cdot(3n+1)}$

$= \frac{1}{3}(\frac{1}{1} - \frac{1}{4} + \frac{1}{4} - \frac{1}{7} \cdots + \frac{1}{3n-2} - \frac{1}{3n+1}) = \frac{1}{3}(\frac{1}{1} - \frac{1}{3n+1}) = \frac{1}{3}(\frac{3n+1-1}{3n+1}) = \frac{n}{3n+1}$

Problem 3. Solution: C.

By $\frac{1}{n(n+1)} = \frac{1}{n} - \frac{1}{n+1}$, $f(1) + f(2) + \ldots f(2011)$

$= = \frac{1}{1} - \frac{1}{3} + \frac{1}{3} - \frac{1}{5} \cdots + \frac{1}{2011} - \frac{1}{2012}) = \frac{1}{1} - \frac{1}{2012}) = \frac{2011}{2012}$.

Problem 4. Solution: D.

$\left(1 - \frac{1}{2}\right)\left(1 - \frac{1}{3}\right)\left(1 - \frac{1}{4}\right)\cdots\left(1 - \frac{1}{2008}\right) = \frac{1}{2} \times \frac{2}{3} \times \frac{3}{4} \cdots \times \frac{2007}{2008} = \frac{1}{2008}$.

Problem 5. Solution: B.

Let $\frac{1}{2} + \frac{1}{3} + \ldots + \frac{1}{2007} = a$ and $\frac{1}{2} + \frac{1}{3} + \ldots + \frac{1}{2006} = b$.

The given expression becomes $a(1 + b) - (a + 1)b = a - b = 1/2007$.

Problem 6. Solution: B.

We know that $\frac{1}{\sqrt{n+1} - \sqrt{n}} = \sqrt{n+1} + \sqrt{n}$. The original expression can be written as

$\sqrt{2} + \sqrt{1} - (\sqrt{3} + \sqrt{2}) + \sqrt{4} + \sqrt{3} - (\sqrt{5} + \sqrt{4}) \cdots + \sqrt{100} + \sqrt{99} = \sqrt{1} + \sqrt{100} = 11$.

Problem 7. Solution: E.
0.35353535... = 35/99.

Problem 8. Solution: B.

$.\overline{518} = \frac{518}{999} = \frac{14}{27}$. The difference of the numerator and denominator is 27 − 14 = 13.

Problem 9. Solution: B.

$2.53535\ldots = 2 + 0.53535\ldots = 2 + \frac{53}{99} = \frac{251}{99}$. The sum of the numerator and denominator is 251 + 99 = 350.

Problem 10. Solution: D.

$3.145454545\ldots + 4.154154154\ldots = 7.29960869960\ldots = 7.2\overline{996086}$

Problem 11. Solution: D.

$1.363636\ldots = 1 + 0.3636\ldots = 1 + \frac{36}{99} = \frac{135}{99} = \frac{15}{11}$.

Problem 12. Solution: E.

Let $\cfrac{1}{2-\cfrac{1}{2-\cfrac{1}{2-\cfrac{1}{2-\cfrac{1}{2-\cdots}}}}} = x$. $\quad \frac{1}{2-x} = x \quad \Rightarrow \quad x^2 - 2x + 1 = 0 \Rightarrow$

$(x-1)^2 = 0 \quad \Rightarrow \quad x = 1$.

Problem 13. Solution: 16.

Let $\cfrac{3}{1+\cfrac{3}{1+\cfrac{3}{1+\ldots}}} = y \Rightarrow \quad \frac{3}{1+y} = y \quad \Rightarrow \quad y^2 + y - 3 = 0 \Rightarrow$

$y^2 + 2 \cdot \frac{1}{2}y + \frac{1}{4} - \frac{1}{4} - 3 = 0 \quad \Rightarrow \quad (y + \frac{1}{2})^2 = \frac{13}{4} \quad \Rightarrow \quad y = \frac{\sqrt{13}}{2} - \frac{1}{2}$.

So $x = 1 + \cfrac{3}{1+\cfrac{3}{1+\ldots}} = 1 + \frac{\sqrt{13}}{2} - \frac{1}{2} = \frac{\sqrt{13}+1}{2}$. $k = 1$, $w = 13$ and $p = 2$. $(k + w + p) = 16$.

Problem 14. Solution: C.

$$\sqrt[x]{\sqrt{\sqrt[3]{25}}} = 25^{\frac{1}{6x}} = (5^2)^{\frac{1}{6x}} = 5^{\frac{1}{3x}}$$

Problem 15. Solution: B.

$$\frac{x^3 - y^3}{x^4 + x^2y^2 + y^4} = \frac{(x-y)(x^2+xy+y^2)}{x^4+x^2y^2+y^4} = \frac{(x-y)(x^2+xy+y^2)}{(x^2+xy+y^2)(x^2-xy+y^2)} = \frac{(x-y)}{(x^2-xy+y^2)}.$$

Problem 16. Solution: C.

$$\frac{\frac{x^2}{y} + \frac{y^2}{x}}{y^2 - xy + x^2} = \frac{\frac{x^3+y^3}{xy}}{y^2-xy+x^2} = \frac{(x+y)(x^2-xy+y^2)}{xy(y^2-xy+x^2)} = \frac{x+y}{xy}.$$

Problem 17. Solution: C.

$$\frac{1 - \frac{1}{m+1}}{1 + \frac{1}{m-1}} = \frac{\frac{m+1-1}{m+1}}{\frac{m-1+1}{m-1}} = \frac{\frac{m}{m+1}}{\frac{m}{m-1}} = \frac{m-1}{m+1}.$$

Problem 18. Solution: D.

$$\frac{\frac{x}{x-2}+1}{\frac{3}{x^2-4}+1} = \frac{\frac{x+x-2}{x-2}}{\frac{3+x^2-4}{x^2-4}} = \frac{\frac{2x-2}{x-2}}{\frac{x^2-1}{(x+2)(x-2)}} = \frac{2(x-1)}{(x-1)(x+1)} \cdot \frac{(x+2)}{1} = \frac{2(x+2)}{x+1}.$$

Problem 19. Solution: B.

$$\frac{\frac{1}{x+2} + \frac{1}{x-5}}{\frac{2x^2-x-3}{x^2-3x-10}} = \frac{\frac{x-5+x+2}{(x+2)(x-5)}}{\frac{2x^2-x-3}{(x+2)(x-5)}} = \frac{2x-3}{(2x-3)(x+1)} = \frac{1}{x-1}.$$

Problem 20. Solution: E.
$((10 + 20\uparrow)\downarrow + 30)\uparrow \Rightarrow ((10 + 23)\downarrow + 30)\uparrow \Rightarrow (31 + 30)\uparrow \Rightarrow 67$

Problem 21. Solution: D.

$$X^2 \oplus (2X \oplus 1) = X^2 \oplus \left(\frac{2X}{2X+1}\right) = \frac{X^2\left(\frac{2X}{2X+1}\right)}{X^2 + \left(\frac{2X}{2X+1}\right)} = \frac{2X^3}{2X^3 + X^2 + 2X},$$ so we need to

solve $\frac{2X^3}{2X^3 + 2X^2 + X} = \frac{1}{4} \Rightarrow 8X^3 = 2X^3 + 2X^2 + X \Leftrightarrow 6X^3 - X^2 - 2X = 0$. Factor this last expression to get $X(6X^2 - X - 2) = X(3X - 2)(2X + 1) = 0$, making $X = 0$, 2/3 or –1/2. Since the operation is only defined for positive reals, 2/3 is the only correct answer.

Problem 22. Solution: C.
Using the operation, we can see the progression of values for $A \oplus A$, $A \oplus A \oplus A$, $A \oplus A \oplus A \oplus A$, …. These values are
$\frac{A^2}{2A} = [\frac{A}{2}]$, $\frac{(A/2)A}{(A/2)+A} = \frac{A^2}{3A} = [\frac{A}{3}]$, $\frac{(A/3)A}{(A/3)+A} = \frac{A^2}{4A} = [\frac{A}{4}]$, … , so it looks like the general term would be $\frac{A}{n}$.

Algebra II Through Competitions   **Chapter 3. Linear Equations & Functions**

1. HANDY FORMULAS

(1.1) Distance formula

Given right triangle ABC, the Pythagorean Theorem yields

$$AB^2 = AC^2 + BC^2$$

Let $AB = d$,

$$d^2 = (x_2 - x_1)^2 + (y_2 - y_1)^2$$

$$d = \sqrt{(x_2 - x_1)^2 + (y_2 - y_1)^2} \qquad (1.1)$$

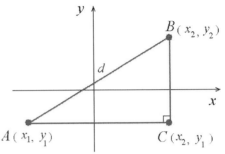

(1) is called the distance formula and is used to calculate the distance between two points $P_1(x_1, y_1)$ and $P_2(x_2, y_2)$.

Note: $d = \sqrt{(x_1 - x_2)^2 + (y_1 - y_2)^2}$ is equivalent to (1).

(1.2) Midpoint formula

The coordinates (x_m, y_m) of the midpoint of the line segment with endpoints (x_1, y_1) and (x_2, y_2) are $x_m = \dfrac{x_1 + x_2}{2}$, and $y_m = \dfrac{y_1 + y_2}{2}$ $\qquad (1.2)$

(1.3) Vertices formula of a parallelogram:

$ABCD$ is a parallelogram. Given three vertices $A(x_A, y_A)$, $B(x_B, y_B)$, and $C(x_C, y_C)$, find the coordinates of D.

$$x_D = x_A + x_C - x_B$$
$$y_D = y_A + y_C - y_B \qquad (1.3)$$

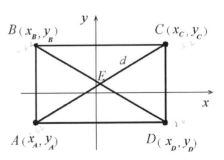

(1.4) Area of polygons in the coordinate system (Shoelace formula)

Given a triangle with three vertices located at (x_1, y_1), (x_2, y_2), and (x_3, y_3), its area is

$$\text{Area} = A = \frac{1}{2}\begin{vmatrix} x_1 & y_1 \\ x_2 & y_2 \\ x_3 & y_3 \\ x_1 & y_1 \end{vmatrix} = \frac{1}{2}|(x_1 y_2 + x_2 y_3 + x_3 y_1) - (y_1 x_2 + y_2 x_3 + y_3 x_1)| \qquad (1.4.1)$$

Given a quadrilateral with four vertices in clockwise order located at (x_1, y_1), (x_2, y_2), (x_3, y_3), and (x_4, y_4), its area is

$$\text{Area} = A = \frac{1}{2}\begin{vmatrix} x_1 & y_1 \\ x_2 & y_2 \\ x_3 & y_3 \\ x_4 & y_4 \\ x_1 & y_1 \end{vmatrix} = \frac{1}{2}|(x_1 y_2 + x_2 y_3 + x_3 y_4 + x_4 y_1) - (y_1 x_2 + y_2 x_3 + y_3 x_4 + y_4 x_1)| \quad (1.4.2).$$

(1.5) Point to Line Distance Formula

Solving problems involving finding the distance from a point (x_1, y_1) to a line $ax + by + c = 0$ might seem simple at first glance, but is very time – consuming, unless you know the following formula: $d = \left|\dfrac{ax_1 + by_1 + c}{\sqrt{a^2 + b^2}}\right|.$

Example 1. Find the area of a triangle whose vertices are $(1, 3)$, $(5, 7)$, and $(9, -1)$.

A. 18   B. 5 17   C. 8 10   D. 24   E. 48

Solution: D.

Method 1:
By the formula (1.4), we have

Algebra II Through Competitions   Chapter 3. Linear Equations & Functions

$$A = \frac{1}{2}\begin{vmatrix} x_1 & y_1 \\ x_2 & y_2 \\ x_3 & y_3 \\ x_1 & y_1 \end{vmatrix} = \frac{1}{2}\begin{vmatrix} 1 & 3 \\ 5 & 7 \\ 9 & -1 \\ 1 & 3 \end{vmatrix} = \frac{1}{2}|(1\times 7 + 5\times(-1) + 9\times 3) - (1\times(-1) + 9\times 7 + 5\times 3)| = 24.$$

Method 2:
Calculate areas of trapezoids under line segments ((1,3), (5,7)) , ((5,7), (9,–1)) and ((1,3), (9,–1)). Triangle = ½(5–1)(7+3) + ½(9–5)(–1+7) – ½(9–1)(–1+3) = 24.

Example 2. Given the line $4x - 3y = 12$, what is the distance between it and the origin?

A. 1.2   B. $\dfrac{7\sqrt{6}}{3}$ C. 2.4   D. 3.5   E. none of these

Solution: C.

The distance from a line $Ax + By + C = 0$ to a point x_0, y_0 is given by the formula $d = \left|\dfrac{Ax_0 + Bx_0 + C}{\sqrt{A^2 + B^2}}\right|$. Plugging in values, $\left|\dfrac{(4)(0) + (-3)(0) - 12}{\sqrt{4^2 + 3^2}}\right| = 2.4$

2. LINES AND THEIR EQUATIONS ON THE COORDINATE PLANE

(2.1). Standard form:

$Ax + By + C = 0$ ($A^2 + B^2 \neq 0$)
or $Ax + By = C$ ($A^2 + B^2 \neq 0$)

(2.2). Point-Slope Form

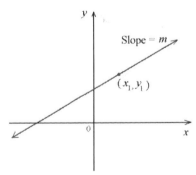

$y - y_1 = m(x - x_1)$
m is the slope of the line and the point (x_1, y_1) is on the line.
If we know the slope of a line, we only need to know a point on the line in order to define the equation of the line.

If we know a point on a line, we only need to know the slope of the line in order to define the equation of the line.

(2.3). Slope-Intercept Form
$y = mx + b$, where m is the slope and b is the y-intercept.

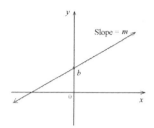

(2.4). Two-Intercept Form:

$$\frac{x}{a} + \frac{y}{b} = 1$$

a is the x-intercept and b is the y-intercept.

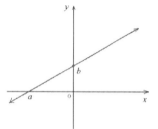

(2.5). Two-point Form:

If two points (x_1, y_1) and (x_2, y_2) are known the equation of the line can be written as: $\frac{y - y_1}{x - x_1} = \frac{y_2 - y_1}{x_2 - x_1}$.

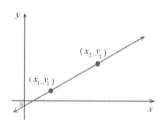

Example 3. A line with slope 3 intersects a line with slope 6 at the point (30, 40). What is the distance between the y-intercepts of these lines?
A. 80   B. 90   C. 100   D. 110   E. 120

Solution (B).

For l_1, we have $\frac{40 - y_1}{30 - 0} = 3$ $\Rightarrow$ $40 - y_1 = 90$ (1)

For l_2, we have $\frac{40 - y_2}{30 - 0} = 6$ $\Rightarrow$ $40 - y_2 = 180$ (2)

(2) – (1): $y_1 - y_2 = 90$.

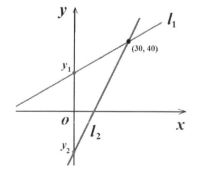

Example 4. Given $A(1, -2)$, $B(5, 1)$, and $C(-2, 2)$, find the equation of the angle bisector at A.
A. $5x - y = 7$ B. $7x - y = 2$ C. $7y - x = 2$ D. $y = x + 3$ E. none of the above

Solution: E

We find that side *AC* and *AB* both have length 5, so the angle bisector is also the median, passing through $A(1, -2)$ and the midpoint $(3/2, 3/2)$, of side *BC*. This makes the slope 7 and the equation $y = 7(x-1) - 2 = 7x - 9$.

Example 5. A line of slope 2/3 passes through the point $(-1, 2)$. What is its *y*-intercept?
A. (0, 3)   B. (0, 11/4)   C. (0, 15/6)   D. (0, 14/5)   E. (0, 8/3)

Solution: (E).

The equation of the line can be written as $\dfrac{y-2}{x+1} = \dfrac{2}{3}$.

We get the *y*-intercept by letting $x = 0$: $\dfrac{y-2}{0+1} = \dfrac{2}{3}$ $\Rightarrow$ $y = \dfrac{2}{3} + 2 = \dfrac{8}{3}$.

Example 6. What is the area of the triangular region in the first quadrant bounded on the left by the y-axis, bounded above by the line $7x + 4y = 168$ and bounded below by the line $5x + 3y = 121$?
A. 16   B. 50/3   C. 17   D. 52/3   E. 53/3

Solution: B.

The triangle has a base along the y-axis of $42 - 121/3 = 5/3$ and an altitude of 20 (the lines intersect at (20, 7). So the area is $(1/2)(5/3)20 = 50/3$.

3. SLOPE OF A LINE

(3.1) The slope formula

Given two points (x_1, y_1) and (x_2, y_2) on a line, the slope of the line is $m = \dfrac{\Delta y}{\Delta x} = \dfrac{y_2 - y_1}{x_2 - x_1}$ (rise over run), where $\Delta y = y_2 - y_1$ is called the change in *y*, and $\Delta x = x_2 - x_1 \neq 0$ is called the change in *x*.

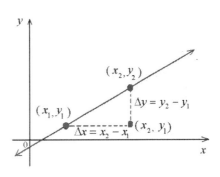

Example 7. What is the slope of the line that goes through points $(-1, 2)$ and $(2, -1)$?
A. 1   B. 0   C. -1   D. 2   E. -2
Solution: (C).

$$m = \frac{y_2 - y_1}{x_2 - x_1} = \frac{2-(-1)}{-1-2} = -1.$$

Example 8. The line through the points $(m, -9)$ and $(7, m)$ has slope m. What is the value of m?
A. 3   B. −7/9   C. 16   D. 5   E. None of these

Solution: A.

By the definition of slope we have $m = (m + 9)/(7 - m)$, which has solution $m = 3$.

(3.2) Positive and negative slopes

θ is called the inclination of the line, and is the angle between the line and the x-axis measured counterclockwise from x-axis.

The slope is positive if $0 < \theta < \frac{\pi}{2}$

The slope is negative if $\frac{\pi}{2} < \theta < \pi$

The slope of a vertical line $\theta = \frac{\pi}{2}$ is undefined.

The slope of a horizontal line $\theta = \pi$ is zero (flat with no inclination).
Note: $m = \tan \theta$

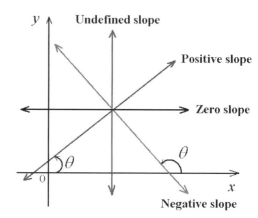

(3.3). Slopes of parallel and perpendicular lines

If line $y_1 = m_1 x_1 + b_1$ and line $y_2 = m_2 x_2 + b_2$ are parallel, then $m_1 = m_2$.
If $m_1 = m_2$, then line $y_1 = m_1 x_1 + b_1$ and line $y_2 = m_2 x_2 + b_2$ are parallel.
If line $y_1 = m_1 x_1 + b_1$ and line $y_2 = m_2 x_2 + b_2$ are perpendicular, then $m_1 \times m_2 = -1$.
If $m_1 \times m_2 = -1$, then line $y_1 = m_1 x_1 + b_1$ and line $y_2 = m_2 x_2 + b_2$ are perpendicular.

Example 9. Given that the points, (1, 4) (6, 12) and (c, 10), are collinear, what is the value of c?
A. 7/2   B. 4   C. 17/4   D. 19/4   E. 5

Solution: D.

The line determined by the points (1, 4) and (6, 12) is the same as the line determined by the points (6, 12) and (c, 10).

So $\dfrac{12-4}{6-1} = \dfrac{12-10}{6-c} \Leftrightarrow \dfrac{8}{5} = \dfrac{2}{6-c} \Rightarrow 8(6-c) = 10 \Rightarrow 6 - c = \dfrac{5}{4} \Rightarrow c = 6 - \dfrac{5}{4} = 19/4$

Example 10. Find the equation of a line perpendicular to $4x - 3y = 12$ that passes through point (2, 3).
A. $y = 0.75x + 1.5$
B. $y = -0.75x + 4.5$
C. $y = -0.25x + 3.5$
D. $y = 1.33x - 4$
E. none of these

Solution: B.

The original equation is $y = \dfrac{4}{3}x - 4$. The new equation will be of the form $y = -\dfrac{3}{4}x + b$.

Since the line passes through point (2, 3), we have $3 = -\dfrac{3}{4}(2) + b \Rightarrow b = \dfrac{9}{2} = 4.5$.

Thus $y = -\dfrac{3}{4}x + b = -0.75x + 4.5$.

Example 11. If the graphs of the lines with equations $3x - 2y + 4 = 0$ and $7x + ay - 1 = 0$ are perpendicular, then a is equal to:
A. $\dfrac{21}{2}$
B. $-\dfrac{2}{21}$
C. $-\dfrac{3}{7}$
D. $\dfrac{7}{3}$
E. none of these

Solution: A.
$3x - 2y + 4 = 0 \Rightarrow y = (3/2)x + 2$.
$7x + ay - 1 = 0 \Rightarrow y = -(7/a)x + 1/a$.
Thus $(3/2) \times (-7/a) = -1 \Rightarrow a = 21/2$.

Algebra II Through Competitions   **Chapter 3. Linear Equations & Functions**

PROBLEMS

Problem 1. What is the slope of the line that goes through points $(-3, 2)$ and $(3, 2)$?
A. Undefined   B. 0   C. 3/2   D. $-2/3$ E. 1

Problem 2. If the graphs of $2y + x + 3 = 0$ and $3y + ax + 2 = 0$ are perpendicular, the value of a is
A. $-2/3$ B. $-3/2$ C. 6   D. -6 E. None of these

Problem 3. Two lines have slopes whose sum is 5 and both go through the point $(1, 1)$. If one line goes through $(5, y_1)$ and the other line goes through $(5, y_2)$ then $y_1 + y_2$ is equal to
A. 22   B. 13   C. 6   D. 25   E. 50

Problem 4. Three points $A = (0, 1)$, $B = (2, a)$ and $C = (3, 7)$ are on a straight line. What is the value of a?
A. 5   B. 3   C. 1   D. 4   E. 2

Problem 5. (Find the point on line $y = 2x - 3$ which lies closest to $(4, 0)$.
A. $(1.5, 0)$ B. $(2, 1)$ C. $(2.5, 2)$ D. $(\sqrt{2}, \sqrt{8} - 3)$ E. $(3, 3)$

Problem 6. Given the line, $2x + 4y = 3$, what is the approximate distance between it and the origin?
A. $\dfrac{3}{\sqrt{5}}$ B. $\dfrac{3}{2}$ C. $\dfrac{3\sqrt{5}}{2}$ D. $\dfrac{3\sqrt{5}}{10}$ E. $\dfrac{2\sqrt{5}}{10}$

Problem 7. For what value of m will the triangle formed by the lines $y = 5$, $y = mx - 6$, and $y = -mx - 6$ be equilateral?
A. $\sqrt{2}$ B. $\sqrt{3}$ C. $\sqrt{5}$ D. $\dfrac{\sqrt{3}}{2}$ E. $\dfrac{\sqrt{5}}{3}$

Problem 8. Determine the value of m such that the three lines $y = 2x + 1$, $x = 3$, and $y = mx + 3$ are concurrent.
A. -1 B. $-3/4$ C. $3/4$ D. 1   E. $4/3$

Problem 9. Find the area of a triangle whose vertices are the intersection of the following three lines: $3x + 4y = 75$, $4x - 3y = 0$, $y = 0$.
A. $\dfrac{675}{8}$ B. 150   C. $\dfrac{375}{2}$ D. $\dfrac{25\sqrt{3}}{2}$ E. $\dfrac{75\sqrt{5}}{2}$

Problem 10. An acute angle is formed by two lines of slope 1 and 3. What is the slope of the line which bisects this angle?
A. 1/2　　　B. 2　　　C. $(1+\sqrt{5})/2$　　　D. $\sqrt{5}$　　　E. $\sqrt{7}$

Problem 11. Find the equation of a line that passes through point, $(2, \sqrt{3})$ and has a 30° angle of inclination.
A. $y = \sqrt{3} \cdot (x-1)$　　B. $y = \sqrt{2} \cdot (x-2) + \sqrt{3}$　　C. $y = 2 \cdot (x-2) + \sqrt{3}$
D. $y = \dfrac{\sqrt{3} \cdot (x-1)}{3}$.　　E. $y = \dfrac{\sqrt{3} \cdot (x+1)}{3}$

Problem 12. Find the y value of the intersection of $y = 2x + 7$ and $4x + 3y = 4$.
A. −1.7　　　B. 2.2　　　C. 3.5　　　D. 3.6　　　E. 5.25

Problem 13. For what value of m will the triangle formed by the lines $y = 4$, $x = 0$ and $y = m \cdot x - 2$ have an area of 42?
A. 1/2　　　B. 11/42　　　C. 4/14　　　D. 4/21　　　E. 3/7

Problem 14. A quadrilateral ABCD has vertices with coordinates $A(0, 0)$, $B(6, 0)$, $C(5, 4)$, $D(3, 6)$. What is its area?
A. 18　　　B. 19　　　C. 20　　　D. 21　　　E. 22

Problem 15. In the xy-plane, how many lines whose x-intercept is a positive prime number and whose y-intercept is a positive integer pass through the point (4, 3)?
A. 0　　　B. 1　　　C. 2　　　D. 3　　　E. 4

Problem 16. Find the distance from the point (3, 2) to the line $y = 3x + 2$.
A. $2/\sqrt{3}$　　　B. 3　　　C. $3/\sqrt{5}$　　　D. 5.　　E. $9/\sqrt{10}$.

Problem 17. Find the value of k if the graph of $y = \dfrac{2kx - 3}{5x + k}$ has a y–intercept of 10. Express your answer as a common fraction reduced to lowest term.

SOLUTIONS

Problem 1. Solution: (B).
$m = \dfrac{y_2 - y_1}{x_2 - x_1} = \dfrac{2-2}{3+3} = 0$.

Problem 2. Solution: D.
The first equation can be written as $y = -\dfrac{1}{2}x - 3$. The second equation will be of the form $y = -\dfrac{a}{3}x - 2$.

Since the lines are perpendicular we have $-\dfrac{1}{2} \times (-\dfrac{a}{3}) = -1 \Rightarrow a = -6$.

Problem 3. Solution: A.
The equation of one line can be written as $m_1 = \dfrac{y_1 - 1}{5 - 1} = \dfrac{y_1 - 1}{4}$.

The equation of another line can be written as $m_2 = \dfrac{y_2 - 1}{5 - 1} = \dfrac{y_2 - 1}{4}$.

We know that $m_1 + m_2 = 5$. So we have: $\dfrac{y_1 - 1}{4} + \dfrac{y_2 - 1}{4} = 5 \Rightarrow$

$\dfrac{y_1 - 1 + y_2 - 1}{4} = 5 \Rightarrow y_1 + y_2 = 22$.

Problem 4. Solution: A.
The slope of AB is the same as that of AC. That is, $(a - 1)/(2 - 0) = (7 - 1)/(3 - 0)$, and it follows that $a - 1 = 4$ so $a = 5$. Alternatively, use A and C to get the slope and then use A to get the slope-intercept equation $y = 2x + 1$.

Then just plug in $x = 2$. Also, there is a graphical approach: draw the line and notice that the line goes through (2, 5).

Problem 5. Solution: B.
The line connecting the point and (4, 0) will be perpendicular to $y = 2x - 3$. It has the equation $y = -\dfrac{1}{2}x + 2$. The point is located at the intersection of the two lines.

$$2x - 3 = -\dfrac{1}{2}x + 2$$
$$x = 2, y = 1$$

Problem 6. Solution: D.
If a line is described as $Ax + By + C = 0$ and there is a point (x_0, y_0), then the distance from the point to the line is $\left|\dfrac{Ax_0 + Bx_0 + C}{\sqrt{A^2 + B^2}}\right|$.

Here, this simplifies to $\left|\dfrac{-3}{\sqrt{2^2 + 4^2}}\right| = \dfrac{3\sqrt{5}}{10}$.

Problem 7. Solution: B.
For the triangle to be equilateral, the sides must all have the same length, so $BC = AC$, or $2a = 2a = \sqrt{a^2 + 11^2}$. Solving for a we get $11/\sqrt{3}$, so the slope of side AC is $m = \dfrac{11}{\frac{11}{\sqrt{3}}} = \sqrt{3}$

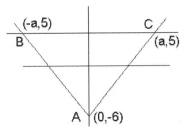

Problem 8. Solution: E.
Since $x = 3$, $y = 2x + 1 \Rightarrow y = 2(3) + 1 = 7$. Thus $7 = m(3) + 3 \Rightarrow 3m = 4 \Rightarrow m = 4/3$.

Problem 9. Solution: B.
A quick sketch of the function shows the triangle in question. Two of the vertices are $(0,0)$ and $(25,0)$ and the third is the intersection of the two oblique lines. By solving the system of equations $3x + 4y = 75$ and $4x - 3y = 0$, we find the solution to be $(9,12)$, which is the final vertex of the triangle. Now the area is $(1/2)bh = (1/2)(25)(12) = 150$.

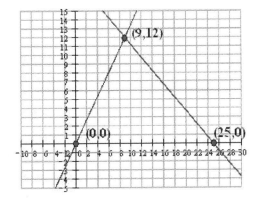

Problem 10. Solution: (C).
By the angle bisector theorem,

$\dfrac{AC}{CD} = \dfrac{AB}{BD} \quad \Rightarrow \quad \dfrac{BD}{CD} = \dfrac{AB}{AC} = \dfrac{\sqrt{10}}{\sqrt{2}} = \sqrt{5}$

We know that $BD + CD = 2$.
Then,

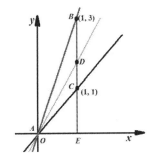

$$CD = \frac{2}{\sqrt{5}+1} = \frac{\sqrt{5}-1}{2}. \qquad DE = CD+1 = \frac{\sqrt{5}-1}{2}+1 = \frac{\sqrt{5}+1}{2}.$$

The slope is $\dfrac{DE}{OE} = \dfrac{DE}{1} = DE = \dfrac{\sqrt{5}+1}{2}$.

Problem 11. Solution: E.

If the angle of inclinatuon is 30°, the tangent of this angle is the slope, so the slope is than $(30°) = \dfrac{\sqrt{3}}{3}$. Using the point – slope equation, we get $y - \sqrt{3} = \dfrac{\sqrt{3}}{3}(x-2)$. This simplifies to $y = \dfrac{\sqrt{3}}{3}(x-2) + \sqrt{3}$, which is equivalent to

$$y = \frac{\sqrt{3}}{3}(x-2) + \sqrt{3} = \frac{\sqrt{3}}{3}x - 2\frac{\sqrt{3}}{3} + \frac{3\sqrt{3}}{3} = \frac{\sqrt{3}}{3}x - 2\frac{\sqrt{3}}{3} + \frac{3\sqrt{3}}{3} = \frac{\sqrt{3}(x+1)}{3}$$

Problem 12. Solution: D.
$y = 2x + 7 \Rightarrow 4x = 2y - 14$
Substituting: $(2y - 14) + 3y = 4 \Rightarrow 5y = 18 \Rightarrow y = 3.6$.

Problem 13. Solution:
Vertices are: $(0, 4)$, $(0, -2)$ and $(6/m, 4)$ with a right angle are $(0, 4)$.
Area $= 1/2 \times 6 \times 6/m = 42 \Rightarrow m = 3/7$.

Problem 14. Solution: D.
Method 1:

$$A = \frac{1}{2}\begin{vmatrix} x_1 & y_1 \\ x_2 & y_2 \\ x_3 & y_3 \\ x_4 & y_4 \\ x_1 & y_1 \end{vmatrix} = \frac{1}{2}\begin{vmatrix} 0 & 0 \\ 6 & 0 \\ 5 & 4 \\ 3 & 6 \\ 0 & 0 \end{vmatrix} = \frac{1}{2}|(0+24+30+0)-(0+12+0+0)| = 21$$

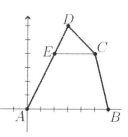

Method 2:

The point $E = (2, 4)$ lies on AD. The line segment from $(2, 4)$ to $(5, 4)$ divides the quadrilateral into a trapezoid with area $(6+3)/2 \times 4$ and a triangle with area $(1/2)(3)(2)$: The total area is 21.

Problem 15. Solution: C.
Let the y-intercept by $(0,k)$ and the x-intercept $(p,0)$.
Then $\dfrac{k-3}{0-4} = \dfrac{3-0}{4-p} \Rightarrow (k-3)(p-4) = 12$. The positive integer factors of 12 are
1, 2, 3, 4, 6, and 12, so
$\begin{matrix} p-4 = 1,2,3,4,6,12 \\ k-3 = 12,6,4,3,2,1 \end{matrix} \Rightarrow \begin{matrix} p = 5,6,7,8,10,16 \\ k = 15,9,7,6,5,4. \end{matrix}$

Of these, only $p = 5$ and 7 are prime.

Problem 16. Solution: E.
The distance from a line $Ax + By + C = 0$ to a point x_0, y_0 is given by the formula
$d = \left|\dfrac{Ax_0 + Bx_0 + C}{\sqrt{A^2 + B^2}}\right|$. Plugging in values, $\left|\dfrac{(3)(3) + (-1)(2) + 2}{\sqrt{3^2 + (-1)^2}}\right| = \dfrac{9}{\sqrt{10}}$.

Problem 17. Solution: $-\dfrac{3}{10}$.

When $x = 0$, $y = 10$. So we have $10 = \dfrac{0x - 3}{0 + k} \Rightarrow k = -\dfrac{3}{10}$.

Chapter 4 Absolute Values

1. DEFINITION

The absolute value of a number x is denoted as $|x|$. It is the distance from the point x to zero on the number line.

As shown in the figure below, $|3|$ indicates that the distance from 3 to 0 is 3 and $|-3|$ indicates that the distance from –3 to 0 is 3.

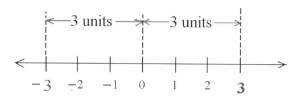

2. PROPERTIES OF ABSOLUTE VALUE:

(1). $|x| = \begin{cases} x & \text{if } x > 0 \\ 0 & \text{if } x = 0 \\ -x & \text{if } x < 0 \end{cases}$ or $|x| = \begin{cases} x & \text{if } x \geq 0 \\ -x & \text{if } x < 0 \end{cases}$ or $|x| = \begin{cases} x & \text{if } x > 0 \\ -x & \text{if } x \leq 0 \end{cases}$

(2). For any real numbers x and y: $|x| = |y|$ If and only if $x = y$ or $x = -y$.

(3). $|x| \geq 0$. The equality holds if and only if $x = 0$.

(4). $|-x| = |x|$

(5). $|x - y| = |y - x|$

(6). $|xy| = |x| \cdot |y|$, and $\left|\dfrac{x}{y}\right| = \dfrac{|x|}{|y|}$ $y \neq 0$.

Example 1. $|b - a| = |b| - |a|$ is true
A. For all real values of a and b.   B. If a and b are positive reals.
C. For no real values of a and b D. If $b = 0$.   E. If $a = 0$ or $a = b$.

Solution: E.
If $a = 0$, then $|b - 0| = |b| - |0| = |b|$. If $a = b$, then $|b - a| = 0 = |b| - |a|$.

3. GRAPHING ABSOLUTE VALUES

(1). To get the graph of $y = |f(x)|$, we first plot $y = f(x)$. Next, we flip the part that is below x-axes up about the x-axes. Note that the part below the x-axes should not be there anymore. The other parts of the graph do not change.

(2). Graph of $y = |kx + b|$: Draw $y_1 = kx + b$ (Figure a). Flip the part of the graph below the x-axes up about the x-axes (Figure b). The solid lines comprise of the graph of $y = |kx + b|$ (Figure c).

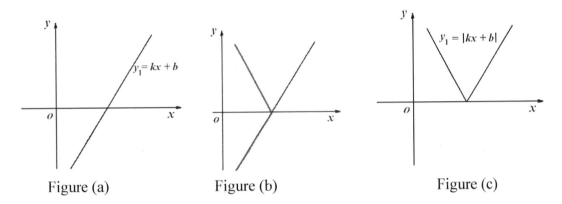

Figure (a)   Figure (b)   Figure (c)

(3). To get the graph of $y = f(|x|)$, we first plot $y = f(x)$ (Figure d). Then we flip the part that is on the right side of the y-axes to the left of the y-axes (Figure e). The solid lines comprise of the graph of $y = f(|x|)$ (Figure f). Note that the part originally on the left of the y-axes is erased while the part on the right of the y-axes is still kept.

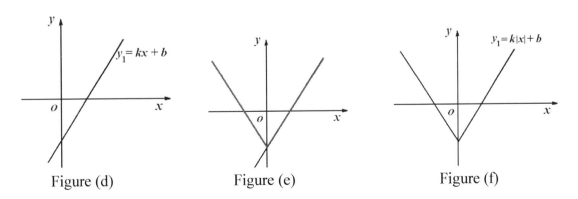

Figure (d)   Figure (e)   Figure (f)

(4). Some graphs of absolute values

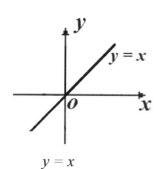
$y = x$

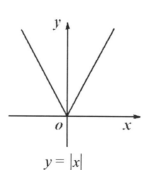
$y = |x|$

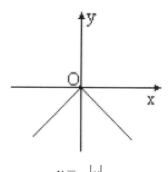
$y = -|x|$

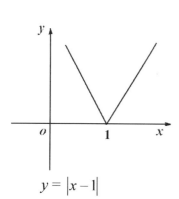
$y = |x - 1|$

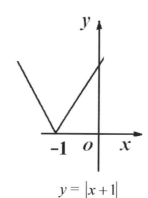
$y = |x + 1|$

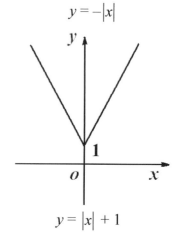
$y = |x| + 1$

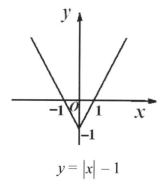
$y = |x| - 1$

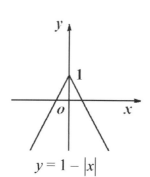
$y = 1 - |x|$

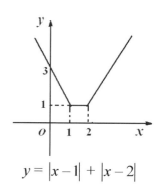
$y = |x - 1| + |x - 2|$

Example 2. The equation $||x + 3| - 2| = p$, where p is a constant integer has exactly three distinct solutions. Find the value of p.

A. 0 B. 1 C. 2 D. 3 E. 4

Solution: C.

The following sequence of graphs gives a progression that leads directly to the answer.

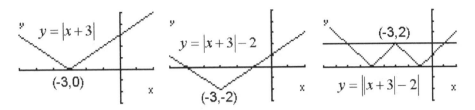

4. FINDING THE ENCLOSED AREA

Theorem: If a and b are real numbers and c is positive integer, the area of the region formed by $|x-a| + |y-b| = c$ is the same as the area formed by $|x| + |y| = c$.

CASE 1: $a|x| + b|y| = c$

Basic procedure:

Let $y = 0$, and our original equation becomes $|x| = \dfrac{c}{a}$. The solutions to this equation are the two points on x-axis: $x_1 = \dfrac{c}{a}$, $x_2 = -\dfrac{c}{a}$

Let $x = 0$, and our original equation becomes $|y| = \dfrac{c}{b}$.

The solutions to this equation are the two points on y-axis: $y_1 = \dfrac{c}{b}$, $y_2 = -\dfrac{c}{b}$.

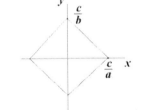

Then, we connect these four points to form a rhombus. The enclosed area will be the area of this rhombus: $A = \dfrac{(2 \times \dfrac{c}{a}) \times (2 \times \dfrac{c}{b})}{2} = 2 \times \dfrac{c}{a} \times \dfrac{c}{b}$ (4.1)

CASE 2. $|x-y| + |x+y| = c$

Basic procedure:
Let $y = 0$, and our original equation becomes
$|x| + |x| = c \Rightarrow |x| = \dfrac{c}{2}$

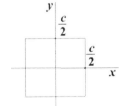

Similarly we also get $|y| = \dfrac{c}{2}$

The enclosed area is then the area of a square of side length $(2 \times \dfrac{c}{2})$.

The area is:

$$A = (2 \times \dfrac{c}{2}) \times (2 \times \dfrac{c}{2}) = c^2 \qquad (4.2)$$

Example 3. Find the number of square units in the area of the region determined by $|x - 3| + |y + 6| = 5$.

Solution: 50.

We can ignore -3 and 6 in the given equation, and re-write the equation as $|x| + |y| = 5$.

(-3 and 6 are only distracters. They shift the figure along the Cartesian plane but do not affect the area of the figure).

We divide both sides by 5 and then use formula (4.1). The enclosed area will be $A = 50$.

Example 4. Find the number of square units in the area of the region determined by $|2x - 3| + |3y + 6| = 12$.

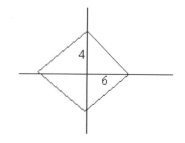

Solution: 48.

Ignoring -3 and 6, we can re-write equation as $|2x| + |3y| = 12$, or $2|x| + 3|y| = 12$.

Divide both sides by 12 to obtain: $\dfrac{|x|}{6} + \dfrac{|y|}{4} = 1$

$A = 2 \times 6 \times 4 = 48$.

Example 5. What is the number of square units in the area of the region determined by $|x - y| + |x + y| = 4$?

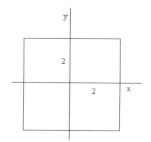

Solution: 16.

Let $y = 0$, and our original equation becomes

$|x| + |x| = 4 \quad \Rightarrow \quad |x| = 2$.

Similarly we get $|y| = 2$.

By formula (4.2), the enclosed area $A = 4 \times 4 = 16$.

5. SOLVING ABSOLUTE VALUE EQUATIONS

$$(|x-a|+|x-b|=c) \quad (5.1)$$

$|x-a|$ is the distance from x to point a, and $|x-b|$ is the distance from x to point b. We call points a and b "critical points". a, b, and c are real numbers.

Case I: If $|b-a| = c$, (we denote $|b-a|$ as the critical distance) equation (5.1) has infinite many solutions. (No matter where we move x along the x-axis between a and b, the distance represented by $|x-a|+|x-b|$ will always equal the value of c).

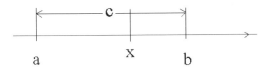

Case II: If $|b-a| < c$, equation (5.1) has two solutions. (One is to the left side of a, and the other solution is to the right side of b).

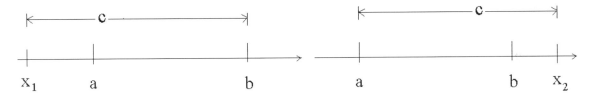

Case III: If $|b-a| > c$, equation (5.1) has no solutions. (No matter where we put x, the equation will not be satisfied).

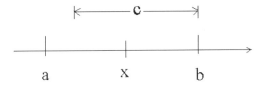

Example 6: Find all integer solutions to $|x-2|+|x-3|=1$.

Solution: 2 and 3.

Method 1:
The critical points are 2 and 3. The critical distance is then $3 - 2 = 1$, which is the same value as the value of the right hand side of the equation. This means that the equation has infinitely many solutions. However, the integer solutions can only be achieved at the two ends, where $x = 2$ and $x = 3$.

Method 2:
We use the basic formula.
We can write the equation as: $|x - 2| = 1 - |x - 3|$. According to the basic formula, we can write:

$x - 2 = 1 - |x - 3|$ (1)
$x - 2 = -(1 - |x - 3|)$ (2)

Re-write equation (1) as $|x - 3| = 3 - x$. Thus, $x - 3 = 3 - x$ or $x - 3 = -(3 - x) \Rightarrow x = 3$.
Re-write equation (2) as $|x - 3| = x - 1$. Thus, $x - 3 = x - 1$ or $x - 3 = -(x - 1) \Rightarrow x = 2$.
The solutions will be $x = 2$ and $x = 3$.

Method 3:
We have the following three intervals:
$\{-\infty, 2], [2, 3], [3, \infty\}$.
On $\{-\infty, 2]$, we have $|x - 2| = 2 - x$ and $|x - 3| = 3 - x$.
Our original equation thus becomes $2 - x + 3 - x = 1$. Solve for x, $x = 2$.

On $[2, 3]$, we have $|x - 2| = x - 2$ and $|x - 3| = 3 - x$.
Our original equation thus becomes $x - 2 + 3 - x = 1$ from where we obtain the identity $1 = 1$. This means that all the numbers on this interval are solutions to this equation.

On $[3, \infty\}$, we have $|x - 2| = x - 2$ and $|x - 3| = x - 3$.
Our original equation thus becomes $x - 2 + x - 3 = 1$ from where $x = 3$.
Upon assembling all this, the integer solutions will be $x = 2$ and $x = 3$.

Example 7. What is the sum of the solutions of $||x + 2| - 5| = 4$?
A. 0 B. –1 C. –2 D. –4 E. –8

Solution: (E).
Method 1:
Case 1: $|x+2| - 5 = 4$ $\Rightarrow$ $|x+2| = 9$.
The solutions are $x = 7$ or $x = -11$.
Case 1: $|x+2| - 5 = -4$ $\Rightarrow$ $|x+2| = 1$.
The solutions are $x = -3$ or $x = -1$.
The sum of them is $7 - 11 - 1 - 3 = -8$.

Method 2:
If $x + 2 > 0$, $x > -2$ and $|x + 2 - 5| = 4$ $\Rightarrow$ $|x - 3| = 4$.
The solutions are $x = 7$ or $x = -1$.

If $x + 2 < 0$, $x < -2$ and $|-x - 2 - 5| = 4$ $\Rightarrow$ $|x + 7| = 4$.
The solutions are $x = -3$ or $x = -11$.

The sum of them is $7 - 1 - 3 - 11 = -8$.

Example 8. Which of following statements is true about the equation $|x^2 - 2x - 3| = x + 2$?
A. There are no solutions. B. There is only one solution.
C. There are exactly two solutions. D. There are exactly three solutions.
E. There are exactly four solutions.

Solution: E.
Draw two graphs of $y = |x^2 - 2x - 3|$ and $y = x + 2$, and count the number of intersection points of the graphs. Alternatively, square both sides of the equation to get $(|x^2 - 2x - 3|)^2 - (x + 2)^2 = 0$, and factor it to get $(x^2 - x - 1)(x^2 - 3x - 5) = 0$. Each quadratic equation has two real solutions, and therefore the equation has four real solutions. Checking to be sure that there are no extraneous roots, we find that there are indeed four solutions.

Example 9. The sum of the solutions of $|3x - 8| = |2x - 7|$ is
A. 1   B. 3   C. 4   D. 7   E. None of these

Solution: C.
Method 1:
Case 1: $|3x - 8| = 2x - 7$
Sub case 1: $3x - 8 = 2x - 7$ $\Rightarrow$ $x = 1$.
Sub case 2: $3x - 8 = -(2x - 7)$ $\Rightarrow$ $x = 3$.

Case 2: $|3x - 8| = -(2x - 7) = 7 - 2x$ (we need to note that $7 - 2x > 0$ or $x < 7/2$.

Sub case 1: $3x - 8 = 7 - 2x$ $\Rightarrow$ $x = 3$.
Sub case 2: $3x - 8 = -(7 - 2x)$ $\Rightarrow$ $x = 1$.

So the solutions are 1 and 3. The sum is 4.

Method 2:
We square both sides of the equation to get rid of the absolute value sign:
$(3x - 8)^2 = (2x - 7)^2$ $\Rightarrow$ $9x^2 + 64 - 48x = 4x^2 + 49 - 48x$
$\Rightarrow$ $5x^2 - 20x + 15 = 0$ $\Rightarrow$ $(5x - 15)(x - 1) = 0$
$\Rightarrow$ $x = 3$ and $x = 1$. We checked and both are the solutions. The sum is 4.

Example 10. Find the sum of all real solutions to the equation $y^2 = |5 - 4y|$.
A. 0　　　　　B. 1　　　　　C. 2　　　　　D. 3　　　　　E. none of these

Solution: E.
If $y^2 = |5 - 4y| \Rightarrow y^2 = 5 - 4y$ or $y^2 = -(5 - 4y) \Rightarrow y^2 + 4y - 5 = 0$ or $y^2 - 4y + 5 = 0$.
The first of these, $y^2 + 4y - 5 = 0 \Rightarrow (y + 5)(y - 1) = 0$ has solutions -5 and 1 with a sum of -4. The second equation has no real solutions.

Algebra II Through Competitions 　　　　　　　　　　　　　　**Chapter 4 Absolute Values**

PROBLEMS

Problem 1. Find the area enclosed by the graph of the following relation: $\dfrac{|x|}{5}+\dfrac{|y|}{3}=1$.

A. 30　　　　B. $\sqrt{34}$　　　　C. 15　　　　D. 60　　　　E. none of these

Problem 2. Find the area enclosed by the graph of the following relation: $\dfrac{|x|}{5}-\dfrac{|y|}{3}=1$

A. 30　　　　B. 34　　　　C. 15　　　　D. 60　　　　E. none of these

Problem 3. The number of real solutions of the equation $|x-2|+|x-3|=3$ is
A. 1　　　　B. 2　　　　C. 3　　　　D. 4　　　　E. many

Problem 4. The sum of the solutions of the equation is $|x-5|=|4x+9|$

A. $\dfrac{11}{4}$　　　　B. 0　　　　C. $-\dfrac{82}{15}$　　　　D. -8　　　　E. none of these.

Problem 5. Find the maximum value of $f(x)=\sqrt{x^2+22x+121}+\sqrt{x^2-26x+169}$ over the interval $-12\leq x \leq 12$.
A. 24　　　　B. 26　　　　C. 28　　　　D. 30　　　　E. 32

Problem 6. The equation $||x+5|-3|=c$, where c is a real number has exactly three distinct solutions. Find the value of c.
A. 0　　　　B. 2　　　　C. 3　　　　D. 5　　　　E. 7

Problem 7. Find the average value of all the solutions to $x^2+x+1=|4x-1|$.

A. $-1/2$　　　B. $\dfrac{-3+\sqrt{2}}{2}$　　　C. 0　　　D. $\dfrac{\sqrt{2}}{4}$　　　E. $1/2$.

Problem 8. Consider the region defined by the inequality $|x|+2|y|\leq 2$. What is the area of this region?
A. 1　　　　B. 2　　　　C. 3　　　　D. 4　　　　E. 5.

Problem 9. The sum of the solutions of $|3x+2|=|2x-7|$ is
A. -8　　　　B. -1　　　　C. 1　　　　D. 8　　　　E. none of these

Problem 10. How many times will the graph of $|x| + |y| = 2$ intersect the graph of $x^2 - y = 2$?
A. 1   B. 2   C. 3   D. 4   E. 5

Problem 11. Compute the area of the region enclosed by the graph $|2x - 10| + |5y - 10| = 20$.
A. 80   B. 85   C. 86   D. 90   E. 96

Problem 12. Which of the following statements is true about the solutions of the equation $|x^2 - 5x| = 6$?
A. The equation has two solutions, both greater than 5.
B. The equation has two solutions, one positive and one negative.
C. The equation has three solutions whose sum is 11.
D. The equation has four solutions whose sum is 10.
E. None of the above statements is true.

Problem 13. For what value of b will $|2x - 1| = x + b$ have a unique solution?
A. $-1/2$ B. 0   C. 1/2   D. 3/2   E. 2

Problem 14. (2008 Arkansas Algebra II) Solve the inequality and write the final answer using interval notation: $|3x - 6| \geq 9$
A. $(-\infty, -1) \cup (5, \infty)$ B. $(-1, 5)$ C. $(-\infty, -1] \cup [5, \infty)$ D. $[-1, 5]$.

Problem 15. (2000 Tennessee Algebra II) Which interval below contains the solution(s) of the equation $3|2x - 3| = x - 1$?
A. $(-2, 9/7]$ B. $(-3/2, 3/2)$ C. $(7/5, 8/5]$ D. $[9/7, 7/5]$ E. $(0, 1)$.

Algebra II Through Competitions   Chapter 4 Absolute Values

SOLUTIONS

Problem 1. Solution: (A).
The area enclosed by the graph is a rhombus (shown in the figure). The length of the diagonals of the rhombus are 10 and 6, so the area is $6 \times 10/2 = 30$.

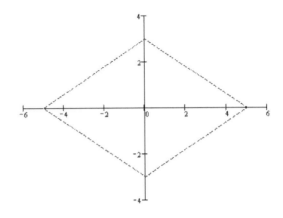

Problem 2. Solution: A.

The given equation can be written as $|y| = \frac{3}{5}|x| - 3$.

This equation represents four equations which can be expressed as $y = \pm\frac{3}{5}x \pm 3$.

The enclosed area is a rhombus whose diagonals have length 10 and 6, so the resulting area is $6 \times 10/2 = 30$.

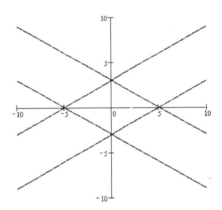

Problem 3. Solution: B.
Method 1: (official solution)
There are just two solutions, $x = 1$ and $x = 4$. We'll condition on three inequalities. If $x < 2$, then $|x - 2| = 2 - x$ and $|x - 3| = 3 - x$, in which case we have $2 - x + 3 - x = 3$ or $-2x = -2$, so $x = 1$. If $2 \le x \le 3$, then $|x - 2| = x - 2$ and $|x - 3| = 3 - x$, in which case we have $x - 2 + 3 - x = 3$, which is impossible. Finally, if $3 < x$, then $|x - 2| = x - 2$ and $|x - 3| = x - 3$, in which case we have $x - 2 + x - 3 = 3$ or $2x = 8$, so $x = 4$.

Method 2:
The critical points are $a = 2$ and $b = 3$. The critical distance is then $3 - 2 = 1 < 3 = c$.
Since $|b - a| < c$, this is the case II of (5.1). The equation has two solutions. (One is on the left side of 2, and the other solution is on the right side of 3).

Problem 4. Solution: C.
We square both sides of the equation to get rid of the absolute value sign:
$(x - 5)^2 = (4x + 9)^2 \Rightarrow x^2 - 10x + 25 = 16x^2 + 72x + 81 \Rightarrow 15x^2 + 82x + 56 = 0$.

By the **Vieta's Theorem**, the sum of the solutions is $x_1 + x_2 = -\dfrac{b}{a} = -\dfrac{82}{15}$.

Problem 5. Solution: B.
Note that $f(x) = |x + 11| + |x - 13|$, whose maximum value over $[-12, 12]$ occurs at (-12); $f(-12) = 26$.

Problem 6. Solution: C.
$||x + 5| - 3| = c \Rightarrow |x + 5| = 3 \pm c \Rightarrow x = -5 \pm (3 \pm c) \Rightarrow x = -2 \pm c$ or $x = -8 \pm c$.
To have three solutions $-2 - c = -8 + c \Rightarrow c = 3$.

Problem 7. Solution: A.
If $x > 1/4$, $4x - 1 = 4x - 1$, otherwise it equals $-(4x - 1)$, so we need to find solutions for $x^2 + x + 1 = 4x - 1$ and $x^2 + x + 1 = -4x + 1$. The solutions to the first equation, $x^2 + x + 1 = 4x - 1 \Leftrightarrow x^2 - 3x + 2 = 0 \Leftrightarrow (x - 2)(x - 1) = 0$ are 2 and 1, and since both are greater than 1/4th, both are valid solutions. The solutions to the second equation $x^2 + x + 1 = -4x + 1 \Leftrightarrow x^2 + 5x = 0 \Leftrightarrow x(x + 5) = 0$ are 0 and -5, and, again, since both are less than 1/4th, both are solutions. The average, then, of the four solutions is $\dfrac{1 + 2 + 0 + (-5)}{4} = \dfrac{-2}{4} = -\dfrac{1}{2}$.

Problem 8. Solution: D.
Divide both sides by 2 to obtain: $\dfrac{|x|}{2} + \dfrac{|y|}{1} = 1$.
$A = 2 \times 2 \times 1 = 4$.

Problem 9. Solution: A.
We square both sides of the equation to get rid of the absolute value sign:
$(3x + 2)^2 = (2x - 7)^2 \Rightarrow 9x^2 + 4 + 12x = 4x^2 + 49 - 28x$
$\Rightarrow 5x^2 + 40x - 44 = 0$
By the **Vieta's Theorem**, the sum of the solutions is
$x_1 + x_2 = -\dfrac{b}{a} = -\dfrac{40}{5} = -8$.

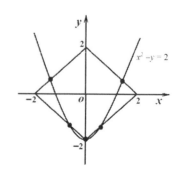

Problem 10. Solution: E.
By graphing we get 5 points of intersection.

Problem 11. Solution: A.

Ignoring two –10's, we can re-write equation as $|2x| + |5y| = 20$, or $2|x| + 5|y| = 20$.

Divide both sides by 20 to obtain: $\dfrac{|x|}{10} + \dfrac{|y|}{4} = 1$.

$A = 2 \times 10 \times 4 = 80$.

Problem 12. Solution: D.
$|x^2 - 5x| = 6 \Rightarrow x^2 - 5x = 6$ or $x^2 - 5x = -6$
$\therefore x^2 - 5x - 6 = 0$ or $x^2 - 5x + 6 = 0$
$\Rightarrow (x - 6)(x + 1) = 0$ or $(x - 2)(x - 3) = 0 \quad \Rightarrow x = 6, -1, 2, 3$
So there are 4 solutions whose sum is 10.

Problem 13. Solution: A.
Method 1:
$|2x - 1| = x + b \Rightarrow 2x - 1 = \pm (x + b) \Rightarrow x = b + 1$ or $3x = 1 - b$
For unique solution: $b + 1 = (1 - b)/3 \Rightarrow 4b = -2 \Rightarrow b = -1/2$.

Method 2:
Since $y = |2x - 1|$ and $y = x + b$ have a unique solution. by graphing we get $b = -1/2$.

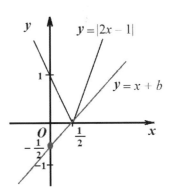

Problem 14. Solution: C.

Case 1: $3x - 6 \geq 9 \quad \Rightarrow \quad 3x \geq 15 \quad \Rightarrow \quad x \geq 5$
Case 2: $3x - 6 \leq -9 \quad \Rightarrow \quad 3x \leq -3 \quad \Rightarrow \quad x \leq -1$.
The solutions are $(-\infty, -1] \cup [5, \infty)$.

Problem 15. Solution: C.
Case 1: $3(2x - 3) = x - 1 \quad \Rightarrow \quad 6x - 9 = x - 1 \quad \Rightarrow \quad 5x = 8 \Rightarrow \quad x = 8/5$.
Case 1: $3(2x - 3) = -(x - 1) \Rightarrow \quad 6x - 9 = -x + 1 \quad \Rightarrow \quad 7x = 10 \Rightarrow \quad x = 10/7$.
The solutions are $x = 10/7$ or $x = 8/5$.
Only the interval C contains both solutions.

Algebra II Through Competitions   Chapter 5 Quadratic Equations and Functions

1. QUADRATIC EQUATIONS

The following equation is called quadratic equation: $ax^2 + bx + c = 0$, where a, b, and c are real numbers with $a \neq 0$.

1.1. Quadratic Formula:

We will derive the quadratic formula below. The method used is called "completing the square" method. The method works for all quadratic equations.
$ax^2 + bx + c = 0$

Since $a \neq 0$, we can divide both sides by a: $x^2 + \frac{b}{a}x + \frac{c}{a} = 0$.

Now we complete the square:

$$x^2 + 2 \times x \times \frac{b}{2a} + \left(\frac{b}{2a}\right)^2 - \left(\frac{b}{2a}\right)^2 + \frac{c}{a} = 0$$

$$x^2 + 2 \quad x \quad y \quad + \quad y^2 = (x+y)^2.$$

$$x^2 + 2 \times x \times \frac{b}{2a} + \left(\frac{b}{2a}\right)^2 = \left(\frac{b}{2a}\right)^2 - \frac{c}{a}$$

We can write the left side as a perfect square, and the right side as a single fraction:

$\left(x + \frac{b}{2a}\right)^2 = \frac{b^2 - 4ac}{4a^2}$. Take the square root of each side: $x + \frac{b}{2a} = \pm\sqrt{\frac{b^2 - 4ac}{4a^2}}$.

Solve for x: $x_{1,2} = \frac{-b}{2a} \pm \sqrt{\frac{b^2 - 4ac}{4a^2}}$.

Simplify: $x_{1,2} = \frac{-b \pm \sqrt{b^2 - 4ac}}{2a}$ \hfill (1.1.1)

Example 1. Solve for x: $x(x - c) = 1 - c$.
A. $x = 1, 1-c$ B. $x = 1, c$ C. $x = -1, c+1$ D. $x = 1, c-1$ E. none of these

Solution: (D).
$x^2 - xc + (c - 1) = 0$

Using the quadratic formula, $x = \frac{c \pm \sqrt{c^2 - 4(c-1)}}{2} = \frac{c \pm (c-2)}{2} = c-1, 1$.

Algebra II Through Competitions   Chapter 5 Quadratic Equations and Functions

Example 2. If the discriminant of $ax^2 + 2bx + c = 0$ is zero, then which of the following statements is true about a, b, and c?
A. they form an arithmetic progression   B. they are unequal   C. only b is negative
D. they are all negative numbers   E. they form a geometric progression

Solution: E.
The discriminant of $ax^2 + 2bx + c = 0$ is $4b^2 - 4ac$, which factors as $4(b^2 - ac)$. If this is zero, then $b^2 = ac$, or $a b/a = c/b$, making b the geometric mean between a and c.

Example 3. For which value of c will the equation $x(x - c) = 1 - c$ have exactly one solution?
A. $c = 0$ B. $c = 1$ C. $c = 2$ D. $c = 0.5$ E. none of these

Solution: C.
In a quadratic Ax^2+Bx+C, only one solution exists if $B^2 - 4AC = 0$. For this equation,
$c^2 - 4(c - 1) = 0$ $\Rightarrow$ $(c - 2)^2 = 0$ $\Rightarrow$ $c = 2$.

Example 4. For which value of b is there only one intersection between the line $y = x + b$ and the parabola $y = x^2 - 3x + 5$?
A. 0   B. 1   C. -2 D. 5   E. none of these

Solution: B.
Intersections occurs when $x + b = x^2 - 3x + 5$ or $x^2 - 4x + (5 - b) = 0$.
There is only one intersection point when the above quadratic has only one solution, i.e. it is a perfect square. This occurs when $(-4)^2 - 4(5 - b) = 0$. $b = 1$.

1.2. Vieta's Theorem

If x_1 and x_2 are two roots of a quadratic equation $ax^2 + bx + c = 0$, $(a \neq 0)$, then

$$x_1 + x_2 = -\frac{b}{a} \qquad (1.2.1)$$

$$x_1 \cdot x_2 = \frac{c}{a} \qquad (1.2.2)$$

Example 5. If m is a positive real number, determine the sum of the roots of the equation $(2x - 3)^2 - m = 0$.
A. 1   B. 3   C. 5   D. 7   E. 9

Solution: B.

$(2x-3)^2 - m = 0 \Rightarrow 4x^2 - 12x + 9 - m = 0 \Rightarrow x^2 - 3x + \dfrac{9-m}{4} = 0$, so the sum of the roots is $-(-3) = 3$.

Example 6. Find the sum of the real roots of the equation, $6x^2 + 11x + k = 0$, if $3x - 2$ is a factor of $6x^2 + 11x + k$.
A. $-7/15$ B. $7/12$ C. $-5/3$ D. $11/3$ E. $-11/6$

Solution: E.
Since $3x - 2$ is a factor, one root is $2/3$. So the second root is also real. By (1.2.1), the sum of the real roots of the equation $6x^2 + 11x + k = 0$ is $-11/6$.

1.2.1. Useful Forms Of Vieta's Theorem

$$x_1^2 + x_2^2 = (x_1 + x_2)^2 - 2x_1 x_2 \tag{1.2.3}$$

$$x_1^2 + x_2^2 = \dfrac{b^2 - 2ac}{a^2} \tag{1.2.4}$$

$$x_1^3 + x_2^3 = (x_1 + x_2)[(x_1 + x_2)^2 - 3x_1 x_2] \tag{1.2.5}$$

$$x_1^3 + x_2^3 = \dfrac{3abc - b^3}{a^3} \tag{1.2.6}$$

$$(x_1 - x_2)^2 = (x_1 + x_2)^2 - 4x_1 x_2 \tag{1.2.7}$$

$$\dfrac{1}{x_1} + \dfrac{1}{x_2} = \dfrac{x_1 + x_2}{x_1 x_2} = -\dfrac{b}{a} \cdot \dfrac{a}{c} = -\dfrac{b}{c} \tag{1.2.8}$$

$$\dfrac{1}{x_1^2} + \dfrac{1}{x_2^2} = \dfrac{b^2 - 2ac}{c^2} \tag{1.2.9}$$

In all the above formulas, x_1 and x_2 represent the two roots of the quadratic equation $ax^2 + bx + c = 0$.

Example 7. Let x_1 and x_2 be the real solutions of the equation $x^2 + bx + c = 0$ with $b \neq 0$. If $x_1 - x_2 = 4$ and $x_1^2 + x_2^2 = 40$, then b must be equal to
A. 8   B. 4   C. 12   D. 8 or 4   E. 8 or -8

Solution: E.

By substitution of the givens, we have $x_1^2 + (x_1 - 4)^2 = 40$ and $x_1 = 6, -2$. Then $x_2 = x_1 - 4 = 2, -6$. Since $-b = x_1 + x_2$ and $b \neq 0$, we must have either $-b = 6 + 2 = 8$ or $-b = -2 - 6 = -8$.

Example 8. Let x_1, x_2 be the two solutions to the equation $2x^2 - x - 2 = 0$. Find the value of $1/x_1 + 1/x_2$.
A. -3 B. -2 C. -1 D. $-1/2$ E. $1/2$.

Solution: (D).
Note that $1/x_1 + 1/x_2 = (x_1 + x_2)/x_1 x_2 = (1/2)/(-1) = -1/2$.

1.2.3. A Different Form Of Vieta Theorem

$$|x_2 - x_1| = \frac{\sqrt{b^2 - 4ac}}{|a|} \tag{1.2.10}$$

If $a > 0$, then (1.2.10) can be simplified as

$$x_2 - x_1 = \frac{\sqrt{b^2 - 4ac}}{a} = \frac{\sqrt{\Delta}}{a} \tag{1.2.11}$$

Example 9. Consider the equation: $x^2 + px + q = 0$. If the roots of this equation differ by 2, then q equals

A. p B. $\dfrac{(p+1)^2}{4}$ C. $\dfrac{p^2+4}{4}$ D. $\dfrac{p^2-4}{4}$ E. $\dfrac{(p-1)^2}{4}$

Solution: (D).

By (1.2.10), $|x_2 - x_1| = 2 = \dfrac{\sqrt{b^2 - 4ac}}{|a|} = \dfrac{\sqrt{p^2 - 4q}}{1} \quad \Rightarrow \quad \sqrt{p^2 - 4q} = 2 \qquad (1)$

Squaring both sides of (1): $p^2 - 4q = 4 \quad \Rightarrow \quad q = \dfrac{p^2 - 4}{4}$.

2. QUADRATIC FUNCTIONS

2.1. The Quadratic Function:

The following function is called the quadratic function:
$$y = f(x) = ax^2 + bx + c \tag{2.1}$$
where a, b, and c are real numbers with $a \neq 0$.

$$y = f(x) = a\left(x + \frac{b}{2a}\right)^2 + \frac{4ac - b^2}{4a} \tag{2.2}$$

2.2. Vertices

$$h = -\frac{b}{2a}, \text{ and } k = \frac{4ac - b^2}{4a} \tag{2.3}$$

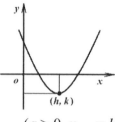
$(a > 0, y_{min} = k)$

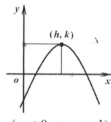
$(a < 0, y_{max} = k)$

Example 10. Find the general expression of a quadratic function that passes through (0, 3) and (8, 3).
A. $f(x) = ax^2 - 8x + 3$ B. $f(x) = a(x^2 - 8x) + 3$ C. $f(x) = a(x^2 - 8x + 3)$
D. $f(x) = x^2 - ax + 3$ E. none of these

Solution: (B).
Initially, let $f(x) = ax^2 + bx + c$. $f(0) = c = 3$. $f(8) = 64a + 8b + c = 3$
$64a + 8b = 0 \Rightarrow b = -8a$.
$f(x) = ax^2 + (-8a)x + 3 = a(x^2 - 8x) + 3$.

Example 11. Find the vertex of a quadratic function that has a for the x^2 coefficient and x intercepts: (2, 0) and (10, 0).
A. $(6, -16a)$ B. $(6, a)$ C. $(a, 0)$ D. $(6/a, -16)$ E. none of these

Solution: A.
Knowing the 2 roots and the constant multiplier, we can say that the form of the quadratic function is $f(x) = a(x - 2)(x - 10)$ This expands to $ax^2 - 12ax + 20a$. The x-value of the vertex is $-b/2a$, which is $-(-12a)/2a = 6$. $f(6) = -16a$, making the vertex $(6, -16a)$.

Example 12. Each spring a 12 meter × 12 meter rectangular garden has its length increased by 2 meters but its width decreased by 50 centimeters. What will be the maximum attainable area of the garden?
A. 144 m²   B. 176 m²   C. 189 m²   D. 200 m²   E. 225 m²

Solution: E.

Algebra II Through Competitions   Chapter 5 Quadratic Equations and Functions

The area will be $(12 + 2x)(12 - 0.5x) = 144 + 24x - 6x - x^2 = 144 + 18x - x^2$, where x is the number of years. This quadratic expression has a maximum at its vertex, which occurs when $x = -b/a = -18/(-2) = 9$. The area when $x = 9$ is $144 + 18(9) - 9^2 = 225$.

2.3. Distance Between Two X-Intercepts Of A Parabola

The quadratic function $y = ax^2 + bx + c$ meets the x-axes at two points $A(x_1, 0)$ and $B(x_2, 0)$. x_1 and x_2 satisfy $ax^2 + bx + c = 0$.

The distance between A and B is: $|AB| = |x_2 - x_1| = \dfrac{\sqrt{b^2 - 4ac}}{|a|} = \dfrac{\sqrt{\Delta}}{|a|}$ (2.4)

Example 13. Find the values of m so that the difference between the roots of $3x^2 + mx - 12$ is 5.

A. 0   B. ± 4 C. ± 7 D. ± 9 E. none of these

Solution: D.
Method 1 (official solution): For this quadratic, we want the following equation to be true

$$\dfrac{-B + \sqrt{B^2 - 4AC}}{2A} - \dfrac{-B - \sqrt{B^2 - 4AC}}{2A} = 5 \quad \Rightarrow \quad \dfrac{\sqrt{B^2 - 4AC}}{A} = 5 \quad \Rightarrow$$

$$\dfrac{\sqrt{m^2 + 144}}{3} = 5 \quad \Rightarrow \quad m = \pm 9.$$

Method 2:
By the formula, we have $5 = \dfrac{\sqrt{b^2 - 4ac}}{|a|} = \dfrac{\sqrt{m^2 - 4 \times 3(-12)}}{3} \quad \Rightarrow$

$15 = \sqrt{m^2 - 4 \times 3(-12)} \quad \Rightarrow \quad m^2 = 81 \Rightarrow m = \pm 9$.

3. SOME USEFUL FORMULAS

3.1. Perfect square trinomial

$(x + y)^2 = x^2 + 2xy + y^2$

$(x - y)^2 = x^2 - 2xy + y^2$

$(x + y)^2 = (x - y)^2 + 4xy$

$(x - y)^2 = (x + y)^2 - 4xy$

$(x + y + z)^2 = x^2 + y^2 + z^2 + 2xy + 2xz + 2yz$

$(x + y + z + w)^2 = x^2 + y^2 + z^2 + w^2 + 2xy + 2xz + 2xw + 2yz + 2yw + 2zw$

Example 14. $(a + b - c)^2 =$
A. $a^2 + b^2 - c^2$ B. $a^2 + b^2 + c^2$ C. $a^2 + b^2 - c^2 + 2ab - 2ac - 2bc$
D. $a^2 + b^2 + c^2 + 2ab - 2ac - 2bc$ E. none of these

Solution: D.

$(a + b - c)^2 = a^2 + b^2 + (-c)^2 + 2ab + 2b(-c) + 2a(-c) = a^2 + b^2 + c^2 + 2ab - 2ac - 2bc.$

3.2. Difference and sum of two squares

$x^2 + y^2 = (x + y)^2 - 2xy$ $x^2 + y^2 = (x - y)^2 + 2xy$
$x^2 - y^2 = (x - y)(x + y)$

3.3. Difference and sum of two cubes

$x^3 + y^3 = (x + y)(x^2 - xy + y^2)$ $x^3 - y^3 = (x - y)(x^2 + xy + y^2)$
$x^n - y^n = (x - y)(x^{n-1} + x^{n-2}y + \cdots + y^{n-1})$ for all n.
$x^n - y^n = (x + y)(x^{n-1} - x^{n-2}y + \ldots - y^{n-1})$ for all even n.
$x^n + y^n = (x + y)(x^{n-1} - x^{n-2}y + \ldots + y^{n-1})$ for all odd n.

Example 15. If $a + b = 3$ and $a^2 + b^2 = 6$, find the numerical value for $a^3 + b^3$.
A. 27/2   B. 18   C. 9   D. 19/2   E. 243

Solution: A.

$(a + b) = 3 \Rightarrow (a + b)^2 = 9$, so $a^2 + 2ab + b^2 = 9$, but we are also told that $a^2 + b^2 = 6$, so $2ab = 3$. $3^3 = (a + b)^3 = a^3 + 3a^2b + 3ab^2 + b^3$, so $27 = a^3 + 3a^2b + 3ab^2 + b^3 = a^3 + b^3 + 3ab(a + b)$, and $a^3 + b^3 = 27 - 3ab(a + b) = 27 - 3\times (3/2) \times 3 = 27/2$.

Algebra II Through Competitions   Chapter 5 Quadratic Equations and Functions

PROBLEMS

Problem 1. For what values of k does the equation $kx^2 + 5x\sqrt{2} - 3 = 0$ have two imaginary roots?

A. $k < \dfrac{-25}{6}$ B. $k < \dfrac{25}{6}$ C. $k \le \dfrac{-25}{6}$ D. $k > \dfrac{25}{6}$ E. $k \ge \dfrac{25}{6}$

Problem 2. Find all real number solutions of $x^4 - 16x^2 - 80 = 0$.

A. $\pm 2\sqrt{5}$ B. $\pm 2\sqrt{5}, \pm 2$ C. ± 2 D. $\pm 2\sqrt{10}, \pm \sqrt{2}$ E. $\pm \sqrt{-2} \pm 2\sqrt{10}$

Problem 3. A quadratic equation $y = ax^2 + bx + c$ is known to pass through the points (0, 5), (2, 11), and (−2, 15). Find the sum of a and b.

A. −7   B. 1   C. 2   D. 3   E. 4

Problem 4. The graph of two parabolas $y = 2x^2$ and $y = x^2 + x + 6$ intersect in two points. An equation for the line that passes through these two points is

A. $x - 2x + 18 = 0$ B. $2x - y - 18 = 0$ C. $2x - y + 12 = 0$
D. $2x - y + 4 = 0$ E. $x - 2y + 12 = 0$

Problem 5. If $1 - \dfrac{4}{x} + \dfrac{4}{x^2} = 0$, $\dfrac{2}{x}$ equals

A. −1   B. 1   C. 2   D. −1 or 2   E. −1 or −2

Problem 6. The product of the values of a, b, and c, that require the graph of $y = ax^2 + bx + c$ to pass through the points (0,−3), (1,6) and (2,9) is

A. 108   B. 56   C. 12   D. −6   E. none of these

Problem 7. Let b be a positive integer and consider $f(x) = 2x^2 + bx + 10$. As b increases, how does the graph of $f(x)$ change?
A. The vertex changes with the x-coordinate decreasing at a slower rate than the y-coordinate.
B. The x-coordinate of the vertex increases by 1 whenever the y-coordinate of the vertex decreases by 6.
C. The vertex changes with the x-coordinate decreasing while the y-coordinate remains constant.
D. The x-coordinate of the vertex decreases by 1 whenever the y-coordinate of the vertex decreases by 4.
E. None of these.

Algebra II Through Competitions   Chapter 5 Quadratic Equations and Functions

Problem 8. A parabola with a vertical axis of symmetry passes through the points (0,7), (4,15), and (12,7). Find the two x-intercepts.
A. {−2, 14}   B. {−1, 13}   C. $\{6-\sqrt{43}, 6+\sqrt{43}\}$ D. $\{6-\sqrt{56}, 6+\sqrt{56}\}$
E. The parabola does not cross the x-axis.

Problem 9. Consider the equation $x^2 + kx + 1 = 0$. A single fair die is rolled to determine the value of the middle coefficient, k. The value of k is the number of dots on the upper face of the die. The probability that the equation will have real, unequal roots is:
A. 1/3   B. 2/3   C. 1/2   D. 3/4   E. None of these.

Problem 10. Jeremy has a bicycle repair shop and needs to know how much to charge for labor. If he charges too much he will lose customers; if he charges too little he won't make much money. At the present, he charges $40 per hour and has 15 hours of work a week. He knows that for every $5 increase in the hourly rate his workload drops by 3 hours, and for every $5 decrease his workload goes up by 3 hours. How much should Jeremy charge per hour to maximize his profits?
A. $42.50   B. $40.00   C. $35.00   D. $32.50   E. $30.00.

Problem 11. Given that (x, y) satisfies $x^2 + y^2 = 9$, what is the largest value of $x^2 + 3y^2 + 4x$?
A. 22   B. 24   C. 36   D. 27   E. 29

Problem 12. If $(mx + 7)(5x + n) = px^2 + 15x + 14$, what is $m(n + p)$?
A. −10   B. −520   C. 480   D. 2   E. 520

Problem 13. What is the largest possible value of y if $y = -x^2 + 12x$ and x is a real number?
A. 0   B. 6   C. 12   D. 36   E. none of these

Problem 14. If the system $\begin{cases} y = x^2 + 2x + 3 \\ y = kx - 1 \end{cases}$ has exactly one solution, then $k =$
A. −6   B. −2   C. −2 or 6   D. 2 or −6   E. none of these

Problem 15. Suppose x is a complex number satisfying the equation $x + 1/x = 1$. What is the value of $x^3 + 1/x^3$?
A. −2   B. −1   C. 0   D. 1   E. 2

Algebra II Through Competitions   Chapter 5 Quadratic Equations and Functions

Problem 16. If $xy = 2$ and $x^2 + y^2 = 5$, then $\dfrac{x}{y} + \dfrac{y}{x}$ is equal to

A. 0   B. $\dfrac{5}{2}$ C. 1   D. $\sqrt{\dfrac{5}{2}}$ E. $\dfrac{2}{5}$

Problem 17. If the equation $x^2 - bx + c = 0$ has two real solutions, $x = p$ and $x = q$, and $p \geq q$, then $p - q = ?$
A. $b^2 - 4c$ B. $b^2 - 2c$ C. $\sqrt{b^2 - 4c}$ D. $\sqrt{b^2 + 4c}$ E. $\sqrt{b^2 - 2c}$

Problem 18. If the vertex of the graph of $y = x^2 + 4x - 7$ is (h, k), then what is $h + k$?
A. −11   B. −12   C. −13   D. −14   E. −15

Problem 19. The line $y = 12 - 2x$ intersects the parabola $y = 5 + 6x - x^2$ at two points. Find the distance between these two points.
A. $6\sqrt{5}$ B. 10   C. $5\sqrt{2}$ D. $8\sqrt{3}$ E. 13

Problem 20. The sum of the cubes of the roots for x of the equation $x^2 - 60x + k = 0$ is 84,420. Find the larger of the two roots for x.

Algebra II Through Competitions   Chapter 5 Quadratic Equations and Functions

SOLUTIONS

Problem 1. Solution: A.
For any quadratic $ax^2 + bx + c = 0$, the roots are imaginary only if $b^2 - 4ac < 0$. Here, $(5\sqrt{2})^2 - 4(k)(-3) < 0$. $k < -25/6$.

Problem 2. Solution: A.
Let $x^2 = b$. Then $b^2 - 16b - 80 = 0$ $\Rightarrow$ $(b - 20)(b + 4) = 0$.
$b = 20$ (–4 is extraneous). So $x = \pm\sqrt{b} = \pm 2\sqrt{5}$.

Problem 3. Solution: B.
Plugging the first point, we get $y = 5 = c$. Then we get from the next two points
$4a + 2b + c = 11$ \qquad (1)
$4a - 2b + c = 15$ \qquad (2)
(1) – (2): $4b = -4$ $\Rightarrow$ $b = -1$
$a = 2$. $a + b = 2 - 1 = 1$

Problem 4. Solution: C.
$y = 2x^2$ and $y = x^2 + x + 6 \Rightarrow 2x^2 = x^2 + x + 6 \Rightarrow x^2 - x - 6 = 0 \Rightarrow (x-3)(x+2) = 0 \Rightarrow$
$x = 3$ or $x = -2$. $y = 2 \cdot 3^2 = 18$ or $y = 2(-2)^2 = 9$.
The parabolas intersect at (3, 18) and (–2, 8). The line through (3, 18) and (–2, 8) has slope $m = 2$ and equation $2x - y + 12 = 0$.

Problem 5. Solution: (B).
$1 - \dfrac{4}{x} + \dfrac{4}{x^2} = 0$ $\Rightarrow$ $x^2 - 4x + 4 = 0$ $\Rightarrow$ $(x-2)^2 = 0$ $\Rightarrow$ $x = 2$ $\Rightarrow$ $\dfrac{2}{x} = 1$.

Problem 6. Solution: A.
All three points satisfy the equation $y = ax^2 + bx + c$, so $-3 = x$, $6 = a + b - 3$, $9 = 4a + 2b - 3 \Rightarrow a + b = 9$, $2a + b = 6 \Rightarrow a = -3$, $b = 12$. $\therefore$ $abc = (-3)(12)(-3) = 108$.

Problem 7. Solution: A.
$f(x) = 2x^2 + bx + 10 = 2(x^2 + \dfrac{b}{2}x + \dfrac{b^2}{4}) + (10 - \dfrac{b^2}{2}) = 2(x + \dfrac{b}{2})^2 + (10 - \dfrac{b^2}{2})$, so when $b = 0$ the vertex is (0, 10) and as b increases, the vertex moves to the left at a rate of $\dfrac{b}{2}$ while it also moves down at a rate of $\dfrac{b^2}{2}$, which, when $b > 1$ is a faster rate than the $\dfrac{b}{2}$.

Algebra II Through Competitions   Chapter 5 Quadratic Equations and Functions

Problem 8. Solution: A.
Since the parabola has a vertical axis, it will be of the form $y = ax^2 + bx + c$.
When we plug the points into this equation, we get the following system of equations:
$7 = a \cdot 0^2 + b \cdot 0 + c$
$15 = a \cdot 4^2 + b \cdot 4 + c$
$7 = a \cdot 12^2 + b \cdot 12 + c$
and the solution to this system is $a = -\frac{1}{4}$, $b = 4$, $c = 7$, so the two x-intercepts are -2 and 14.

Problem 9. Solution: B.
For real, unequal roots $b^2 - 4ac > 0$. So $k^2 - 4 > 0$ $\Rightarrow$ $k = 3, 4, 5, 6$. So the probability that $x^2 + kx + 1 = 0$ will have real, unequal roots is 4/6 or 2/3.

Problem 10. Solution: D.
Let x be the amount he charges per hour. Hours worked = $15 - (3/5) \times (x - 40)$. Thus Profits = $x(15 - 0.6(x - 40)) = x(39 - 0.6x)$. Its maximum is at $x = (1/2) \times (39/0.6) = \$32.5$.

Problem 11. Solution: E.
Replace y^2 with $9 - x^2$ to get $x^2 - 3x^2 + 27 + 4x = 29 - 2x^2 + 4x - 2 = 29 - 2(x-1)^2$, which must be at least 29. It's value at $x = 1$ is 29.

Problem 12. Solution: C.
Since $(mx + 7)(5x + n) = 5mx^2 + 35x + mnx + 7n$, it follows that $n = 2$ and $35x + 2m = 15$. Thus $m = -10$ and $5m = p$ so $p = -50$. Hence $m(n + p) = -10(2 + -50) = 480$.

Problem 13. Solution: D.
By (2.3), the largest possible value of y is $k = \dfrac{4ac - b^2}{4a} = \dfrac{4 \cdot (-1) \cdot 0 - 12^2}{4 \cdot (-1)} = 36$.

Problem 14. Solution: C.

$x^2 + 2x + 3 = kx - 1$ $\Rightarrow$ $x^2 + (2-k)x + 4 = 0$.
Since the sysyem has exactly one solution, $\Delta = 0$ $\Rightarrow$ $(2-k)^2 - 4 \cdot 1 \cdot 4 = 0$ $\Rightarrow$ $k = -2$ or $k = 6$.

Problem 15. Solution: A.
$(x + \dfrac{1}{x}) = 1$ $\Rightarrow$ $(x + \dfrac{1}{x})^3 = 1^3$ $\Rightarrow$ $x^3 + 3x^2 \dfrac{1}{x} + 3x \cdot \dfrac{1}{x^2} + \dfrac{1}{x^3} = 1$

$\Rightarrow \quad x^3 + 3x + 3\dfrac{1}{x} + \dfrac{1}{x^3} = 1 \quad \Rightarrow \quad x^3 + 3(x + \dfrac{1}{x}) + \dfrac{1}{x^3} = 1 \quad \Rightarrow$

$x^3 + 3 \cdot 1 + \dfrac{1}{x^3} = 1 \quad \Rightarrow \quad x^3 + \dfrac{1}{x^3} = -2$.

Problem 16. Solution: B.

$\dfrac{x}{y} + \dfrac{y}{x} = \dfrac{x^2 + y^2}{xy} = \dfrac{5}{2}$.

Problem 17. Solution: C.

By (1.2.11), $x_2 - x_1 = \dfrac{\sqrt{b^2 - 4ac}}{a} = \dfrac{\sqrt{\Delta}}{a} \quad \Rightarrow \quad p - q = \dfrac{\sqrt{(-b)^2 - 4c}}{1} = \sqrt{b^2 - 4c}$.

Problem 18. Solution: C.

By (2.3), $h = -\dfrac{b}{2a}$, and $k = \dfrac{4ac - b^2}{4a}$.

$h + k = -\dfrac{b}{2a} + \dfrac{4ac - b^2}{4a} = -\dfrac{4}{2} + \dfrac{4 \cdot (-7) - 4^2}{4} = -2 - 11 = -13$.

Problem 19. Solution: A.

First we should find the two points. So $12 - 2x = 5 + 6x - x^2$, which simplifies to $x^2 - 8x + 7 = 0$, which factors to $(x-1)(x-7) = 0 \Rightarrow x = 1, 7$ and the points of intersection are $(1, 10)$ and $(7, -2)$. So the distance between these points is

$\sqrt{(7-1)^2 + (-2-10)^2} = \sqrt{6^2 + (-12)^2} = \sqrt{36 + 144} = \sqrt{180} = 6\sqrt{5}$.

Problem 20. Solution: 43.

By the Vieta's Theorem, we have $x_1^3 + x_2^3 = \dfrac{3abc - b^3}{a^3}$, or

$84420 = \dfrac{3a \cdot 1 \cdot (-60) \cdot k - (-60)^3}{1^3} \quad \Rightarrow \quad k = 731$.

Solving $x^2 - 60x + k = 0 \quad \Rightarrow \quad x^2 - 60x + 731 = 0 \quad \Rightarrow \quad (x - 43)(x - 17) = 0$
$x_1 = 43, x_2 = 17$. The larger root is 43.

Algebra II Through Competitions **Chapter 6 Systems of Equations**

1. SYSTEM OF LINEAR EQUATIONS

A group (two or more) of equations is called a system of equations. The solutions of a system of equations should satisfy all the equations.

A system of two equations is as follows:

$$\begin{cases} a_1x + b_1y = c_1 \quad (1)\\ a_2x + b_2y = c_2 \quad (2) \end{cases}$$

Case I: When $\dfrac{a_1}{a_2} \neq \dfrac{b_1}{b_2}$, the system of equations has one unique solution.

Case II: When $\dfrac{a_1}{a_2} = \dfrac{b_1}{b_2} = \dfrac{c_1}{c_2}$, the system of equations has infinitely many solutions.

Case III: When $\dfrac{a_1}{a_2} = \dfrac{b_1}{b_2} \neq \dfrac{c_1}{c_2}$, the system of equations has no solution.

Example 1. Let b be a positive number such that the system
$\begin{cases} ax + 3y = 1\\ 5x + ay = b \end{cases}$ has an infinite number of solutions. Then b equals:

A. 3/5.   B. $\dfrac{\sqrt{15}}{3}$.   C. $\sqrt{15}$.   D. $\dfrac{\sqrt{15}}{2}$ E. no unique answer.

Solution: B.

Two lines with infinite solutions are coincident, so $\dfrac{b}{1} = \dfrac{a}{3} = \dfrac{5}{a} \Rightarrow a = \sqrt{15}$, $b = \dfrac{\sqrt{15}}{3}$.

1.2. Methods to solve system of equations

(1). Substitution method

First solve either equation for one variable. Let's solve (2) for x:

$$x = \dfrac{c_2 - b_2 y}{a_2} \quad (3)$$

Now substitute the result for x in equation (1):

73

Algebra II Through Competitions   **Chapter 6 Systems of Equations**

$$a_1 \times \frac{c_2 - b_2 y}{a_2} + b_1 y = c_1$$

To solve for y, first multiply both sides of the equation by a_2 to eliminate the denominator:

$$a_1 \times (c_2 - b_2 y) + a_2 b_1 y = c_1 a_2 \Rightarrow a_1 c_2 - a_1 b_2 y + b_1 a_2 y = c_1 a_2 \Rightarrow a_1 b_2 y - a_2 b_1 y = a_1 c_2 - c_1 a_2$$

$$\Rightarrow y(a_1 b_2 - a_2 b_1) = a_1 c_2 - c_1 a_2 \Rightarrow y = \frac{a_1 c_2 - c_1 a_2}{a_1 b_2 - a_2 b_1} \qquad (4)$$

Find x by substituting (4) in equation (3):

$$x = \frac{c_2 - b_2 y}{a_2} = \frac{c_2 - b_2 \times \left(\frac{a_1 c_2 - c_1 a_2}{a_1 b_2 - a_2 b_1} \right)}{a_2} \Rightarrow x = \frac{c_1 b_2 - c_2 b_1}{a_1 b_2 - a_2 b_1}.$$

(2). Elimination method

We eliminate x and solve for y by first multiplying both sides of equation (2) by b_1 and both sides of equation (1) by b_2

$$\begin{cases} a_1 b_2 x + b_1 b_2 y = c_1 b_2 & (5) \\ a_2 b_2 x + b_1 b_2 y = c_2 b_1 & (6) \end{cases}$$

(5) – (6):

$$x(a_1 b_2 - a_2 b_2) = c_1 b_2 - c_2 b_1 \Rightarrow x = \frac{c_1 b_2 - c_2 b_1}{a_1 b_2 - a_2 b_1} \qquad (7)$$

Then we get y: $y = \dfrac{a_1 c_2 - c_1 a_2}{a_1 b_2 - a_2 b_1}$

Example 2. Given $x + 3y = a$ and $2x + 5y = b$, solve for x.
A. $x = -5a + 3b$ B. $x = 2a - b$ C. $x = 3a - 5b$ D. $x = 5a - 3b$ E. none of these

Solution: (A).
$$\begin{cases} x + 3y = a & (1) \\ 2x + 5y = b & (2) \end{cases}$$

74

Algebra II Through Competitions **Chapter 6 Systems of Equations**

(1) × 5: $5x + 15y = 5^a$ (3)
$-3 \times (2)$: $-6x - 15y = -3b$ (4)
(3) + (4): $-x = 5a - 3b$ $\Rightarrow$ $x = -5a + 3b$.

Example 3. Solve simultaneously:
$x + 2y + z = 14$
$2x + y + z = 12$
$x + y + 2z = 18$
(A) $x = 1, y = 3, z = 7$ (B) $x = -2, y = 4, z = 8$ (C) $x = -3, y = 12, z = 0$
(D) $x = 6, y = -4, z = 4$ (E) $x = 4, y = 4, z = 5$

Solution: A.

Add the three equations together to get $4x + 4y + 4z = 44$, so $x + y + z = 11$. Subtract this from each of the given equations, one at a time, to get $x = 1, y = 3$, and $z = 7$.

Example 4. Given the following system of equations
$$\begin{cases} \dfrac{1}{x} + \dfrac{1}{y} = \dfrac{1}{3} \\ \dfrac{1}{x} + \dfrac{1}{z} = \dfrac{1}{5} \\ \dfrac{1}{y} + \dfrac{1}{z} = \dfrac{1}{7} \end{cases}$$
What is the value of the ratio z/y ?
A. 17   B. 23   C. 29   D. 31   E. 36

Solution: C.

Add the three equations together to get: $2(\dfrac{1}{x} + \dfrac{1}{y} + \dfrac{1}{z}) = \dfrac{1}{3} + \dfrac{1}{5} + \dfrac{1}{7} = \dfrac{35 + 21 + 15}{105} = \dfrac{71}{105}$

(1)

Then subtract the $2(\frac{1}{x}+\frac{1}{y})=\frac{2}{3}$ from both sides to get $\frac{2}{z}=\frac{71}{105}-\frac{70}{105}=\frac{1}{105}$. So $z = 210$.

Subtracting $2(\frac{1}{x}+\frac{1}{z})=\frac{2}{5}$ from both sides of (1) yields $\frac{2}{y}=\frac{71}{105}-\frac{42}{105}=\frac{29}{105}$. So $y = 210/29$. It follows that $\frac{z}{y}=\frac{210}{\frac{210}{29}}=29$.

Example 5. A collection of nickels, dimes and quarters contains 56 coins altogether. The total value of the coins is nine dollars, and the value of the quarters alone is twice the value of nickels and dimes taken together. The number of nickels in the collection is:
A. less than 5. B. 5, 6 or 7.   C. 7, 8 or 9.   D. 10, 11 or 12.   E. more than 12.

Solution: A.

Let n be the number of nickels, d the number of dimes, and q the number of quarters.
Then $\begin{cases} n + d + q = 56 \\ 0.05n + 0.10d + 0.25q = 9 \\ 0.25q = 2(0.05n + 0.10d). \end{cases}$
$\Rightarrow q = 24, n = 4$.

2. SYSTEM OF NONLINEAR EQUATIONS

A system of equations is called nonlinear system of equations if at least one equation is nonlinear.

Example 6: Solve the system:
$\begin{cases} x + y = 5 \\ xy = 4 \end{cases}$

Solution:

Method 1:
First solve either equation for one variable. Let's solve $x + y = 5$ for x:
$x = 5 - y$
Now substitute the result for x in equation $xy = 4$.
$(5 - y)y = 4 \Rightarrow 5y - y^2 = 4 \Rightarrow y^2 - 5y - 4 = 0 \Rightarrow (y-1)(y-4) = 0$

$y = 1$ or $y = 4$.
The solutions are

$$\begin{cases} x = 1 \\ y = 4 \end{cases} \qquad \begin{cases} x = 4 \\ y = 1 \end{cases}$$

Method 2:
In the quadratic equation $ax^2 + bx + c = 0$, the sum of the root is $-b/a$, and the product of the roots is c/a. Therefore, x and y are two roots of the quadratic equation:
$t^2 - 5t + 4 = 0 \Rightarrow (t-1)(t-4) = 0 \qquad t = 1$ or $t = 4$.
The solutions are

$$\begin{cases} x = 1 \\ y = 4 \end{cases} \qquad \begin{cases} x = 4 \\ y = 1 \end{cases}$$

Example 7. If $x + y = 5$ and $x^2 + 3xy + 2y^2 = 40$, find the value of $2x + 4y$.
A. 16   B. 17   C. 19   D. 18   E. none of the above

Solution: A.

Factoring, we get $x^2 + 3xy + 2y^2 = 40 \Rightarrow (x + y)(x + 2y) = 40$, but since $x + y = 5$, we know that $(x + 2y) = 8$ and $2x + 4y = 16$.

Example 8. Solve the following system of equations simultaneously for x where $x > 0$.
$y = 11x + 3x^2$
$y = 11x^2 - 3x$
What is the value of x?
A. 1   B. 2   C. $1 + \dfrac{\sqrt{7}}{2}$ D. 3/11   E. 7/4

Solution: E.
Setting $11x + 3x^2 = 11x^2 - 3x \Rightarrow 8x^2 - 14x = 0 \Rightarrow x = 0, \dfrac{7}{4}$. Since we are told that x is greater than zero, we accept only the 7/4.

Example 9. Let $x + y = A$ and $x^2 + y^2 = B$. Express $x^3 + y^3$ in terms of A and B.
A. $A \times B$ B. ½ $A(B - A)$ C. $A(B - 3A^2)$ D. ½$A(3B - A^2)$ E. ½$A(B - A^2)$

Solution: D.

$A^3 = x^3 + 3x^2y + 3xy^2 + y^3$ and $A \cdot B = x^3 + x^2y + xy^2 + y^3$
Thus $½A(3B - A^2) = x^3 + y^3$.

Example 10. If x, y, and z satisfy
$$yz = 1$$
$$zx = 2$$
$$xy = 3$$
then what is the value of $x^2 + y^2 + z^2$?
A. 17/2   B. 25/3   C. 33/4   D. 41/5   E. 49/6.

Solution: E.
Multiplying the equations together we get:
$x^2 y^2 z^2 = 6$ (1)

We also can have
$y^2 z^2 = 1$ (2)
$z^2 x^2 = 4$ (3)
$x^2 y^2 = 9$ (4)

(1) ÷ (2): $x^2 = 6$
(1) ÷ (3): $y^2 = 3/2$
(1) ÷ (4): $z^2 = 2/3$
$x^2 + y^2 + z^2 = 6 + 3/2 + 2/3 = 49/6$.

Example 11. Suppose a, b, and c are integers satisfying
$a + b^2 + 2ac = 22$
$b + c^2 + 2ab = 36$
$c + a^2 + 2bc = -2$
What is $a + b + c$?
A. −6   B. −2   C. 4   D. 7   E. 9.

Solution: D.

Add the three equations and note that $a + b + c + (a + b + c)^2 = 22 + 36 - 2 = 56$, so $a + b + c$ must be 7.

Algebra II Through Competitions   Chapter 6 Systems of Equations

PROBLEMS

Problem 1. If $\begin{cases} 2x+3y=-1 \\ x-3y=13 \end{cases}$ is a system of simultaneous equations, then the product of the coordinates of the solution (x, y) is
A. −12   B. −8   C. 8   D. 12   E. none of these

Problem 2. Find the intersection of the lines $2x + 3y = A$ and $x + 2y = B$.
A. $(3B - 2A, A - 2B)$ B. $(3A - 2B, B - 2A)$ C. $(2A - 3B, 2B - A)$
D. $(1/2\,A, 1/2\,B)$ E. $(3B - 2A, 2B - A)$

Problem 3. If the following system of equations
$y\ =\ -x\ +\ 5$
$kx\ +\ y\ =\ 17$
$x\ +\ ky\ =\ -2$
is consistent and k is a constant, then the value of $2k + 1$ is:
A. 2   B. 6   C. 4   D. 3   E. 5

Problem 4. Given: $ax + by = (a-b)^2$ and $ax - by = a^2 - b^2$.
Determine the difference in x and y.
A. $3(a+b)$ B. $3(a-b)$ C. $2(a+b)$ D. $2(a-b)$ E. $(a-b)$

Problem 5. Let a, b, c, and d represent four distinct positive integers where $a^2 - b^2 = c^2 - d^2 = 81$. Find the value of $a + b + c + d$.
A. 27   B. 65   C. 98   D. 108   E. 111

Problem 6. Solve the following system of linear equations and determine the value of x.
$x + y = 7$
$x - 2z = 8$
$y + 3z = 5$
A. 6   B. −17   C. 20   D. 7.5   E. −3

Problem 7. The sum of three numbers is 20. The first is four times the sum of the other two. The second is seven times the third. What is the product of all three?
A. 28   B. 32   C. 60   D. 84   E. 140

Problem 8. If $x + y + z = 7^2$ and $x^2 + y^2 + z^2 = 21$, what is $xy + yz + zx$.

Problem 9. In a course "Leadership in Mathematics" there are several tests. Each test is worth 100 points. After the last test John realized that if he had received 97 points for the last test, his average score for the course would have been a 90, and that if he had made a 73, his average score would have been an 87. How many tests are there in the course?
A. 4 B. 5 C. 6 D. 7 E. 8

Problem 10. Given that $x^2 + y^2 = 10$, $\sqrt[4]{xy} + \sqrt{xy} + 27 = 29$, $x > 0$, and $y > 0$. What is $x + y$?

A) $2\sqrt{3}$ B) $3\sqrt{2}$ C) $\dfrac{1}{2}$ D) $\dfrac{\sqrt{3}}{2}$ E) 8

Problem 11. If $xy = 2$ and $x^2 + y^2 = 5$, then $\dfrac{x}{y} + \dfrac{y}{x}$ is equal to
A. 0 B. 5/2 C. 1 D. $\sqrt{5}/2$ E. 2/5.

Problem 12. If $x^2 + y^2 = 10$ and $x^2 - y^2 = 1$, then what is the value of $|xy|$?

A. $\dfrac{3\sqrt{11}}{2}$ B. 5 C. $3\sqrt{3}$ D. 21/4 E. $\dfrac{5\sqrt{10}}{3}$

Problem 13. If $a = 2b^2$, $b = 4c^3$, and $c = 8d^4$, then which of the following must be true?
A. $a = 2^{23}d^{36}$ B. $a = 2^{23}d^{24}$ C. $a = 2^{11}d^{36}$ D. $a = 2^{11}d^{24}$ E. $a = 2^{11}d^{12}$

Problem 14. In five years Vic will be half as old as his dad. Twenty five years ago he was 1/8 as old as his dad. How old was his dad on the day Vic was born?
A. 20 B. 25 C. 30 D. 35 E. 40

Problem 15. An airplane, flying with a tail wind, travels 1200 miles in 5 hours; the return trip, against the wind, takes 6 hours. Find the cruising speed of the plane and the speed of the wind (assume that both are constant).
A. 220 mph, 20 mph B. 220 mph, 40 mph C. 230 mph, 20 mph
D. 230 mph, 40 mph E. 240 mph, 20 mph

Problem 16. If the length and width of a rectangle were increased by 1, the area would be 84. The area would be 48 if the length and width were diminished by 1. Find the perimeter P of the original rectangle.
A. $10 < P < 20$ B. $20 < P < 30$ C. $30 < P < 40$ D. $40 < P < 50$ E. none of them.

Problem 17.   Find the solution to the system
$2x + 4y - 10z = -2$
$3x + 9y - 21z = 0$
$x + 5y - 12z = 1$
A. (1, 2, –3)   B. (3, –1, –2)   C. 2, –1, 3)   D. (–2, 3, 1)

Problem 18.   Solve the system $\begin{array}{l} x + y = -4 \\ x^2 + y = 2 \end{array}$.

Problem 19. (2004 Tennessee Algebra II) The solution set for the following system is:
$y - 5x = -6$
$16x^2 - 19x - y = 1$
A. {(5/4, 1/4), (1/4, -19/4}   B. {(2, 4)}   C. {(2, 4), (1, -4)}   D. {(4, 25)}
E. {(5/4, -1/4), (1, -4)}

Problem 20.   If the following system is solved, one of the ordered pairs (x, y) in the solution set has a negative value for y. Find that value of y.
$x = -7y$
$x^2 - 3xy = 70$.

SOLUTIONS

Problem 1. Solution: A.
Adding two equations:
$3x = 12 \Rightarrow x = 4$.
Substituting x = 4 into the first equation: $8 + 3y = -1 \Rightarrow y = -3$.
So $xy = 4(-3) = -12$.

Problem 2. Solution: C.
$$\begin{cases} 2x+3y = A \\ x+2y = B \end{cases} \Leftrightarrow \begin{cases} 2x+3y = A \\ 2x+4y = 2B \end{cases} \Rightarrow y = 2B - A.$$
But $x = B - 2y \Rightarrow x = B - 2(2B - A) = 2A - 3B$.

Problem 3. Solution: E.
Using substitution with the first two equations and then the first and third equations results in the system of two equations in two variables.

$$\begin{cases} kx + (-x+5) = 17 \\ x + k(-x+5) = -12 \end{cases} \Rightarrow \begin{cases} kx - x = 12 \\ -kx + x = -2 - 5k \end{cases} \Rightarrow 0 = 10 - 5k \Rightarrow k = 2k+1 = 5$$

Problem 4. Solution: D.
$ax + by = (a-b)^2$ and $ax - by = a^2 - b^2 \Rightarrow 2ax = a^2 - 2ab + a^2$.
$\Rightarrow ax = a^2 - ab \Rightarrow x = a - b$ and $y = b - a \Rightarrow x - y = a - b - (b-a) = 2(a-b)$.

Problem 5. Solution: D.
$a^2 - b^2 = c^2 - d^2 = 81 \Rightarrow (a-b)(a+b) = (c-d)(c+d) = 81$. If a, b, c, d are distinct positive integers, then $1 \times 81 = 81$ or $3 \times 27 = 81$.
Let $c - d = 1$ and $c + d = 81$. Then $c = 41$ and $d = 40$. Let $a - b = 3$ and $a + b = 27$.
Then $a = 15$ and $b = 12$. So $a + b + c + d = 108$.

Problem 6. Solution: C.
Subtracting the second equation from the first yields $y + 2z = -1$. This equation subtracted from the third equation yields $z = 6$, $y = -13$, and $x = 20$.

Problem 7. Solution: A.
Let the three numbers be x, y, and z. The given information tells us that
$y = 7z$
$x = 4(y + z) = 4(7z + z) = 32z$

Hence, $x + y + z = 32z + 7z + z = 40z = 20$. Hence, $z = 1/2$. Substitution gives $y = 7/2$, $x = 16$. Therefore the product of all three numbers is $(16)(7/2)(1/2) = 28$.

Problem 8. Solution:
We know that $(x + y + z)^2 = x^2 + y^2 + z^2 + 2(xy + yz + zx)$.
$2(xy + yz + zx) = (x + y + z)^2 - (x^2 + y^2 + z^2) = 7^2 - 21 = 28$.
$xy + yz + zx = 28/2 = 14$.

Problem 9. Solution: E.
Denote by N the number of tests in the course and by S the sum of John's scores for all but the last test. Then we can write the following system of equations:
$(S + 97)/N = 90$
$(S + 73)/N = 87$
Solving this system for N we obtain $N = 8$.
Alternatively, Since the difference in the averages is 3 and the difference in the last tests is 24, $3 = 24=N$ where n is the total number of exams. So $N = 8$.

Problem 10. Solution: A.
$\sqrt[4]{xy} + \sqrt{xy} + 27 = 29$ can be written as $\sqrt[4]{xy} + \sqrt{xy} - 2 = 0$ \hfill (1)
Let $\sqrt{xy} = a$.
(1) becomes: $a + a^2 - 2 = 0 \Rightarrow (a - 1)(a + 2) = 0$.
Solving we get $\Rightarrow$ ($a = 1$ or $a = -2$ (extraneous).
Thus $\sqrt{xy} = 1 \Rightarrow xy = 1 \Rightarrow 2xy = 2$.
From $x^2 + y^2 = 10$ and $2xy = 2$, we get $(x + y)^2 = 12$.
So is $x + y = 2\sqrt{3}$.

Problem 11. Solution: B.
$\dfrac{x}{y} + \dfrac{y}{x} = \dfrac{x^2 + y^2}{xy} = \dfrac{5}{2}$.

Problem 12. Solution: A.
Adding two equations we get: $2x^2 = 11$ \hfill (1)
Subtracting one from another we get $2y^2 = 9$ \hfill (2)
Multiplying (1) by (2): $4(xy)^2 = 99 \Rightarrow (xy)^2 = 99/4 \Rightarrow |xy| = \dfrac{3\sqrt{11}}{2}$.

Problem 13. Solution: B.
$a = 2b^2 = 2(4c^3)^2 = 2^5 c^6 = 2^5 (8d^4)^6 = 2^{23} d^{24}$.

Problem 14. Solution: D.
Let x denote Vic's current age and y Vic's dads current age. The problem information yields the following linear system: $x + 5 = (1/2)(y + 5)$ and $x - 25 = (1/8)(y - 25)$. The system can be solved for $x = 30$, $y = 65$. So, Vic's dad was $65 - 30 = 35$ on the day Vic was born.

Problem 15. Solution: A.
This is a $d = rt$ problem. Let c = cruising speed and w = wind speed
$1200/(c + w) = 5$
$1200/(c - w) = 6$
$1200/5 = c + w = 240$
$1200/6 = c - w = 200$
From here, it's a system of equations. $c = 220$, $w = 20$.

Problem 16. Solution: C.
$(L + 1)(W + 1) = 84$ and $(L - 1)(W - 1) = 48$ so we get $LW + L + W = 84$ and $LW - L - W = 48$.
Subtracting these last two equations yields $2L + 2W = 36$, so we could go on and solve for L and W, however, we are looking for the perimeter, so we are done.

Problem 17. Solution: D.
The original system of equations can be written as
$x + 2y - 5z = -1$ (1)
$x + 3y - 7z = 0$ (2)
$x + 5y - 12z = 1$ (3)
(1) + (2): $2x + 5y - 12z = -1$ (4)
(4) - (3): $x = -2$ (5)
(1) becomes: $2y - 5z = 1$ (6)
(2) becomes: $3y - 7z = 2$ (7)
Solving (6) and (7) we get $y = 3$ and $z = 1$.

Problem 18. Solution: $(3, -7)$ and $(-2, -2)$.
We subtract the first equation from the second one: $x^2 - x = 6$ $\Rightarrow$ $x^2 - x - 6 = 0$.
Factoring we get $(x - 3)(x + 2) = 0$
So $x = 3$ or $x = -2$. Then $y = -7$ or $y = -2$.
The solutions are $(3, -7)$ and $(-2, -2)$.

Problem 19. Solution: A.

We add two equations together: $16x^2 - 24x + 5 = 0$ $\Rightarrow$ $(4x-5)(4x-1) = 0$

We get $x = 5/4$ or $x = 1/4$.

$y - 5(1/4) = -6$ $\Rightarrow$ $y - 5 \times \dfrac{1}{4} = -6 + \dfrac{5}{4} = -\dfrac{19}{4}$, or

$y - 5(5/4) = -6$ $\Rightarrow$ $y = -6 + \dfrac{25}{4} = -\dfrac{1}{4}$.

The solutions are $\{(5/4, 1/4), (1/4, -19/4)\}$.

Problem 20. Solution: -1.

Substituting $x = -7y$ to the equation $x^2 - 3xy = 70$: $(-7y)^2 - 3(-7y)y = 70$ $\Rightarrow$
$49y^2 + 21y^2 = 70$ $\Rightarrow$ $70y^2 = 70$ $\Rightarrow$ $y^2 = 1$.

Since y is negative, $y = -1$.

1. PROPERTIES OF INEQUALITY

a, b, and c are real numbers.

Transitive:
$$\text{If } a > b \text{ and } b > c, \text{ then } a > c.$$

Addition:
$$\text{If } a > b, \text{ then } a + c > b + c.$$

Multiplication:
$$\text{If } a > b \text{ and if } c > 0, \text{ then } ac > bc.$$
$$\text{If } a > b \text{ and if } c < 0, \text{ then } ac < bc.$$
$$\text{If } a > b > 0, \text{ and } c > d > 0, \text{ then } ac > bd.$$

Division:
$$\text{If } a > b \text{ and } ab > 0, \text{ then } \frac{1}{a} < \frac{1}{b}$$
$$\text{If } a > b > 0, \text{ and } d > c > 0, \text{ then } \frac{a}{c} > \frac{b}{d}$$

Always remember to change the direction of the inequality sign when multiplying or dividing by a negative number.

Example 1. If $a + b > 0$ and $c + d > 0$, which of the following must be true?
A. $a + b + c > 0$ B. $ac + bd > 0$ C. $a + b > c + d$
D. $a^2 + b^2 > c^2 + d^2$ E. $a^2 + b^2 > 0$

Solution: E.
Obviously E should be true no matter what. All others are not certain.

2. LINEAR INEQUALITIES

A linear inequality with one variable is in the form: $ax > b$.
(1). When $a > 0$, the solution is $x > \frac{b}{a}$.
(2). When $a < 0$, the solution is $x < \frac{b}{a}$.
When $a = 0$, if $b < 0$, there are infinite number of solutions.

if $b > 0$, there are no solutions.

The solution to the inequality $a < x < b$ is the same as the solution to the system of inequalities:
$$\begin{cases} x < b \\ x > a \end{cases}$$

Example 2. What is the sum of all integers x that satisfy $-5 \leq x/\pi \leq 10$?
A. 312   B. 324   C. 346   D. 376   E. 412

Solution: D.
The sum is $-15 - 14 - 13 - \cdots 0 + 1 + 2 \cdots + 31$, which is just the sum of 16 consecutive integers $16 + 17 + \cdots + 31 = 16(16+31)/2 = 376$.

Example 3. What is the length of the interval of solutions to the inequality $1 \leq 3 - 4x \leq 9$?
A. 1.75   B. 2.00   C. 2.25   D. 2.50   E. 3.25

Solution: B.
Subtract 3 from all parts to get $-2 \leq -4x \leq 6$, then divide all by -4 to get $1/2 \geq x \geq -3/2$, so the length of the interval is $1/2 - (-3/2) = 2$.

3. POLYNOMIAL INEQUALITIES

Theorem 1:
$(x-a)^{2k\pm1}\varphi(x) > 0$ has the same solution as $(x-a)\varphi(x) > 0$.
$(x-a)^{2k\pm1}$ will have the same sign as $x - a$.

Theorem 2:
$(x-a)^{2k\pm1}\varphi(x) \geq 0$ has the same solution as $(x-a)\varphi(x) \geq 0$.
$(x-a)^{2k\pm1}$ will have the same sign as $x - a$.

Theorem 3:
$(x-a)^{2k}\varphi(x) > 0$ has the same solution as the system of inequalities $\begin{cases} \varphi(x) > 0 \\ x \neq a \end{cases}$.

$(x-a)^{2k}$ will always be greater than zero, if $x \neq a$.

Theorem 4:

$(x-a)^{2k}\varphi(x) \geq 0$ has the same solution as the system of inequalities $\begin{cases}\varphi(x) \geq 0 \\ x = a\end{cases}$.

$(x-a)^{2k}$ will always be greater than zero, if $x \neq a$, and equal to zero if $x = a$.

Steps in solving high degree polynomial inequalities:

(1) Factor the inequality into: $f(x) = (x-x_1)^{a_1}(x-x_2)^{a_2}\cdots(x-x_m)^{a_m}$ where $m \leq n$, $x_1 < x_2 < \cdots < x_m, a_1, a_2, \cdots, a_m \in N$.

(2) Divide a number line into $m + 1$ sections using the roots $x_1, x_2, \cdots, x_m$

(3) Draw a curve, always starting from the upper right region of the number line above x_m and going down below the number line through x_m. Continue drawing the curve below the number line and going up through x_{m-1}, if a_{m-1} is odd. Otherwise, ignore x_{m-1} and go up through x_{m-2}, if a_{m-2} is odd.

(4) Mark each region with "+" or "–" sign. If the region is above the number line, mark the region with a "+" sign. Otherwise mark the region with a "–" sign.

Example 4. What is the greatest rational number n such that $n^2 - 11n + 24 \leq 0$?

Solution: 8.
Step 1: Find the roots of the equation $n^2 - 11n + 24 = 0$: $n = 3$ and $n = 8$.
Step 2: Mark these points on the number line:

Step 3: Draw a curve in this fashion:
Start from the upper right of the number line above the maximum root (in this case, 8), and go down below the number line through 8. Continue the curve below the number line and go up through 3. See below figure for a clearer image.

Algebra II Through Competitions **Chapter 7 Inequalities**

Step 4: Mark the regions with "+" and "–" signs. If the region is above the number line, mark the region with a "+" sign. Otherwise mark a "–" sign.

Since $n^2 - 11n + 24 \leq 0$, the solutions will be: $3 \leq x \leq 8$. The greatest number n is 8.

Example 5. Find the complete solution to the inequality: $x^3 - 8 \leq 7x - 14$.
a) $x \leq -3$ or $x \geq 1$ b) $-3 \leq x \leq 1$ c) $x \leq -1$ or $2 \leq x \leq 3$
d) $-3 \leq x \leq 1$ or $x \geq 2$ e) $x \leq -3$ or $1 \leq x \leq 2$

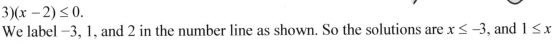

Solution: (E).
We write $x^3 - 8 \leq 7x - 14$ as $x^3 - 7x + 6 \leq 0$, or $(x-1)(x+3)(x-2) \leq 0$.
We label -3, 1, and 2 in the number line as shown. So the solutions are $x \leq -3$, and $1 \leq x \leq 2$.

Example 6. Solve: $(6x - x^2 - x^3)(x^2 - 7x + 10) > 0$

Solution:
The inequality can be factored into $x(x^2 + x - 6)(x^2 - 7x + 10) < 0$, which can be further factored into $x(x+3)(x-5)(x-2)^2 < 0$.
We know that the given inequality is equivalent to the following system of inequalities:
$\begin{cases} x(x+3)(x-5) < 0 \\ x \neq 2 \end{cases}$

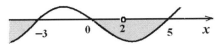

Therefore the set of solutions is $(-\infty, -3) \cup (0, 2) \cup (2, 5)$.

Example 7. If it is true that $a < a^5 < a^4$, then which of the following must be also true?
A. $a < 1$ B. $-1 < a < 0$ C. $0 < a < 1$ D. $a > 1$ E. None of these

Solution: B.
We have two cases:

Case 1: $a < a^5$ $\Rightarrow$ $a^5 - a > 0$ $\Rightarrow$ $a(a^4 - 1) > 0$ $\Rightarrow$ $a(a^2 - 1)(a^2 + 1) > 0$ $\Rightarrow$ $a(a - 1)(a + 1)(a^2 + 1) > 0$

The solutions are $-1 < a < 0 \cup (1, \infty)$

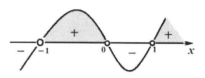

Case 2: $a^5 < a^4$ $\Rightarrow$ $a^5 - a^4 < 0$ $\Rightarrow$ $a^4(a - 1) < 0$.

The solution is $a < 1$.

We plot these solutions as follows: (red: $-1 < a < 0 \cup (1, \infty)$; blue: $a < 1$).

The shaded area is our answer.

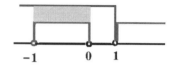

4. ABSOLUTE VALUE INEQUALITIES

(1). $|a| = a$ if $a \geq 0$

(2). $|a| = -a$ if $a < 0$

(3). $|a| < b$ if and only if $-b < a < b$;

(4). $|a| > b$ if and only if $a < -b$ or $a > b$.

Example 8. The number of integral values of x that satisfy the inequality $8 < |3x + 4| < 32$ is
A. 7 B. 8 C. 15 D. 17 E. 22

Solution: C.
Since $8 < |3x + 4| < 32$ $\Rightarrow$ $8 < |3x + 4|$ and $|3x + 4| < 32$

$8 < |3x + 4|$ $\Rightarrow$ $3x + 4 > 8$ or $3x + 4 < -8$

$3x > 4$ or $3x < -12$

$x > 4/3$ or $x < -4$

$|3x + 4| < 32$ $\Rightarrow$ $3x + 4 < 32$ and $3x + 4 > -32$

$3x < 28$ and $3x > -36$

$x < 28/3$ and $x > -12$

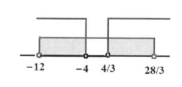

When these are plotted on a number line we see that there are 15 integral values in this set ($-11, -10, -9, -8, -7, -6, -5, 2, 3, 4, 5, 6, 7, 8, 9$).

5. RATIONAL INEQUALITIES

For rational inequalities, it is usually not a good idea to multiply both sides of the inequality by the denominator to get rid of the denominator unless you are sure that the denominator is always greater than zero.

Theorem 1: $\dfrac{f(x)}{g(x)} \geq 0$ and $\begin{cases} f(x) \cdot g(x) \geq 0 \\ g(x) \neq 0; \end{cases}$ have the same solutions.

Theorem 2: $\dfrac{f(x)}{g(x)} \leq 0$ and $\begin{cases} f(x) \cdot g(x) \leq 0 \\ g(x) \neq 0. \end{cases}$ have the same solutions.

Example 9. Solve for x: $\dfrac{x+3}{x-1} > 0$.

A. $x < -3$ or $x > 1$ B. $-3 < x < 1$ C. $x \leq -3$ or $x > 1$
D. $-3 \leq x < 1$ E. none of these.

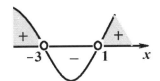

Solution: A.
The given inequality is equivalent to the inequality
$\begin{cases} (x+3)(x-1) > 0 \\ x-1 \neq 0. \end{cases}$
We label -3 and 1 in the number line as shown. The solutions are: $x < -3$ or $x > 1$.

6. SYSTEM OF INEQUALITIES

Example 10: Solve $-4 < x^2 - 5x + 2 < 26$.

Solution:
The given inequality may be split into the following two inequalities:
$\begin{cases} x^2 - 5x + 2 < 26 & (1) \\ x^2 - 5x + 2 > -4 & (2). \end{cases}$

$\begin{array}{l} x^2 - 5x - 24 < 0, \\ x^2 - 5x + 6 > 0, \end{array} \Rightarrow \begin{array}{l} (x-8)(x+3) < 0, \\ (x-2)(x-3) > 0. \end{array} \Rightarrow \begin{array}{l} -3 < x < 8; \\ x > 3,\ x < 2 \end{array}$

The final set of solutions is $-3 < x < 2$, $3 < x < 8$.

PROBLEMS

Problem 1. How many integers are in the solution set of $|4x + 3| < 8$?
A. Zero   B. Two   C. Three   D. Four   E. Infinitely many

Problem 2. For how many integer values of n is $\dfrac{3}{17} < \dfrac{n}{68} < \dfrac{32}{51}$?
A. 28   B. 29   C. 30   D. 32   E. 34.

Problem 3. What is the largest integer k such that $\dfrac{3}{2} \cdot \dfrac{2}{1} \cdot \dfrac{1}{2} \cdot \dfrac{2}{3} \cdot \dfrac{3}{4} \cdots \dfrac{k}{k+1} \geq \dfrac{1}{8}$
A. 20   B. 21   C. 23   D. 24   E. 26

Problem 4. Given $x > y$, and $z \neq 0$, the inequality which is always correct is
A. $x + z > y + z$ B. $xz > yz$ C. $x/z > y/z$ D. $x^2z > y^2z$ E. None of these

Problem 5. If $3 \leq 3t - 18 < 18$, then which of the following must be true?
A. $15 \leq 2t + 1 \leq 20$ B. $8 \leq t < 12$ C. $8 \leq t + 1 < 13$
D. $21 \leq 3t < 24$ E. $t \leq 7$ or $t > 12$

Problem 6. The solution of the inequality $(x + 5)/(2x - 1) > 3$ is an interval. What is the length of the interval?
A. 7/6   B. 8/7   C. 9/8   D. 10/9   E. 11/10

Problem 7. If $x^2 + 2x + n > 10$ for all real numbers x, then which of the following conditions must be true?
A. $n > 11$ B. $n < 11$ C. $n = 10$ D. $n = 1$ E. $n > -11$

Problem 8. Find the values of x for which $x^2 + 3x - 4 > 0$.
A. $x < 1$ and $x > -4$ B. $x > 1$ or $x < -4$ C. $x < -1$ or $x > 4$
D. $x > -1$ and $x < 4$ E. none of A, B, C, or D

Problem 9. Determine the approximate percentage of the interval $[-5,15]$ for which the inequality $x > 4 - \dfrac{7}{x+4}$ is satisfied.

A. 55 %   B. 65 %   C. 60 %   D. 45 %   E. none of the above

Problem 10. The set of real numbers satisfying $\dfrac{1}{x+1} > \dfrac{1}{x-2}$ is:

A. $\{x \mid x > 2\}$ B. $\{x \mid -1 < x < 2\}$ C. $\{x \mid x < 2\}$ D. $\{x \mid x < -1\}$ E. $\{x \mid x > -1\}$.

Problem 11. Solve the following inequality $\dfrac{(x^2 - x - 6)}{x - 5} \geq 0$.

A. $(5, +\infty)$ B. $[-2, 3] \cup [5, +\infty)$ C. $[-2, 3]$
D. $(-\infty, -2] \cup [3, 5)$ E. $[-2, 3] \cup (5, +\infty)$

Problem 12. Determine the solution set to $\dfrac{x^2+1}{x+2} < \dfrac{x+5}{2}$.

A. $\{x \mid -1 < x < 8\}$ B. $\{x \mid (x < -1) \cup (x > 8)\}$ C. $\{x \mid (x < -2) \cup (x > 8)\}$
D. $\{x \mid (x < -2) \cup (-1 < x < 8)\}$ E. $\{x \mid (-2 < x < -1) \cup (x > 8)\}$

Problem 13. Solve for x: $\dfrac{x-5}{x+2} \leq 0$.

A. $x \geq 5$ or $x \leq -2$ B. $x \geq 5$ or $x < -2$ C. $-2 \leq x \leq 5$ D. $-2 < x \leq 5$
E. none of these.

Problem 14. Solve for x. $\dfrac{x^2(x+4)}{x-2} \leq 0$.

A. $-4 \leq x < 2$ B. $-x \leq -4$ or $x > 2$ C. $-4 \leq x \leq 0$ or $x > 2$ D. $x \neq 2$ E. none of these

Problem 15. Let k represent a positive integer. If $k/12$ is a member of the solution set for the inequality $\dfrac{1}{x-2} > 3$, find the sum of all possible distinct values of k.

Problem 16. Anita attends a baseball game in Atlanta and estimates that there are 50,000 fans in attendance. Bob attends a baseball game in Boston and estimates there are 60,000 fans in attendance. A league official who knows the actual numbers attending the two games notes that:
i. The actual attendance in Atlanta is within 10% of Anita's estimate.
ii. Bob's estimate is within 10% of the actual attendance in Boston.

To the nearest 1000, the largest possible difference between the numbers attending the two games is
A. 10,000.   B. 11,000.   C. 20,000.   D. 21,000.   E. 22,000.

Problem 17. the solution of the inequality $\dfrac{2}{x-3} < \dfrac{3}{x+4}$ is
A. $(-\infty, -4) \cup \{3, \infty)$ B. $(-4, \infty)$ C. $(-4, 3) \cup (17, \infty)$ D. $(-\infty, \infty)$ E. $\emptyset$.

Problem 18. Solve the equation for x: $\dfrac{2x-7}{x-5} \le 3$.
A. $(-\infty, 5) \cup [8, \infty)$ B. $[8, \infty)$ C. $(-\infty, 8)$ D. $(5, 8]$

Problem 19. Find the positive number that is **not** a member of the solution set of the inequality $9x^3 + 21x^2 - 17x + 3 > 0$.

Problem 20. Let $p(x) = 3(x-2)^3 (x-1)^4 (x+1)(x+2)^5 (x+5)^2$. Find all x +for which $p(x) < 0$.
A. $(-\infty, -5) \cup (-1, 1) \cup (1, \infty)$ B. $(-\infty, -2) \cup (-1, 2)$ C. $(-5, -2) \cup (-1, 1) \cup (1, \infty)$
D. $(-5, -1) \cup (1, 2))$ E. $(-\infty, -5) \cup (-5, -2) \cup (-1, 1) \cup (1, 2)$.

Algebra II Through Competitions **Chapter 7 Inequalities**

SOLUTIONS

Problem 1. Solution: D.
We have two cases:
Case 1: $4x + 3 < 8$ $\Rightarrow$ $x < 5/4$
Case 2: $4x + 3 > -8$ $\Rightarrow$ $x > -3$.
The integer solutions are $-2, -1, 0,$ and 1.

Problem 2. Solution: C.
For positive numbers a, b, c, d, $a/b < c/d$ if and only if $ad < bc$, it follows that $3 \cdot 17 \cdot 4 < 17 \cdot n$ and $3 \cdot 17 \cdot n < 32 \cdot 4 \cdot 17$. Solving these simultaneously for n yields $13 \leq n \leq 42$ which is 30 values.

Problem 3. Solution: C.
The inequality is equivalent to $\dfrac{3}{k+1} \geq \dfrac{1}{8}$ which is true if and only if $k + 1 \leq 24$. Thus $k = 23$ is the largest integer satisfying the inequality.

Problem 4. Solution: A.
By the addition property, $x + z > y + z$ is always true.

Problem 5. Solution: C.
$3 \leq 3t - 18 < 18$ $\Rightarrow$ $3 + 18 \leq 3t < 18 + 18$ $\Rightarrow$ $21 \leq 3t < 36$ $\Rightarrow$ $7 \leq t < 12$

$\Rightarrow$ $8 = 7 + 1 \leq t + 1 < 12 + 1 = 13$ $\Rightarrow$ $8 \leq t + 1 < 13$.
So the answer is C.

Problem 6. Solution: E.
$\dfrac{x+5}{2x-1} - 3 > 0$ $\Rightarrow$ $\dfrac{x+5}{2x-1} - \dfrac{3(2x-1)}{2x-1} > 0$

$\Rightarrow$ $\dfrac{x+5-6x+3}{2x-1} > 0$ $\Rightarrow$ $\dfrac{-5x+8}{2x-1} > 0$ $\Rightarrow$

$\dfrac{5x-8}{2x-1} < 0$.

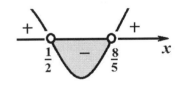

The length of the interval is $\dfrac{8}{5} - \dfrac{1}{2} = \dfrac{11}{10}$.

Problem 7. Solution: A.
Complete the square to find that $f(x) = x^2 + 2x + n = (x + 1)^2 - 1 + n > 10$ if and only if $n > 11$. Alternatively, the minimum value of $x^2 + 2x + n$ occurs at the vertex of the parabola,

whose x-coordinate is given by $-b/2a = -2/2 = -1$. Thus $f(-1) = (-1)^2 + 2(-1) + n > 10$ if and only if $n > 11$.

Problem 8. Solution: B.
Factor to find the zeros of $x^2 + 3x - 4 = (x + 4)(x - 1)$. As shown in the figure, $x > 1$ or $x < -4$.

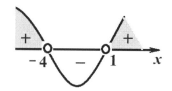

Problem 9. Solution: B.
$x > 4 - \dfrac{7}{x+4} \Leftrightarrow x - 4 + \dfrac{7}{x+4} > 0$. Now simplify the last inequality to $\dfrac{(x-4)(x+4)+7}{(x+4)} > 0$, or $\dfrac{x^2-9}{(x+4)} > 0$. Now factor:

$\dfrac{(x-3)(x+3)}{(x+4)} > 0$ \hfill (1)

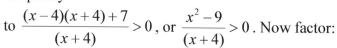

The inequality (1) is equivalent to the inequality $\begin{cases}(x+3)(x-3)(x+4) > 0 \\ x+4 \neq 0.\end{cases}$

From here we see that on the interval $[-5, 15]$, the expression on the left is positive on the intervals $(-4, -3)$ and $(3, 15)$. This represents 13/20ths of the entire interval, or 65%.

Problem 10. Solution: B.
$\dfrac{1}{x+1} > \dfrac{1}{x-2} \Rightarrow \dfrac{x-2}{(x+1)(x-2)} - \dfrac{x+1}{(x+1)(x-2)} > 0 \Rightarrow \dfrac{-3}{(x+1)(x-2)} > 0 \Rightarrow$
$\dfrac{3}{(x+1)(x-2)} < 0$.

The inequality is equivalent to the inequality $\begin{cases}(x+1)(x-2) < 0 \\ x+1 \neq 0, \ x-2 \neq 0\end{cases}$.

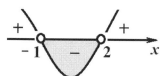

The solution is $-1 < x < 2$.

Problem 11. Solution: E.

$\dfrac{(x^2-x-6)}{x-5} \geq 0 \Rightarrow \dfrac{(x-3)(x+2)}{(x-5)} \geq 0$

$\Rightarrow -2 \leq x \leq 3$ or $x > 5$.

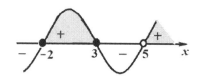

Problem 12. Solution: D.
The best way to look at this is to get an expression which is less than zero, so
$$\frac{x^2+1}{x+2} < \frac{x+5}{2} \Leftrightarrow \frac{x^2+1}{x+2} - \frac{x+5}{2} < 0 \Leftrightarrow \frac{2(x^2+1)-(x+2)(x+5)}{2(x+2)} < 0.$$

Simplify the numerator to get $\frac{2x^2+2-(x^2+7x+10)}{2(x+2)} < 0 \Leftrightarrow \frac{x^2-7x-8}{2(x+2)} < 0$

$\Leftrightarrow \frac{(x-8)(x+1)}{2(x+2)} < 0.$

This last expression is equivalent to $\begin{cases}(x-8)(x+1)(x+2) < 0 \\ x+2 \neq 0\end{cases}$.

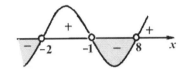

The solution is $\{x \mid (x < -2) \cup (-1 < x < 8)\}$.

Problem 13. Solution: D.
The points we are going to put in the number line is -2 and 5. So the solutions are $-5 \leq x \leq 5$. Note that we include 5 in our solutions since $x = 5$ will make the inequality true.

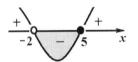

Problem 14. Solution: A.
The points we are going to put in the number line is -4, 2 and 0. Note that any value for x in x^2 does not change the sign of the inequality. So the solutions are $-4 \leq x < 2$.

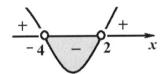

Problem 15. Solution: 78.
$\frac{1}{x-2} > 3 \Rightarrow \frac{1}{x-2} - 3 > 0 \Rightarrow \frac{1}{x-2} - \frac{3(x-2)}{x-2} > 0 \Rightarrow \frac{-(3x-7)}{x-2} > 0$

$\Rightarrow \frac{3x-7}{x-2} < 0.$

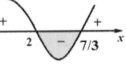

The solution is $2 < x < \frac{7}{3}$, or $2 < \frac{k}{12} < \frac{7}{3} \Rightarrow 24 < k < 28$.

The possible values are 25, 25, and 27. The sum is $25 + 26 + 27 = 78$.

Problem 16. Solution: E.
Let A denote the number is attendance in Atlanta and let B denote the number is attendance in Boston. We are given $45,000 \leq A \leq 55,000$, and $0.9B \leq 60,000 \leq 1.1B$, so $54,546 \leq B \leq 66,666$. Hence the largest possible difference between A and B is $66,666 - 45,000 = 21,666$, so the correct choice is E.

Problem 17. Solution: C.

$$\frac{2}{x-3} - \frac{3}{x+4} < 0 \Rightarrow \frac{2(x+4) - 3(x-3)}{(x-3)(x+4)} < 0 \Rightarrow \frac{-x+17}{(x-3)(x+4)} < 0$$

$$\Rightarrow \frac{x-17}{(x-3)(x+4)} > 0.$$

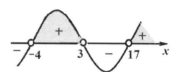

The solutions are $(-4, 3) \cup (17, \infty)$.

Problem 18. Solution: A.

$$\frac{2x-7}{x-5} - 3 \le 0 \Rightarrow \frac{2x-7}{x-5} - \frac{3(x-5)}{x-5} \le 0$$

$$\Rightarrow \frac{2x-7-3x+15}{x-5} \le 0 \Rightarrow \frac{-x+8}{x-5} \le 0 \Rightarrow \frac{x-8}{x-5} \ge 0.$$

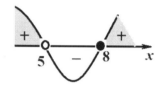

The solutions are $(-\infty, 5) \cup [8, \infty)$.

Problem 19. Solution: 1/3.
$9x^3 + 21x^2 - 17x + 3 = 9x^3 + 27x^2 - 6x^2 - 18x + x + 3 = (9x^3 + 27x^2) - (6x^2 - 18x) + x + 3 = 9x^2(x+3) - 6x(x+3) + (x+3) = (x+3)(9x^2 - 6x + 1) = (x+3)(3x-1)^2$.
The solutions to the inequality $(x+3)(3x-1)^2 > 0$ are: $x > -3$ with $x \ne 1/3$.

Problem 20. Solution: E.
The inequality $3(x-2)^3(x-1)^4(x+1)(x+2)^5(x+5)^2 < 0$ is equivalent to
$(x-2)(x+1)(x+2) < 0$ with $x \ne 1$ and $x \ne -5$.

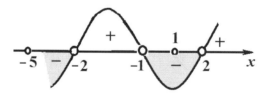

The solutions are: $(-\infty, -5) \cup (-5, -2) \cup (-1, 1) \cup (1, 2)$.

Algebra II Through Competitions Chapter 8 Function Composition and Operations

1. FUNCTIONS

1.1 Definition
A function is a relationship between the independent variable x and dependent variable y. Each value of x corresponds exactly one value of y.

Note two different values of x can have the same value of y, but one value of x cannot have two different values of y.

For example, the price of fruits in a store is a function of fruit kind. One pound of apple (x_1) and one pound of orange (x_2) can have the same price $1.99 per pound ($y$).

But one pound of apple (x) cannot have two different prices at the same time (one price tag says $1.99 per pound ($y_1$) and price tag says $0.99 per pound ($y_2$)).

Example 1. Let $f(x) = \begin{cases} x, & x < 1 \\ x+1, & 1 \leq x < 3 \\ x+3, & x \geq 3 \end{cases}$. Then $f(0) + f(2) + f(4)$ is equal to

A. 0   B. 9   C. 10   D. 21   E. None of these

Solution: C.
$f(0) = 0$.
$f(2) = 2 + 1 = 3$.
$f(4) = 4 + 3 = 7$.
$f(0) + f(2) + f(4) = 0 + 3 + 7 = 10$.

Example 2. If $f(x^2) = 4 \cdot x \cdot f(x + 2) + 3$, what is the value of $f(4)$?

A. -1 B. $-\dfrac{3}{7}$ C. 2   D. $\dfrac{7}{3}$ E. $\dfrac{22}{7}$

Solution: B.
$f(2^2) = 4 \cdot 2 \cdot f(2 + 2) + 3 \Rightarrow f(4) = 4 \cdot 2 \cdot f(2 + 2) + 3 \Rightarrow f(4) = 4 \cdot 2 \cdot f(4) + 3 \Rightarrow f(4) = -3/7$.

Example 3. Let $g(x) = x^2 + b \cdot x + c$ and $g(2) = -6$. Determine $g(5)$.

A. -15 B. $2c - 9$ C. $-1.5c$ D. $2.5c - 10$ E. $-4c$

Solution: C.

$g(2) = -6 = (2)^2 + b\cdot 2 + c \quad \Rightarrow \quad -10 - c = 2b \quad \Rightarrow \quad b = -5 - c/2$.
$g(5) = (5)^2 + b\cdot 5 + c = 25 + 5(-5 - c/2) + c = -5c/2 + c = -1.5c$.

Example 4. (2007 NC Algebra II) If $f(1) = 2$ and $f(n + 1) = (f(n))^2$, what is the value of $f(4)$?

A. 4. B. 16. C. 64. D. 256. E. 65,536.

Solution: D.
$f(2) = f(1 + 1) = (f(1))^2 = 4$ and $f(3) = f(2 + 1) = (f(2))^2 = 16$, and $f(4) = f(3 + 1) = (f(3))^2 = 16^2 = 256$.

1.2. Domain and range

A domain is all the possible values of x. A range is all the possible values of y.

Example 5. Determine the range of the function $f(x) = \dfrac{x+3}{x}$.

Solution: E.
Let $y = F(x) = (x + 3)/x = 1 + 3/x \quad \Rightarrow \quad y - 1 = \dfrac{3}{x}$

Solving for x we get: $x = \dfrac{3}{y-1}$.

We see that y can never equal 1, but can equal anything else, the range of this function is $\{y \in \text{Reals}, y \neq 1\}$.

1.3. Vertical line test

If each vertical line intersects a graph at no more than one point, the graph is the graph of a function.

Figure (a) is the graph of $y^2 = x$ and it is not a function.
Figure (b) is the graph of $y = \sqrt{x}$ and it is a function.

Algebra II Through Competitions Chapter 8 Function Composition and Operations

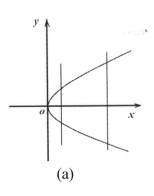

(a)

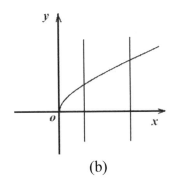

(b)

2. SYMMETRY

2.1. Properties

Symmetric with respect to		
y-axis	x-axis	Origin
$y = f(x) = f(-x)$	$y = f(x) = -f(x)$	$y = f(x) = -f(-x)$
Even Function	Nothing	Odd function

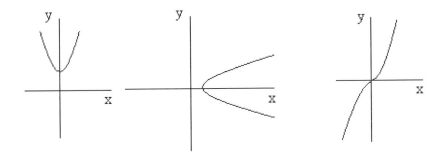

Example 6. A function f is even if for each x in the domain of f, $f(x) = f(-x)$. A function f is odd if for each x in the domain of f, $f(-x) = -f(x)$. Which of the following statement(s) is (are) true?
I. The product of two odd functions is odd.
II. The sum of two even functions is even.
III. The product of an even function and an odd function is odd.
IV. If f is any function and the function F is defined by $F(x) = [f(x) + f(-x)]/2$, then F is even.
A. All statements are true.   B. Only I, II and III are true.   C. Only II, III and IV are true.

Algebra II Through Competitions Chapter 8 Function Composition and Operations

D. Only II and III are true.   E. Only III and IV are true.

Solution: C.
First note that I is false since the product of two odd functions, like $f(x) = x$ and $g(x) = x^3$ is clearly an even function. Next note that the next two are clearly true. The final option, that $F(x) = \dfrac{f(x) + f(-x)}{2}$ is always even bears checking. First, if $f(x)$ is even, then $F(x) = \dfrac{f(x) + f(-x)}{2} = \dfrac{f(x) + f(x)}{2} = f(x)$, which is even. If $f(x)$ is odd, then $F(x) = \dfrac{f(x) + f(-x)}{2} = \dfrac{f(x) - f(x)}{2} = 0$, which is a constant function, hence it is also even.

2.2 Graph of functions

To graph	Shift $y = f(x)$ by c
$y = f(x) + c$	Upward
$y = f(x) - c$	Downward
$y = f(x + c)$	Left
$y = f(x - c)$	Right

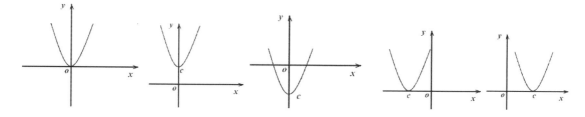

Example 7. Suppose we are given the graph of $y = f(x)$. Which functional expression below describes a graph that is reflected across the x-axis then shifted 3 units to the right and 5 units up?
A. $3 - f(x - 5)$ B. $f(x + 3) - 5$ C. $f(3 - x) + 5$ D. $f(x - 3) + 5$ E. $5 - f(x - 3)$.

Solution: E.
When a graph is flipped over the x-axis, $f(x)$ turns into $-f(x)$. To move a graph to the right, we must subtract from the x-value. To move a graph up we add to the final value, so the desired formula is $-f(x - 3) + 5 = 5 - f(x - 3)$.

Algebra II Through Competitions Chapter 8 Function Composition and Operations

Example 8. If $(-1, 5)$ is a point of the graph of $y = f(x)$, then the graph of $y = f(x - 3)$ contains the point $(2, c)$ where c equals
A. 2   B. 3   C. 4   D. 5   E. 6

Solution: D.
If $(-1, 5)$ is a point on the graph of $y = f(x)$ then the point $(2, 5)$ will be the corresponding point on the graph of $y = f(x - 3)$, since this graph is the result of sliding the original graph to the right 3 units.

3. OPERATIONS ON FUNCTIONS

Given two functions $f(x)$ and $g(x)$, we have the following operations
Sum: $(f + g)(x) = f(x) + g(x)$
Difference: $(f - g)(x) = f(x) - g(x)$
Product: $(fg)(x) = f(x) \cdot g(x)$
Quotient: $(f/g)(x) = f(x)/g(x)$ $g(x) \neq 0$.

The domains of $f + g$, $f - g$, fg include all real numbers of the intersection of the domains of f and g.
The domain of f/g includes all real numbers of the intersection of the domains of f and g for which $g(x) \neq 0$.

4. COMPOSITE FUNCTIONS

The composite function of f and g) is $(f \circ g)(x) = f(x) \circ g(x) = f(g(x))$.

The domain of $f \circ g$ is the set of all numbers x in the domain g such that $g(x)$ is in the domain of f.

Example 9. For $f(x) = x^2 - 1$ and $g(x) = |2x + 3|$, which has the greatest value?
A. $f(g(3))$.   B. $g(f(2))$.   C. $g(f(-1))$.   D. $f(g(-5))$.   E. $g(10)$.

Solution: A.
$f(g(x)) = f(|2x+3|) = |2x+3|^2 - 1 = |4x^2 + 12x + 9| - 1$ and
$g(f(x)) = g(x^2 - 1) = |2(x^2 - 1) + 3| = 2x^2 + 1$. So $f(g(3)) = 80$, $g(f(2)) = 9$, $g(f(-1)) = 3$, $f(g(-5)) = 48$, and $g(10) = 23$, so $f(g(3))$ has the largest value.

Example 10. Let $\otimes$ be an operation defined on functions such that: $f \otimes g(x) = f(g(x)) - g(f(x))$. If $f(x) = x^2 - 1$ and $g(x) = 2x + 1$, find $f \otimes g(x)$.

103

A. $x^2 - 2x - 2$ B. $2x^2 - 4x + 1$ C. $2x^2 + 4x - 2$ D. $2x^2 + 4x + 1$ E. none of the above.

Solution: (D).
$f \otimes g(x) = (2x + 1)^2 - 1 - (2(x^2 - 1) + 1) = 4x^2 + 4x + 1 - 1 - (2x^2 - 2 + 1) = 2x^2 + 4x + 1.$

5. INVERSE FUNCTIONS

5.1 One-to-one function

Horizontal line test: If each horizontal line intersects the graph of a function in no more than one point, the function is one-to-one.

Figure (a) is the graph of $y = \sqrt{r^2 - x^2}$ and it is not one-to-one.
Figure (b) is the graph of $y = \sqrt{r^2 - x^2}$, $x \geq 0$ and it is one-to-one.

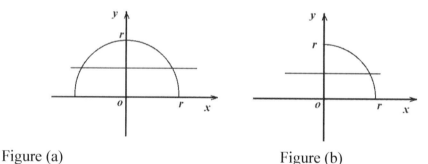

Figure (a)   Figure (b)

5.2 Inverse function

f is a one-to-one function. g is the inverse function of f if $(f \circ g)(x) = x$ for every x in the domain of g and $(g \circ f)(x) = x$ for every x in the domain of f. g is written as f^{-1}.

Notes: (1) Do not confuse -1 in f^{-1} with a negative exponent.
(2) A function f has the inverse function if and only if f is one-to-one.
(3) The graph of f and f^{-1} are symmetric with respect to the line $y = x$, that is, the graph of f^{-1} is the image obtained by reflecting the graph of f about the line $y = x$.

5.3. Steps in find an equation for f^{-1}

(1) Check to make sure that f defined by $y = f(x)$ is a one-to-one function.

Algebra II Through Competitions Chapter 8 Function Composition and Operations

(2) Solve for x. Let $x = f^{-1}$ and then replace y by x.
(3) Check that $(f \circ f^{-1})(x) = x$ and $(f^{-1} \circ f)(x) = x$.

Note: If it is hard to explicitly solve for x, we just exchange x and y to get an expression that define the inverse function implicitly.

Example 11. Suppose f and g are both linear functions, with $y = -2x + 1$ and $f(g(x)) = x$. Find the sum of the slope and the y-intercept of g.

A. -2   B. -1   C. 0   D. 1   E. none of these

Solution: C.
If $f(g(x)) = x$, then $f(x)$ and $g(x)$ are inverse functions. To find the inverse of $y = -2x + 1$, switch x and y and solve for y. The result is that $y = g(x) = -(1/2)x + 1/2$. The sum of the slope and y-intercept is $-1/2 + 1/2 = 0$

Example 12. Find, if possible, the inverse of $f(x) = \dfrac{x+2}{x-3}$.

A. $\dfrac{6x+2}{x-1}, x \neq 1$ B. $\dfrac{3x-2}{x+2}, x \neq -2$ C. $\dfrac{3x+2}{x-1}, x \neq 1$ D. $\dfrac{6x-2}{x+2}, x \neq -2$

E. no inverse exists

Solution: C.
To find the inverse, exchange x and y in the equation $y = \dfrac{x+2}{x-3}$ and solve for y. Solving for y, $xy - 3x = y + 2 \Rightarrow xy - y = 3x + 2 \Rightarrow y(x-1) = 3x + 2 \Rightarrow y = \dfrac{3x+2}{x-1}, x \neq 1$ or

$f^{-1}(x) = \dfrac{3x+2}{x-1}, x \neq 1$.

Example 13. If $f^{-1}(\dfrac{1}{x+1}) = 2x - 3$ then

A. $f(x) = 2x - 4$.   B. $f(x) = 2x - 5$.   C. $f(x) = \dfrac{x-5}{2}$.   D. $f(x) = \dfrac{x+2}{2x+2}$.

E. $f(x) = \dfrac{2}{x+5}$.

Solution: E.

Algebra II Through Competitions Chapter 8 Function Composition and Operations

If $f^{-1}\left(\dfrac{1}{x+1}\right) = 2x-3$, then $f(2x-3) = \dfrac{1}{x+1}$. Now let $u = 2x-3$, so $x = \dfrac{u+3}{2}$

and $f(u) = \dfrac{1}{\dfrac{u+3}{2}+1} = \dfrac{1}{\dfrac{u+5}{2}} = \dfrac{2}{u+5}$, or $f(x) = \dfrac{2}{x+5}$.

Example 14. If $f(x)$ is a linear function and the slope of $y = f(x)$ is $\dfrac{1}{2}$, what is the slope of $y = f^{-1}(x)$?

A. –2   B. $-\dfrac{1}{2}$ C. $\dfrac{1}{2}$ D. 2   E. None of these

Solution: D.

Let $y = f(x) = \dfrac{1}{2}x + b$ $\Rightarrow$ $2y = x + 2b$ $\Rightarrow$ $x = 2y - 2b$

$\Rightarrow$ $f^{-1}(x) = 2x - 2b$.

The slope is 2.

Algebra II Through Competitions Chapter 8 Function Composition and Operations

PROBLEMS

Problem 1. Given: $f(t) = \dfrac{t-1}{t+1}$, evaluate $f^{-1}\left(\dfrac{1}{c}+1\right)$.

A. $1/(1+2c)$ B. $1/(2+c)$ C. $2c-1$ D. $c-2$ E. none of these

Problem 2. (1999 NC Algebra II) Given $f(x) = x^3 + x - 5$, find t so that $f^{-1}(t) = 0.2$
A. -4.792 B. $-\sqrt[3]{4.8}$ C. 2 D. 1.5437 E. none of these

Problem 3. If $f(2x) = \dfrac{2}{2+x}$ for all $x > 0$, then $2f(x) =$

A. $\dfrac{2}{1+x}$ B. $\dfrac{2}{2+x}$ C. $\dfrac{4}{1+x}$ D. $\dfrac{4}{2+x}$ E. $\dfrac{8}{4+x}$

Problem 4. (2001 NC Algebra II) Let $h(x+1) = \dfrac{3h(x)+4}{3}$ for $x = 1, 2, 3,\ldots$ and, $h(1) = -\dfrac{2}{3}$, find $h(3)$.

A. 1   B. 0   C. 5   D. 2   E. 6

Problem 5. Given $f(x) = \dfrac{x+1}{x-1}$. Solve $f^{-1}=\left(\dfrac{1}{x}\right)=3$.

A. $x = 1$ B. $x = 2$ C. $-1/2$ D. $x = \dfrac{1}{2}$ E. $x = \dfrac{1}{3}$

Problem 6. (2003 NC Algebra II) If $f(x) = f(x-2) + x$, and $f(7) = 11$, find $f(5)$.
A. 10   B. 8   C. 6   D. 8   E. 4

Problem 7. Given $f(x) = x + \dfrac{1}{4}$ and $h(x) = x^{\frac{3}{2}}$, find $(h \circ f^{-1})(\dfrac{1}{2})$.

A. $\dfrac{1}{2}$ B. $-\dfrac{1}{4}$ C. $\dfrac{1}{8}$ D. $\dfrac{1}{4}$ E. $-\dfrac{1}{8}$

Problem 8. Let f be a linear function with the properties that $f(1) \le f(2)$, $f(3) \ge f(4)$ and $f(5) = 5$. Which of the following statements is true?
A. $f(0) < 0$.   B. $f(0) = 0$.   C. $f(1) < f(0) < f(-1)$ D. $f(0) = 5$ E. $f(0) > 5$

107

Algebra II Through Competitions Chapter 8 Function Composition and Operations

Problem 9. Let $f(x) = x^2 - 12x + 5$, $g(x) = x + a$ and $f(g(x)) = x^2 + c$. Find the value of c.
A. –31   B. –7   C. $a^2 - 12a$ D. 5 – 12a   E. 5.

Problem 10. A function of the form $f(x) = \dfrac{a}{x+b}$ has the following properties $f(1) = 3$ and $f^{-1}(5) = -1$. What is the value of $f(0)$?
A. $\dfrac{15}{4}$ B. $-\dfrac{3}{5}$ C. 4   D. $\dfrac{15}{7}$ E. None of these

Problem 11. If $f(x) = (x+5)^2 + 8$, then what is the sum of the values of x for which $f(x) = 12$?
A. -10.   B. -7.   C. 10.   D. 20.   E. 297.

Problem 12. Let $f(x)$ be defined as the least integer greater than $x/5$. Let $g(x)$ be defined as the greatest integer less than $x/5$. What is the value of $g(18) + f(102)$?
A. 21.   B. 22.   C. 23.   D. 24.   E. 25.

Problem 13. A function f from the integers to the integers is defined as follows:
$$f(n) = \begin{cases} n+3 & \text{if } n \text{ is odd} \\ n/2 & \text{if } n \text{ is even} \end{cases}.$$
Suppose k is odd and $f(f(f(k))) = 27$. What is the sum of the digits of k?
A. 3.   B. 6.   C. 9.   D. 12.   E. 15.

Problem 14. If $f(1 - t^{-1}) = (5t+1)/t$ then $f(t) = $?
A. $6 - t$ B. $\dfrac{6t-1}{t-1}$ C. $\dfrac{t+1}{4t+1}$ D. $5 + \dfrac{1}{t}$ E. $\dfrac{1-4t}{5t+1}$

Problem 15. If $f(x) = x^2 - 3$ and $g(x) = 5 - x$, then $f(g^{-1}(x))$ is equal to:
A. $x^2 + 10x + 22$ B. $x^2 - 10x + 22$ C. $5 - \sqrt{x+3}$ D. $5 + \sqrt{x+3}$ E. none of these

Problem 16. If $f(x) = 1 - 4x$ and $f^{-1}(x)$ is the inverse function of $f(x)$, then $f(-3) \cdot f^{-1}(-3)$ is equal to
A. 1   B. 3   C. 4   D. 10   E. 13

Problem 17. If $f(x) = 3x + 2$ and $g(x) = 3x + 1$ then $g(f^{-1}(x)) = $
A. $x - 1$ B. $x - 5$ C. $9x + 3$ D. $9x + 7$ E. $9x + 5$

Algebra II Through Competitions Chapter 8 Function Composition and Operations

Problem 18. Suppose $f(0) = 3$ and $f(n) = f(n - 1) + 2$. Let $T = f(f(f(f(5))))$. What is the sum of the digits of T?
A. 6 B. 7 C. 8 D. 9 E. 10.

Problem 19. Let the function f be defined by $f(x) = x^2 + 40$. If m is a positive number such that $f(2m) = 2f(m)$ which of the following is true?

A. $0 < m \le 4$ B. $4 < m \le 8$ C. $8 < m \le 12$ D. $12 < m \le 16$ E. $16 < m$

Problem 20. Function f satisfies $f(x) + 2f(5 - x) = x$ for all real numbers x. The value of $f(1)$ is
A. 7/3 B. 3/7 C. 5/2 D. 2/5 E. None of these

Problem 21. Suppose that $f(n + 1) = f(n) + f(n - 1)$ for $n = 1, 2, \ldots$. Given that $f(6) = 2$ and $f(4) = 8$, what is $f(3) + f(5)$?
A. -18 B. -19 C. -20 D. -21 E. -22

Problem 22. If $f(x) = x^2 - c^2$, solve for x when $(f \circ f)(x) = 0$.
A. $x = \pm c$ B. $x = \pm(c + \sqrt{c}), x = \pm(c - \sqrt{c})$ C. $x = \pm\sqrt{c^2 + c}, x = \pm\sqrt{c^2 - c}$
D. $x = \pm c^2 + c, x = \pm c^2 - c$ E. none of the above.

Problem 23. Which of the following functions is an odd function?
A. $f(x) = 3x^4 - 2x^2 + 5$ B. $f(x) = |x|$ C. $f(\theta) = \cos\theta$
D. $f(x) = 2x^5 - x^3 + x$ E. $f(x) = x^2\sqrt{1 - x^2}$

Problem 24. Find the inverse of: $f(x) = 4x^3 - 8$.

A. $f^{-1}(x) = \sqrt[3]{x + 2}$ B. $f^{-1}(x) = \sqrt[3]{\dfrac{x - 8}{4}}$ C. $f^{-1}(x) = \sqrt[3]{\dfrac{x + 8}{4}}$

D. The inverse does not exist E. None of these.

Problem 25. Let $f(x + 2) = \dfrac{\sqrt[3]{9x - 18}}{3}$. Then $f^{-1}(x) = kx^w + f$ where k, w, and f are positive integers. Find the value of $(2k + 3w + 4f)$.

Algebra II Through Competitions Chapter 8 Function Composition and Operations

SOLUTIONS:

Problem 1. Solution: E.

Let $y = \dfrac{t-1}{t+1}$. If the y's and the t's are switched, then solving for y arrives at the inverse.

$$t = \dfrac{y'-1}{y'+1}$$

$$f^{-1}(t) = y' = \dfrac{t+1}{1-t} \quad \Rightarrow \quad f^{-1}\left(\dfrac{1}{c}+1\right) = \dfrac{\dfrac{1}{c}+2}{-\dfrac{1}{c}} = -1 - 2c.$$

Problem 2. Solution: A.

The input of $f(x)$ will be the output of $f^{-1}(x)$ and the output of $f(x)$ will become the input of f^{-1}. This means that $x = f^{-1}(t) = 0.2$ and $t = f(x) = f(0.2) = -4.792$.

Problem 3. Solution: E.

The x in the denominator of $f(2x) = 2/(2+x)$ represents half of the function's input. So $f(x) = 2/(2+1/2\ x) = 4/(4+x)$. $2f(x) = 8/(4+x)$.

Problem 4. Solution: D.

$h(x+1) = \dfrac{3h(x)+4}{3} = h(x) + \dfrac{4}{3}$, so each term is just four-thirds greater than the one before it. So $h(1) = -\dfrac{2}{3}$, $h(2) = -\dfrac{2}{3} + \dfrac{4}{3} = \dfrac{2}{3}$, and $h(3) = \dfrac{2}{3} + \dfrac{4}{3} = 2$.

Problem 5. Solution: D.

To find the inverse, exchange the x and y and solve for y. Thus $x = \dfrac{y+1}{y-1} \Leftrightarrow xy - x = y + 1 \Leftrightarrow xy - y = x + 1 \Leftrightarrow y = \dfrac{x+1}{x-1}$. From this we see that the function is its own inverse. If we had graphed it, and noticed the symmetry about the line $y = x$, we could have drawn the same conclusion. Thus $f^{-1} = \left(\dfrac{1}{x}\right) = f = \left(\dfrac{1}{x}\right) = \dfrac{\dfrac{1}{x}+1}{\dfrac{1}{x}-1} = \dfrac{x+1}{x-1}$. Setting this equal to 3, we have $\dfrac{x+1}{x-1} = 3 \Leftrightarrow x+1 = 3(x-1) \Rightarrow 2x = 4$, so $x = \dfrac{1}{2}$.

Algebra II Through Competitions Chapter 8 Function Composition and Operations

Problem 6. Solution: E.
Since $f(x) = f(x - 2) + x, f(7) = f(5) + 7$ $\Rightarrow$ $11 = f(5) + 7 \Rightarrow f(5) = 4$.

Problem 7. Solution: C.
To find f^{-1}, $x = y + \dfrac{1}{4} \Rightarrow 4x = 4y + 1 \Rightarrow y = \dfrac{4x-1}{4} \Rightarrow f^{-1}(x) = x - \dfrac{1}{4}$. Then
$(h \circ f^{-1})(\dfrac{1}{2}) = h\left(f^{-1}\left(\dfrac{1}{2}\right)\right) = h\left(\dfrac{1}{2} - \dfrac{1}{4}\right) = h\left(\dfrac{1}{4}\right) = \left(\dfrac{1}{4}\right)^{3/2} = \dfrac{1}{8}$.

Problem 8. Solution: D.
Since f is a linear function and $f(1) \le f(2)$, we know that the slope is greater than or equal to zero. Likewise, since $f(3) \ge f(4)$, the slope is less than or equal to zero. Taken together, this means that the slope must be zero, making the function $f(x) = 0x + 5$, so $f(0) = 5$.

Problem 9. Solution: A.
$f(g(x)) = f(x + a) = (x + a)^2 - 12(x + a) + 5 = x^2 + (2a - 12)x + (a^2 - 12a + 5)$.
Since we are told that $f(g(x)) = x^2 + c$, we know that $2a - 12 = 0$ and $a^2 - 12a + 5 = c$,
so $a = 6$ and $a^2 - 12a + 5 = 36 - 72 + 5 = -31 = c$.

Problem 10. Solution: A.
$f(1) = \dfrac{a}{1+b} = 3$ and $f^{-1}(5) = 1 \Rightarrow f(-1) = \dfrac{a}{-1+b} = 5$, so $a = 3 + 3b$ and $a = -5 + 5b$ and
$3 + 3b = -5 + 5b \Rightarrow 8 = 2b \Rightarrow b = 4$ and $a = 3 + 3(4) = 15$, making $f(0) = \dfrac{15}{0+4}$.

Problem 11. Solution: A.
$12 = (x + 5)^2 + 8$ $\Rightarrow$ $4 = (x+5)^2$ $\Rightarrow$ $\pm 2 = (x + 5)$, so $x = -3$ or $x = -7$.
So $-3 + (-7) = -10$.

Problem 12. Solution: D.
$g(18) = 4$, since 4 is the least integer greater than 18/5. Similarly $f(102) = 20$, since 20 is the greatest integer less than 102/5. So $g(18) + g(20) = 4 + 20 = 24$.

Problem 13. Solution: B.
Since k is odd, $f(k) = k + 3$. Since $k + 3$ is even, $f(k + 3) = f(f(k)) = \dfrac{k+3}{2}$. If $\dfrac{k+3}{2}$ is
odd, then $27 = f(f(f(k))) = f\left(\dfrac{k+3}{2}\right) = \dfrac{k+3}{2} + 3$, which implies that $k = 45$. This is not

Algebra II Through Competitions Chapter 8 Function Composition and Operations

possible because $f((f(45))) = f(f(48)) = f(24) = 12$. Hence $\dfrac{k+3}{2}$ must be even, and $27 = f(f(f(k))) = f\left(\dfrac{k+3}{2}\right) = \dfrac{k+3}{4}$, which implies that $k = 105$. Checking, we find that $f((f(105))) = f(f(108)) = f(54) = 27$. Hence the sum of the digits of k is $1 + 0 + 5 = 6$.

Problem 14. Solution: A.
Let $x = 1 - t^{-1}$ then $t^{-1} = 1 - x$.
$f(1 - t^{-1}) = 5 + t^{-1}$ $\Rightarrow$ $f(x) = 5 + (1-x) = 6 - x$. $\Rightarrow$ $f(t) = 6 - t$.

Problem 15. Solution: B.
$g^{-1}(x) = -x + 5$.
We know that If $f(x) = x^2 - 3$, so $f(g^{-1}(x)) = (g^{-1}(x))^2 - 3 = = (-x + 5)^2 - 3 = x^2 + 25 - 10x - 3 = x^2 - 10x + 22$.

Problem 16. Solution: E.
Since $f(x) = 1 - 4x$, $f(-3) = 1 - 4 \times (-3) = 13$.
Since $f^{-1}(x) = -\dfrac{1}{4}x + \dfrac{1}{4}$, $f^{-1}(-3) = -\dfrac{1}{4} \times (-3) + \dfrac{1}{4} = 1$.
$f(-3) \cdot f^{-1}(-3) = 13 \times 1 = 13$.

Problem 17. Solution: A.
Since $f(x) = 3x + 2$, $f^{-1}(x) = \dfrac{1}{3}x - \dfrac{2}{3}$.
Since $g(x) = 3x + 1$, $g(f^{-1}(x)) = 3 \times f^{-1}(x) + 1 = 3 \times (\dfrac{1}{3}x - \dfrac{2}{3}) + 1 = x - 2 + 1 = x - 1$.

Problem 18. Solution: C.
In fact, $f(n) = 2n + 3$, and $T = f(f(f(f(5)))) = f(f(f(13))) = f(f(29)) = f(61) = 125$.

Problem 19. Solution: B.
Note that $f(2m) = (2m)^2 + 40 = 4m^2 + 40$ and $2f(m) = 2(m^2 + 40) = 2m^2 + 80$. It follows that $2m^2 = 40$, so $4 < m \le 8$.

Problem 20. Solution: A.
We know that $f(x) + 2f(5 - x) = x$.
$f(1) + 2f(5 - 1) = 1$ (1)
$f(4) + 2f(5 - 4) = 4$ (2)
$(2) \times 2 - (1)$: $3f(1) = 7$ $\Rightarrow$ $f(1) = 7/3$.

Algebra II Through Competitions Chapter 8 Function Composition and Operations

Problem 21. Solution: C.
$f(6) = f(5) + f(4) = f(4) + f(3) + f(3) + f(2) = 5f(2) + 3f(1) = 2$ (1)
$f(4) = f(3) + f(2) = f(2) + f(1) + f(2) = 2f(2) + f(1) = 8$ (2)
$(2) \times 3 - (1)$: $f(2) = 22$. Thus $f(1) = 8 - 44 = -36$.
$f(3) + f(5) = f(2) + f(1) + f(4) + f(3) = f(2) + f(1) + f(3) + f(2) + f(2) + f(1)$
$= 4f(2) + 3f(1) = 4 \times 22 + 3 \times (-36) = -20$.

Problem 22. Solution: (C).
$f(f(x)) = f(x^2 - c^2) = (x^2 - c^2)^2 - c^2 = 0$
Let $y = x^2$ and $b = c^2$. $y^2 - 2yb + (b^2 - b) = 0$
$$y = \frac{2b \pm \sqrt{(2b)^2 - 4(b^2 - b)}}{2} = b \pm \sqrt{b}.$$
$x^2 = c^2 \pm \sqrt{c^2} \quad\Rightarrow\quad x = \pm\sqrt{c^2 \pm c}$.

Problem 23. Solution: D.
For odd function, we have $y = f(x) = -f(-x)$. We test and we know that D is the answer.

Problem 24. Solution: C.
Let $f(x) = y$.

$f(x) = 4x^3 - 8 \quad\Rightarrow\quad y = 4x^3 - 8 \quad\Rightarrow\quad y + 8 = 4x^3 \quad\Rightarrow\quad \frac{y+8}{4} = x^3$

$\Rightarrow \quad x = \sqrt[3]{\frac{y+8}{4}}$.

We switch the position of x and y: $f^{-1}(x) = \sqrt[3]{\frac{x+8}{4}}$.

Problem 25. Solution: 31.
Let $x + 2 = X$.
$f(x+2) = \frac{\sqrt[3]{9x-18}}{3} \quad\Rightarrow\quad f(X) = \frac{\sqrt[3]{9(X-4)}}{3}$.
Let $f(X) = y$.
$f(X) = \frac{\sqrt[3]{9(X-4)}}{3} \quad\Rightarrow\quad y = \frac{\sqrt[3]{9(X-4)}}{3} \quad\Rightarrow\quad 3y = \sqrt[3]{9(X-4)} \quad\Rightarrow\quad X = 3y^3 + 4$.
We switch the position of X and y: $f^{-1}(X) = 3X^3 + 4$.
So $k = 3$, $w = 3$ and $f = 4$. $2k + 3w + 4f = 6 + 9 + 16 = 31$.

1. POLYNOMIAL FUNCTION

The general form for a polynomial with degree of n is

$$P(x) = a_n x^n + a_{n-1} x^{n-1} + a_{n-2} x^{n-2} + \cdots + a_2 x^2 + a_1 x + a_0 \qquad (1.1)$$

In this equation, a_n, a_{n-1}, ..., a_1, a_0 are real numbers, where n is a nonnegative integer and $a_n \neq 0$. a_n is called the leading coefficient, and the degree of a polynomial is the value n.

$p(x)$ can also be written as $f(x)$.

2. DEGREE OF POLYNOMIALS:

(1). When $p(x)$ with degree of m and $g(x)$ with degree of n are added together ($m \geq n$), the degree of the resulting polynomial has the degree $k = m$.

(2). When $p(x)$ of the degree m and $g(x)$ with degree of n are subtracted from one another ($m \geq n$), the degree of the resulting polynomial has the degree $k \leq m$.

(3). When $p(x)$ of the degree of m and $g(x)$ of degree of n are multiplied, the degree of the resulting polynomial has the degree of $k = m + n$.

Example 1: Find the degree of the following polynomial: $x^3 + 5x^4 - 13x^2 + 15x - 78$.

Solution: 4.

Rearrange the terms as follows: $5x^4 + x^3 - 13x^2 + 15x - 78$.
The degree of a polynomial is the highest value of the exponents.

Number of Zeros Theorem

A polynomial of degree n has at most n distinct zeros.

Example 2: If a polynomial function with real coefficients has 3 distinct x-intercepts, what is the maximal degree of the polynomial?
A. 3   B. 4   C. 5   D. 6   E. None of these

Solution: E.

By the Number of Zeros Theorem, we know that the polynomial has at least 3 distinct roots. We do not know at most how many distinct zeros of the polynomial.

3. DIFFERENT FORMS OF POLYNOMIALS

There are many different expressions of a polynomial. The following five forms are commonly used:

1. $f(x) = a_0 + a_1 x + \cdots + a_n x^n$.   (3.1)

2. $f(x) = (ax + b)q(x) + r$.   (3.2)
   $f(x) = (ax^2 + bx + c)q(x) + (cx + d)$.

3. $f(x) = a_n(x - x_1)(x - x_2)\cdots(x - x_n)$.   (3.3)

Example 3: A cubic polynomial $p(x)$ with leading coefficient 1 has three zeros, $x = 1$, $x = -1$, and $x = 3$. What is the value $p(2)$?
A. −3   B. −1   C. 1   D. 2   E. 3

Solution: A.

The polynomial is $p(x) = a(x - 1)(x + 1)(x - 3)$, and since the leading coefficient of p is 1, it follows that $a = 1$. Thus $p(2) = 1 \cdot 3 \cdot (-1) = -3$.

Example 4: For the function $f(x) = x^4 + ax^3 + bx^2 + cx + 2$ where −1 is the triple root of $f(x)$, what is a?
A. −2   B. −3   C. 2   D. 5   E. 6

Solution: D.

$f(x) = x^4 + ax^3 + bx^2 + cx + 2 = (x + 1)^3 (x + 2)$.
$f(1) = 1 + a + b + c + 2 = (1 + 1)^3(1 + 2) = 24$ ⇒ $a + b + c = 21$ (1)
$f(-1) = 1 - a + b - c + 2 = 0$ ⇒ $a - b + c = 3$ (2)
$f(-2) = 16 - 8a + 4b - 2c + 2 = 0$ ⇒ $4a - 2b + c = 9$ (3)

(1) − (2): $2b = 18$ ⇒ $b = 9$.
(3) − (2): $3a - b = 6$ ⇒ $3a - 9 = 6$ ⇒ $a = 5$.

4. DIVISION ALGORITHM

When $p(x)$ is divided by $g(x)$, where $g(x)$ has a lower degree than $p(x)$, there exists unique polynomials $q(x)$ and $r(x)$ such that

$$p(x) = g(x) \cdot q(x) + r(x) \tag{4.1}$$

where either $r(x) = 0$ or the degree of $r(x)$ is less than the degree of $g(x)$.

$q(x)$ is called the quotient polynomial and $r(x)$ is called the remainder polynomial or the remainder.

When $p(x)$ is divided by a n-degree polynomial, the remainder is a $(n-1)$-degree polynomial.

When $p(x)$ is divided by a first-degree polynomial, the remainder is a constant c.

When $p(x)$ is divided by a second-degree polynomial, the remainder is in the linear form of $ax + b$.

When $p(x)$ is divided by a third-degree polynomial, the remainder is in the quadratic form of $ax^2 + bx + c$.

When $p(x)$ is divided by a fourth-degree polynomial, the remainder is in the cubic form of $ax^3 + bx^2 + cx + d$.

Example 5: Find the remainder when $x^{100} - 4x^{98} + 5x + 6$ is divided by $x^3 - 2x^2 - x + 2$. Record the product of the coefficients.

Solution: 10.

When $p(x)$ is divided by a third-degree polynomial, the remainder is in the quadratic form of $ax^2 + bx + c$.

$$P(x) = x^{100} - 4x^{98} + 5x + 6 = (x^3 - 2x^2 - x + 2) \cdot Q(x) + (ax^2 + bx + c)$$
$$P(1) = 1^{100} - 4 \times 1^{98} + 5 \times 1 + 6 = 8 = a + b + c$$
$$P(-1) = (-1)^{100} - 4 \times (-1)^{98} + 5 \times (-1) + 6 = -2 = a - b + c$$
$$P(2) = 2^{100} - 4 \times 2^{98} + 5 \times 2 + 6 = 16 = 4a + 2b + c$$

Solving the system of equations, we get $a = 1$, $b = 5$, $c = 2$.
The product of the coefficients is 10.

Algebra II Through Competitions Chapter 9 Polynomials

5. REMAINDER THEOREM:

In (4.1), $p(x) = g(x) \cdot q(x) + r(x)$, if we set the polynomial $g(u)$ as 0 then we have:
$p(u) = r(u)$

$$p(x) = g(x) \cdot q(x) + r(x) \quad \Rightarrow \quad p(u) = r(u) \qquad (5.1)$$

In other words, when the polynomial $p(x)$ is divided by $(x - k)$, if we set $x - k$ to be 0, we get $x = k$ and the remainder as $p(k)$.

$$p(x) = (x-k)q(x) + r \quad \Rightarrow \quad p(k) = r \qquad (5.2)$$

When the polynomial $p(x)$ is divided by $(ax - b)$, if we set $ax - b$ to be 0, we get $x = \dfrac{b}{a}$ and the remainder as $p(\dfrac{b}{a})$:

$$p(x) = (ax-b)q(x) + r \quad \Rightarrow \quad p\left(\dfrac{b}{a}\right) = r \qquad (5.3)$$

Let's try some examples:

Example 6: Let $p(x) = x^7 - 3x^5 + x^3 - 7x^2 + 5$ and $q(x) = x - 2$. Find the remainder of $p(x)/q(x)$.
a) 5/2   b) 5   c) 17   d) -3   e) none of these

Solution: (C).

Method 1 (official solution):
Long division will find the answer, but since $q(x)$ is a linear term, we can use a shortcut. Let $p(x) = q(x)h(x) + r$, where r is the remainder and $h(x)$ is the quotient polynomial produced by long division. Let $x = 2$. $p(2) = q(2)h(2) + r$.
Since $q(2) = 0$, $r = p(2) = 17$.

Method 2: By the Remainder Theorem, the remainder is $p(2) = 2^7 - 3\cdot 2^5 + 2^3 - 7\cdot 2^2 + 5 = 17$.

Example 7: Find the remainder when $x^{100} - 2x^{51} + 1$ is divided by $x^2 - 1$.

Solution:

Method 1:
$x^{100} - 2x^{51} + 1 = (x^2 - 1)q(x) + ax + b$.

Setting *x* equal to 1, we get $a + b = 0$ (1)
Setting *x* equal to –1, we get $-a + b = 4$ (2)
Solving (1) and (2), we get $a = -2$ and $b = 2$. The remainder is $-2x + 2$.

Method 2:
$p(x) = x^{100} - 2x^{51} + 1 = (x^2 - 1)q(x) + r(x)$.
Setting $x^2 - 1 = 0$, we get $x^2 = 1$.
We know that the remainder will be $x^{100} - 2x^{51} + 1$ where $x^2 = 1$:
$(x^2)^{50} - 2(x^2)^{25} \cdot x + 1 = 1 - 2x + 1 = -2x + 2$.

Example 8: If $x \ne 5$, then $\dfrac{x^3 - 2x + 20}{x - 5} = x^2 + 5x + 23 + \dfrac{k}{x - 5}$. Find the value of *k*.

Solution:
k is the remainder when $x^3 - 2x + 20$ is divided by $x - 5$.
By the Remainder Theorem, $k = (5)^3 - 2 \cdot 5 + 20 = 135$.

Example 9: A polynomial $f(x)$ has remainder *c* when divided by $x - a$ and remainder *d* when divided by $x - b$. Find the remainder when $f(x)$ is divided by $(x - a)(x - b)$.

Solution:

Let $f(x) = (x - a)(x - b)q(x) + mx + n$.
By the Remainder Theorem, we have $f(a) = c$ and $f(b) = d$.
Substituting in *a* and *b* as *x* into the equation for $f(x)$, we get
$ma + n = c$ and $mb + n = d$.
Solving the system of equations, we get $m = \dfrac{c - d}{a - b}$ and $n = \dfrac{ad - bc}{a - b}$.
Therefore the remainder is $\dfrac{c - d}{a - b}x + \dfrac{ad - bc}{a - b}$.

Example 10: A polynomial $P(x)$ has remainder – 5 when divided by $x + 1$ and remainder 7 when divided by $x - 5$. What is the remainder when $P(x)$ is divided by the product of $x + 1$ and $x - 5$?
A. $2x + 3$ B. $2x - 3$ C. $3x - 2$ D. 35   E. – 35

Solution: B.

The remainder is $\dfrac{c-d}{a-b}x + \dfrac{ad-bc}{a-b} = \dfrac{-5-7}{-1-5}x + \dfrac{(-1)7 - 5(-5)}{-1-5} = 2x - 3$.

6. FACTOR THEOREM:

$(x - k)$ is a factor of the polynomial $p(x)$ if and only if $p(k) = 0$.

In other words, k is a root of the polynomial $p(x)$ when $x - k$ is a factor of $p(x)$.

Example 11: What are the factors of a polynomial function if its graph has x-intercepts at -1, 0, and 4 ?
A. x, $(x - 1)$, $(x + 4)$ B. x, $(x + 1)$, $(x - 4)$ C. $(x - 1)$, $(x - 4)$ D. $(x - 1)$, $(x + 4)$
E. x, $(x + 4)$

Solution: B.

Since $x = 0$, $x = -1$, and $x = 4$ are the solutions, by the factor theorem, x, $(x + 1)$, and $(x - 4)$ are the factors of the polynomial. The answer is B.

Example 12: Find k if $x + 2$ is a factor of $x^3 + kx + 6$.

Solution:

Let $f(x) = x^3 + kx + 6$.
Since we are given that $x + 2$ is a factor of $f(x)$, when $x + 2 = 0$, or $x = -2$, $f(x) = 0$.
Thus $f(-2) = 0$, or $(-2)^3 + k(-2) + 6 = 0 \implies -2k - 2 = 0$.
Solving for k yields $k = -1$.

Example 13: Find a and b if $f(x) = x^4 - 5x^3 + 11x^2 + ax + b$ is divisible by $g(x) = (x - 1)^2$.

Solution:

Since the polynomial $f(x)$ has the degree of 4 and $g(x)$ has the degree of 2, the quotient of $f(x)$ divided by $g(x)$ must have the degree of 2. The quotient must also have the leading coefficient of 1 and constant term of b since the leading coefficient of $g(x)$ is 1 and constant term is 1.
Let the quotient be denoted as $h(x) = x^2 + cx + b$
So $f(x) = g(x)h(x)$.

Therefore we have

$$x^4 - 5x^3 + 11x^2 + ax + b = (x-1)^2(x^2 + cx + b)$$
$$= x^4 + (c-2)x^3 + (b-2c+1)x^2 + (-2b+c)x + b$$

Matching the coefficients of like powers of x, we get the following system of equations:
$$\begin{cases} c - 2 = -5, \\ b - 2c + 1 = 11, \\ -2b + c = a. \end{cases} \Rightarrow \begin{cases} a = -11, \\ b = 4, \\ c = -3. \end{cases}$$

Example 14: The polynomial $p(x) = 2x^4 - x^3 - 7x^2 + ax + b$ is divisible by $x^2 - 2x - 3$ for certain values of a and b. What is the sum of a and b?
A. −34   B. −30   C. −26   D. −18   E. 30

Solution: A.

Because $x^2 - 2x - 3 = (x - 3)(x + 1)$, $p(x)$ has zeros of 3 and −1, $p(3) = 72 + 3a + b = 0$ and $p(-1) = -4 - a + b = 0$ which we can solve simultaneously to get $a = -19$ and $b = -15$.

7. LONG DIVISION AND SYNTHETIC DIVISION

Long division Long division is useful in finding the value of a polynomial for a given x as well as the factors of the polynomial.

Example 15: The polynomial $p(x) = 2x^4 - x^3 - 7x^2 + ax + b$ is divisible by $x^2 - 2x - 3$ for certain values of a and b. What is the sum of a and b?
A. −34   B. −30   C. −26   D. −18   E. 30

Solution: A.

Dividing $2x^4 - x^3 - 7x^2 + ax + b$ by $x^2 - 2x - 3$ by long division. The "last step" is to subtract $5x^2 - 10x - 15$ from $5x^2 + (9 + a)x + b$. The difference must be 0. So $b = -15$ and $9 + a = -10$. Thus $a = -19$ and $a + b = -34$.

Example 16: For $x^2 + 2x + 5$ to be a factor of $x^4 + px^2 + q$, the values of p and q must be, respectively:
(A) −2, 5   (B) 5, 25   (C) 10, 20   (D) 6, 25   (E) 14, 25

Solution:

Long division:

$$\require{enclose}
\begin{array}{r}
x^2 \quad - \quad 2x \quad\quad +(p-1) \\
x^2+2x+5 \enclose{longdiv}{x^4 \quad + \quad px^2 \quad\quad +q}
\end{array}$$

$$
\begin{array}{r}
x^4 + 2x^3 + 5x^2 \\ \hline
-2x^3 + (p-5)x^2 \\
-2x^3 - 4x^2 - 10x \\ \hline
(p-1)x^2 + 10x + q \\
(p-1)x^2 + 2(p-1)x + 5(p-1) \\ \hline
(12-2p)x + (q-5p+5)
\end{array}
$$

Since the remainder must be zero, $12 - 2p = 0$: $p = 6$ and $q - 5p + 5 = 0$: $q = 25$.

Synthetic Division

Example 17: Is $x - 2$ a factor of $p(x) = x^3 + x^2 + x - 14$?
Solution:

In Synthetic Division, write the number being checked (as a possible zero perhaps) and then the coefficients as follows:

2| 1   1   1 -14

Now leave a row blank and draw a horizontal line below the coefficients:

2| 1   1   1 -14

Bring down the leading coefficient and then multiply it by the zero candidate, placing this product in the space below the next coefficient:

2| 1   1   1 -14
               2
     1

Now add, write the total below, and repeat.

```
2 | 1    1    1    -14
        2    6     14
    ─────────────────────
    1    3    7     0
```

Notice that the final sum is zero, telling us that 2 is a zero (or root) and $x - 2$ is a factor. Synthetic division also tells us that the other coefficients are the coefficients of the quotient polynomial.

$$x^3 + x^2 + x - 14 = (x-2)(x^2 + 3x + 7).$$

Example 18: Determine A and B such that $Ax^4 + Bx^3 + 1$ is divisible by $(x-1)^2$.

Solution:

Method 1:
By synthetic division:

```
1 |  A    B      0       0       1
              A      A+B     A+B     A+B
     ──────────────────────────────────
1 |  A   A+B    A+B     A+B    |A+B+1|
              A      2A+B    3A+2B
     ──────────────────────────────
     A   2A+B   3A+2B  |4A+3B|
```

So $A + B + 1 = 0$, $4A + 3B = 0$.
Solving we get: $A = 3$, $B = -4$.

Method 2:
Let $f(x) = Ax^4 + Bx^3 + 1$.
By the remainder theorem, we have $f(1) = A + B + 1 = 0$ (1),
$B = -(1 + A)$.
Substituting the value of B into $f(x)$, we get
$f(x) = Ax^4 - (1+A)x^3 + 1 = Ax^3(x-1) - (x^3 - 1)$
$= (x-1)(Ax^3 - x^2 - x - 1)$.
Let $f_1(x) = Ax^3 - x^2 - x - 1$.
Since $f(x)$ is divisible by $(x-1)^2$, $f_1(x) = Ax^3 - x^2 - x - 1$ contains the factor $(x-1)$.
Then $f_1(1) = A - 3 = 0 \quad \Rightarrow \quad A = 3$.
Substituting $A = 3$ into (1), we get $B = -4$.

Method 3:

Algebra II Through Competitions   **Chapter 9 Polynomials**

Since $f(x)$ is divisible by $(x-1)^2$, we can write $Ax^4 + Bx^3 + 1 = (x^2 - 2x + 1)(Ax^2 + cx + 1)$.
$Ax^4 + Bx^3 + 1 = (x^2 - 2x + 1)(Ax^2 + cx + 1) =$
$Ax^4 - 2Ax^3 + Ax^2 + cx^3 - 2cx^2 + cx + x^2 - 2x + 1$.

Matching the confidents of the like powers of x:
$$\begin{cases} -2A + c = B \\ A - 2c + 1 = 0 \\ c - 2 = 0 \end{cases} \Rightarrow \begin{cases} c = 2 \\ A = 3 \\ B = -4 \end{cases}$$

Example 19: Find the other two zeros of $p(x) = 6x^3 + 19x^2 + 2x - 3$ if one root is -3.

Solution:

Since -3 is a zero of $p(x)$, by the Factor Theorem, $x + 3$ is a factor of $p(x)$.

Use synthetic division to divide $p(x)$ by $x + 3$.

$$\begin{array}{r|rrrr} -3 & 6 & 19 & 2 & -3 \\ & & -18 & -3 & 3 \\ \hline & 6 & 1 & -1 & 0 \end{array}$$

The quotient is $6x^2 + x - 1$, so
$p(x) = 6x^3 + 19x^2 + 2x - 3 = (x + 3)(6x^2 + x - 1)$. This can be further factored into:
$(x + 3)(2x + 1)(3x - 1)$.
Therefore, the other two zeros are $x = -1/2$ and $x = 1/3$.

Example 20: The polynomial $p(x) = 2x^4 - x^3 - 7x^2 + ax + b$ is divisible by $x^2 - 2x - 3$ for certain values of a and b. What is the sum of a and b?
A. -34 B. -30 C. -26 D. -18 E. 30

Solution: A.
Factor $x^2 - 2x - 3$ and see that it has two zeros, $x = 3$ and $x = -1$. Use synthetic division (starting with either zero–here starting with $x = -1$) to see that (1) $b - a - 4 = 0$, and (2) $2x^4 - x^3 - 7x^2 + ax + b = (x + 1)(2x^3 - 3x^2 - 4x + a + 4)$. Now use synthetic division with $x = 3$ on the quotient $2x^3 - 3x^2 - 4x + a + 4$ to see that $a + 19 = 0$. Thus $a = -19$. Put this into (1) to see that $b = -15$, etc.

8. SOME SPECIAL VALUES OF $f(x)$

$$f_1(0) = a_0, \tag{8.1}$$

(It is obvious in any polynomial that $f(0)$ is the y-intercept).

$$f(1) = a_0 + a_1 + a_2 + a_3 + \cdots + a_n, \tag{8.2}$$

($f(1)$ is the sum of all the coefficients).

$$f(-1) = a_0 - a_1 + a_2 - a_3 + \cdots + (-1)^n a_n \tag{8.3}$$

(The difference of the sums of the coefficients of all even terms and all odd terms).

$$\frac{f(1) + f(-1)}{2} = a_0 + a_2 + a_4 + \cdots \tag{8.4}$$

(The sum of the coefficients of all even terms).

$$\frac{f(1) - f(-1)}{2} = a_1 + a_3 + a_5 + \cdots \tag{8.5}$$

(The sum of the coefficients of all odd terms).

Example 21: Given the function $f(x) = 2x^4 - 3x^3 + 4x^2 - 5x + 6$, what is the sum of A, B, C, D, and E if $f(x) = A(x-1)^4 + B(x-1)^3 + C(x-1)^2 + D(x-1) + E$?
A. 16   B. 17   C. 18   D. 19   E. 20

Solution: E.
Let $x = 2$, we get:
$f(2) = A(2-1)^4 + B(2-1)^3 + C(2-1)^2 + D(2-1) + E = 2 \cdot 2^4 - 3 \cdot 2^3 + 4 \cdot 2^2 - 5 \cdot 2 + 6$
$A + B + C + D + E = 2 \cdot 2^4 - 3 \cdot 2^3 + 4 \cdot 2^2 - 5 \cdot 2 + 6 = 20$.

9. ROOTS OF A POLYNOMIAL FUNCTION

The values of x that satisfy $p(x) = 0$ is called the zeros of $p(x)$. The zeros have the same meaning as the roots of a polynomial. The real roots are the x–intercepts of the graph of the polynomial.

Example 22: If $P(x)/(x-a) = Q(x)$, where $P(x)$ and $Q(x)$ are polynomials, then
A. $P(a) = 0$.   B. $x - a$ is a factor of $P(x)$.   C. a is an x-intercept of the graph of $y = P(x)$.
D. All of the above.   E. Insufficient information to answer.

Solution: D.

By the Remainder Theorem, we know that A is true. By the Factor Theorem, we know that B is true. By the definition of the roots of a polynomial function, we know C is true the answer is then D.

Example 23: Find the sum of the x-intercepts of the function $g(x) = 3(2x + 7)^2(x - 1)^2 - (2x + 7)(x - 1)^3$.
A. −69/10   B. −67/10   C. −33/5   D. −5/2   E. 5/2

Solution: A.
Factor out the common terms to get $g(x) = (2x + 7)(x - 1)^2[3(2x + 7) - (x - 1)] = (2x + 7)(x - 1)2[5x + 22]$. Setting each factor equal to zero, we find the zeros are $x = -7/2$, $x = 1$, and $x = -22/5$. So the sum is −69/10.

10. RATIONAL ROOT THEOREM:

If the coefficients of a polynomial are integers, then the rational roots, if any, will have to be of the form $\pm p/\pm q$, where p is an integer factor of the constant term and q is an integer factor of the leading coefficient of the polynomial.

Example 24: The roots of the equation $x^3 + kx^2 - 54x + 216 = 0$ can be arranged to form a geometric progression. Find the value of k.
A. − 3   B. − 6   C. − 9   D. 8   E. NOTA

Solution: C.
Notice that the equation has a cubed x term and a cubed constant term (216). This information can aid us in finding a factor of $x^3 + kx^2 - 54x + 216 = 0$.

We know that the rational roots, if any, of this cubic will have to be of the form $\pm p/\pm q$, where p is an integer factor of the constant term, 216, and q is an integer factor of the leading coefficient, 1.

Using this information, we can assume that -6 is a root. Using synthetic division, we get the quadratic $x^2 - 15x + 36$. Factoring this quadratic, we get $(x - 3)(x - 12)$. So, now we have $(x - 3)(x + 6)(x - 12) = 0$. The roots are in a geometric progression with a common ratio of − 2.

By the Factor Theorem, we can plug one of the roots, 3, into the equation to get:
$(3)^3 + k \times (3)^2 - 54 \times 3 + 216 = 0 \Rightarrow k = -9$.

11. DESCARTES' RULE OF SIGN CHANGES

In a polynomial with real coefficients, the number of positive real roots equals the number of sign changes in the coefficients subtracted by 0 or a multiple of 2. The number of negative real roots is the number of sign changes subtracted by 0 or a multiple of 2 in the polynomial $p(-x)$.

Example 25: The equation $x^3 + 6x^2 + 11x + 6 = 0$ has:
(A) no negative real roots   (B) no positive real roots   (C) no real roots
(D) 1 positive and 2 negative roots   (E) 1 negative and 2 positive roots

Solution: (B).
By Descartes' Rule of Signs, there are no sign changes, so (B) is correct. We can also find the roots through factoring to be: $-1, -2, -3$.

12. ROOTS AND COEFFICIENTS

If the coefficients of a polynomial are rational, there can be individual rational roots, but irrational roots will always occur in conjugate pairs. If the irrational root $a + bi$ was a root of a polynomial, then so would its conjugate: $a - bi$.

For example, suppose you were told that $1, -2,$ and $2 - \sqrt{3}$ are zeros of $P(x)$ and also that the coefficients are rational. We then know that $2 + \sqrt{3}$ is also a zero.

Vieta's Theorem:

(1). Let x_1, x_2, and x_3 be the roots for a 3-degree polynomial $a_0x^3 + a_1x^2 + a_2x^1 + a_3 = 0$.
Then

$$x_1 + x_2 + x_3 = -\frac{a_1}{a_0} \tag{12.1}$$

$$x_1x_2 + x_2x_3 + x_3x_1 = \frac{a_2}{a_0} \tag{12.2}$$

$$x_1x_2x_3 = -\frac{a_3}{a_0} \tag{12.3}$$

(2). Let x_1, x_2, x_3, and x_4 be the roots for a 4-degree polynomial
$a_0x^4 + a_1x^3 + a_2x^2 + a_3x^1 + a_4 = 0$.

Then $x_1 + x_2 + x_3 + x_4 = -\dfrac{a_1}{a_0}$ \hfill (12.4)

$x_1x_2 + x_1x_3 + x_1x_4 + x_2x_3 + x_2x_4 + x_3x_4 = \dfrac{a_2}{a_0}$ \hfill (12.5)

$x_1x_2x_3 + x_1x_2x_4 + x_1x_3x_4 + x_2x_3x_4 = -\dfrac{a_3}{a_0}$ \hfill (12.6)

$x_1x_2x_3x_4 = \dfrac{a_4}{a_0}$ \hfill (12.7)

Example 26: If $p(x)$ is a polynomial with rational coefficients and roots at 0, 1, $\sqrt{2}$ and $1-\sqrt{3}$, then the degree of $p(x)$ is at least?

Solution: 6.
Since irrational roots will always occur in conjugate pairs, we know that $-\sqrt{2}$ and $1 + 1+\sqrt{3}$ will also be the roots of the polynomial. Since we know that there must be at least 6 roots, the degree of $p(x)$ is at least 6.

Example 27: The sum of the zeros of $f(x) = x(2x + 3)(4x + 5) + (6x + 7)(8x + 9)$ is
A. $-35/4$ \qquad B. $35/4$ \qquad C. -70 \qquad D. 70 \qquad E. 0

Solution: A.
Since $f(x) = 8x^3 + 70x^2 + 125x + 63$, the sum of zeros is $-70/8 = -35/4$.

Example 28 (same as Example 19): Find the other two zeros of $p(x) = 6x^3 + 19x^2 + 2x - 3$ if one root is -3.

Solution:
Let the other two roots be r_1 and r_2.
By Vieta's Theorem, we have

$r_1 + r_2 + (-3) = -\dfrac{19}{6}$ \hfill (1)

$r_1 r_2 (-3) = -\dfrac{-3}{6}$ \hfill (2)

Solving the system of equations of (1) and (2), we get $r_1 = -\dfrac{1}{2}$ and $r_2 = \dfrac{1}{3}$.

Algebra II Through Competitions Chapter 9 Polynomials

PROBLEMS

Problem 1. Find all real number solutions of $x^4 - 16x^2 - 80 = 0$
A. $\pm 2\sqrt{5}$ B. $\pm 2\sqrt{5}, \pm 2$ C. ± 2 D. $\pm 2\sqrt{10}, \pm\sqrt{2}$ E. $\pm\sqrt{-2}\pm\sqrt{10}$

Problem 2. Find the product of all real solutions of $x^4 - 5x^2 - 24 = 0$.
A. –24 B. –8 C. –3 D. 24 E. None of these

Problem 3. Let $g(x) = ax^7 + bx^3 + cx - 7$, where a, b, and c are constants. If $g(-5) = 5$, then $g(5)$ is
A. – 7 B. – 5 C. –15 D. –19 E. not enough information

Problem 4. When $x^4 - 6x^2 + 5$ is factored completely with integer coefficients, then the sum of the factors is:
A. $x^2 + 2x - 5$ B. $x^2 + 2x + 3$ C. $x^2 - 7$ D. $x^3 - 5$ E. none of these

Problem 5. What is the sum of the squares of the real and complex solutions of $x^4 + 2x^3 + 9x^2 + 18x = 0$
A. -14 B. 8 C. 5 D. - 7 E. - 2

Problem 6. Given $p^2 - 2p + 3 = 0$, determine the value of $p^4 - 4p^3 + 8p - 2$.
A. – 2 B. 3 C. 19 D. 21 E. 23

Problem 7. Find the difference between the largest and smallest root of $p(x) = x^4 - 6x^3 - 2x^2 + 6x + 1$.

a) 6 b) $4 + \sqrt{10}$ c) $4 - \sqrt{10}$ d) $1 + \sqrt{38}$ e) $\dfrac{7 + \sqrt{23}}{2}$.

Problem 8. If $x^2 = x + 1$ then which of the following expressions is equal to x^6?
A. $6x + 4$ B. $8x + 5$ C. $x^5 + 1$ D. $5x + 3$ E. $5x + 8$

Problem 9. Which of the following expression is NOT a factor of $x^6 - 64$.
A. $x - 2$ B. $x + 2$ C. $x^2 + 2x + 4$ D. $x^2 + 4$ E. $x^2 - 2x + 4$

Problem 10. The polynomial $P(x) = (x^6 - 1)(x - 1) - (x^3 - 1)(x^2 - 1)$ has potentially 7 real zeros. Which of the following is a zero of multiplicity greater than 1?
A. –2 B. –1 C. 0 D. 1 E. 2

Problem 11. How many distinct real number solutions does $(3x^2 + 2x)^2 = (x^2 + 2x + 1)^2$ have?
A. 0   B. 1   C. 2   D. 3   E. 4

Problem 12. Suppose a, b, c are integers such that
1. $0 < a < b$,
2. The polynomial $x(x - a)(x - b) - 17$ is divisible by $(x - c)$.
What is $a + b + c$?
A. 14   B. 17   C. 21   D. 24   E. 27

Problem 13. Let $P(x)$ be a polynomial with $P(1) = 1$ and $P(x) = P(x - 1) + x^3$ for all real x. Calculate $P(-3)$.
A. -36 B. -9 C. 1   D. 9   E. 36

Problem 14. (2004 UNC- Charlotte Algebra II Competition) What is the remainder when $x^4 - x^2 + 1$ is divided by $x^2 + 1$?
A. -3 B. -1 C. 0   D. 3   E. 4

Problem 15. If x^6 is divided by $(x + 1/3)$, the quotient is $Q(x)$ and the remainder is r. If $Q(x)$ is then divided by $(x + 1/3)$, the remainder is f. The value of f can be expressed as $-k/w$ where k and w are relatively prime positive integers. Find the value of $(k + w)$.

Problem 16. It is known that $x = 1$ is a solution to the equation $2x^3 - 3x^2 - 4x + 5 = 0$. What are the two other solutions?
A. $\dfrac{-1 \pm \sqrt{41}}{4}$ B. $\dfrac{1 \pm \sqrt{41}}{4}$ C. $\dfrac{1 \pm \sqrt{39}}{4}$ D. $\dfrac{1 \pm \sqrt{41}}{2}$ E. ± 1

Problem 17. What is the remainder when $x^2 + 3x - 5$ is divided by $x - 1$?
A. -5 B. -2 C. -1 D. 0   E. 1

Problem 18. Determine m such that $x^3 - 5x^2 + 7x + (m - 5)$ is divisible by $(x - 4)$
A. -7 B. 0   C. 5   D. 7   E. 17

Problem 19. How many polynomials are there of the form $x^3 - 9x^2 + cx + d$ such that c and d are real numbers and the three roots of the polynomial are distinct positive integers?
A. 1   B. 2   C. 3   D. 4   E. 5

Problem 20. Given the function $f(x) = 5x^3 - 7x^2 + 11x + 13$, what is the sum of A, B, C and D where $f(x) = A(x + 1)^3 + B(x + 1)^2 + C(x + 1) + D$?
A. 10   B. 11   C. 12   D. 13   E. 14

Problem 21. The polynomial $x^3 + 2x^2 - 4x - 4$ has three different real number roots a, b, and c. If the polynomial $x^3 + qx^2 + rx + s$ has roots $1/a$, $1/b$, and $1/c$, what is $q + r + s$?
A. 3/4   B. 1/2   C. 1/4   D. 1/6   E. 3/8

Problem 22. The polynomial $x^3 - 2x^2 - 5x + 6$ has three real roots. What is the sum of these roots?
A. ≤2   B. −1   C. 0   D. 1   E. 2

Problem 23. The polynomial $P(x) = x^3 - 5x^2 + x - 5$ has
A. 3 integer zeros
B. 1 integer and 2 non-integer rational zeros
C. 1 integer and 2 complex (non-real) zeros
D. 3 irrational zeros
E. none of these

Problem 24. Solve for x: $(x + 1)(x + 2)(x + 3) = 2(x + 1)(x + 2)(x + 4)$.
A. $\{-1, -2, -5\}$ B. $\{-5\}$ C. $\{-1, -2\}$ D. $\{-1, -2, -3, -4\}$ E. None of these

Problem 25. (Alabama 2009 State Math Contest Algebra II) The equation $e^{5x} + e^{4x} - 3e^{3x} - 17e^{2x} - 30e^x = 0$ has how many real solutions?
(A) 1   (B) 2   (C) 3   (D) 4   (E) 5

Problem 26. If $f(x) = \dfrac{3x^3 - 31x^2 + 62x + 56}{x^2 + x - 20}$, then exactly one of the zeros of $f(x)$ is an integer. Find that integer.

Problem 27. Suppose that $p(x)$ is a polynomial of the form $x^{120} + a_{119}x^{119} + a_{118}x^{118} + \ldots + a_1 x + 7200$ where the coefficients $a_1, a_2, \ldots a_{119}$ are all integers. What is the maximum possible number of distinct rational roots of p?
A. 36   B. 54   C. 72   D. 108   E 120.

Algebra II Through Competitions **Chapter 9 Polynomials**

SOLUTIONS

Problem 1. Solution: A.
Let $x^2 = b$. Then $b^2 - 16b - 80 = 0$

$b = \dfrac{16 \pm \sqrt{16^2 - 4 \cdot 1 \cdot (-80)}}{2} = 20$ (–4 is extraneous)

$x = \pm\sqrt{b} = \pm 2\sqrt{5}$

Problem 2. Solution: B.
$x^4 - 5x^2 - 24 = 0 \Rightarrow (x^2 - 8)(x^2 + 3) = 0$
So the only real roots come from $x^2 - 8 = 0 \Rightarrow x_1 = \sqrt{8}$, and $x_2 = -\sqrt{8}$.
The product of the roots is -8.

Problem 3. Solution: D.
Let $g(x) = h(x) - 7$, where $h(x) = ax^7 + bx^3 + cx$. Notice that $h(x)$ is odd,
so $h(5) = -h(-5) = -12$. Thus $g(5) = h(5) - 7 = -12 - 7 = -19$.

Problem 4. Solution: A.
$x^4 - 6x^2 + 5 = (x^2 - 5)(x^2 - 1) = (x^2 - 5)(x + 1)(x - 1)$, so the sum of the factors is $(x^2 - 5) + (x + 1) + (x - 1) = x^2 + 2x - 5$.

Problem 5. Solution: A.
Method 1:
$x^4 + 2x^3 + 9x^2 + 18x = 0 \Rightarrow x(x + 2)(x^2 + 9) = 0 \Rightarrow x = 0, -2,$ or $\pm 3i$. The sum of the squares of all solutions is $0^2 + (-2)^2 + (3i)^2 + (-3i)^2 = 4 - 9 - 9 = -14$.
Method 2:
We know that
$x_1^2 + x_2^2 + x_3^2 + x_4^2 = (x_1 + x_2 + x_3 + x_4)^2 - 2(x_1x_2 + x_1x_3 + x_1x_4 + x_2x_3 + x_2x_4 + x_3x_4)$ (1)
By Vieta's Theorem, $(x_1 + x_2 + x_3 + x_4)^2 = (-2)^2 = 4$, and
$2(x_1x_2 + x_1x_3 + x_1x_4 + x_2x_3 + x_2x_4 + x_3x_4) = 2(-9) = -18$. Thus
$x_1^2 + x_2^2 + x_3^2 + x_4^2 = 4 - 18 = -14$.

Problem 6. Solution: C.
Since $(p^2 - 2p + 3) = 0$, we see that $(p^2 - 2p + 3)^2 = p^4 - 4p^3 + 10p^2 - 12p + 9 = 0$, but we can rearrange the second expression to get $p^4 - 4p^3 + 10p^2 - 12p + 9 = p^4 - 4p^3 + 8p - 2 +$

$(10p^2 - 20p + 11)$, but this too can be rewritten as $[p^4 - 4p^3 + 8p - 2] + 10(p^2 - 2p + 3) - 19 = 0$, so the desired quantity $[p^4 - 4p^3 + 8p - 2] = 19 - 10(p^2 - 2p + 3) = 19$.

Problem 7. Solution: B.
The only rational possibilities for roots are 1 and -1, and it is easy to show that both work. When we divide or factor these out, we have

$f(x) = (x - 1)(x + 1)(x^2 - 6x - 1)$. The remaining solutions are $x = \dfrac{6 \pm \sqrt{40}}{2} = 3 \pm \sqrt{10}$.

The larger of these is the largest root of the function, but -1 is the smallest, so the difference is $3 + \sqrt{10} - (-1) = 4 + \sqrt{10}$.

Problem 8. Solution: B.
$x^2 = x + 1 \Rightarrow x^3 = x^2 + x \Rightarrow x^3 = (x + 1) + x \Rightarrow x^3 = 2x + 1$; similarly $x^4 = 2x^2 + x = 3x + 2$, $x^5 = 3x^2 + 2x = 5x + 3$, and $x^6 = 5x^2 + 3x = 8x + 5$.

Problem 9. Solution: D.
$x^6 - 64 = (x^3 - 8)(x^3 + 8) = (x - 2)(x^2 + 2x + 4)(x + 2)(x^2 - 2x + 4)$ Thus $x^2 + 4$ not a factor.

Problem 10. Solution: D.
One can factor the polynomial as $(x^6 - 1)(x - 1) - (x^3 - 1)(x^2 - 1) = (x^3 - 1)(x^3 + 1)(x - 1) - (x^3 - 1)(x - 1)(x + 1) = (x^3 - 1)(x - 1)[(x^3 + 1) - (x + 1)] = (x - 1)(x^2 + x + 1)(x - 1)x(x - 1)(x + 1) = x(x - 1)^3(x + 1)(x^2 + x + 1)$
Hence $x = 1$ is the only multiple zero.

Problem 11. Solution: D.
Using the formula $a^2 - b^2 = (a - b)(a + b)$ we obtain:
$(3x^2 + 2x)^2 - (x^2 + 2x + 1)^2 = (2x^2 - 1)(4x^2 + 4x + 1) = (2x^2 - 1)(2x + 1)^2$,
so the only zeros of this polynomial are $\pm 1/\sqrt{2}$ and -1. So there are a total of 3 real solutions.

Problem 12. Solution: C.
Since it is divisible by $(x - c)$, we have $c(c - a)(c - b) = 17$.
Since $c(c - a)(c - b) = 17 > 0$, it follows that $c > 0$ and we have the following two cases:
Case 1: $0 < (c - b) < (c - a) < c$
Case 2: $(c - b) < (c - a) < 0 < c$:
Since 17 is a prime number, case 1 does not occur. In case 2, $c = 1$, $c - a = -1$, $c - b = -17$. Hence $a = 2$, $b = 18$, $c = 1$. Thus, $a + b + c = 21$.

Problem 13. Solution: D.

Algebra II Through Competitions   **Chapter 9 Polynomials**

Note that $P(x-1) = P(x) - x^3$ for all x. Hence $P(0) = P(1)-1 = 0$ and $P(-1) = P(0) - 0^3 = 0$. Now one can proceed to $P(-3) = P(-2) - (-2)^3 = P(-1) - (-1)^3 - (-2)^3 = 1^3 + 2^3 = 9$.

Problem 14. Solution: D.
By long division $x^4 - x^2 + 1 = (x^2 + 1)(x^2 - 2) + 3$, so the remainder is 3. Alternatively, since all of the exponents are even, the remainder obtained from dividing $x^4 - x^2 + 1$ by $x^2 + 1$ is the same as the remainder obtained from dividing $y^2 - y + 1$ by $y + 1$. For the latter pair, the remainder can be found easily by simply calculating $(-1)^2 - (-1) + 1 = 3$ since $y + 1 = 0$ for $y = -1$.

Problem 15. Solution:
We know that when x^6 is divided by $(x + 1/3)$, the quotient is $Q(x)$ and the remainder is r. Thus we have:
$$\frac{x^6}{x+\frac{1}{3}} = Q(x) + \frac{r}{x+\frac{1}{3}}.$$
By the Remainder Theorem, $r = (-1/3)^6 = (1/3)^6$.

$$Q(x) = \frac{x^6}{x+\frac{1}{3}} - \frac{r}{x+\frac{1}{3}} = \frac{x^6}{x+\frac{1}{3}} - \frac{(\frac{1}{3})^6}{x+\frac{1}{3}} = \frac{x^6 - (\frac{1}{3})^6}{x+\frac{1}{3}}.$$

We know that $x^n - y^n = (x+y)(x^{n-1} - x^{n-2}y + \ldots - y^{n-1})$ for all even n.

$$Q(x) = \frac{x^6 - (\frac{1}{3})^6}{x+\frac{1}{3}} = \frac{(x+\frac{1}{3})(x^5 - x^4 \cdot \frac{1}{3} + x^3 \cdot \frac{1}{3^2} - x^2 \cdot \frac{1}{3^3} + x \cdot \frac{1}{3^4} - \frac{1}{3^5})}{x+\frac{1}{3}}$$

$$= x^5 - x^4 \cdot \frac{1}{3} + x^3 \cdot \frac{1}{3^2} - x^2 \cdot \frac{1}{3^3} + x \cdot \frac{1}{3^4} - \frac{1}{3^5}.$$

The remainder f is
$$Q(-\frac{1}{3}) = (-\frac{1}{3})^5 - (-\frac{1}{3})^4 \cdot \frac{1}{3} + (-\frac{1}{3})^3 \cdot \frac{1}{3^2} - (-\frac{1}{3})^2 \cdot \frac{1}{3^3} + (-\frac{1}{3}) \cdot \frac{1}{3^4} - \frac{1}{3^5} = -\frac{2}{3^4}.$$

So $k = 2$ and $w = 3^4 = 81$. The sum is $2 + 81 = 83$.

Problem 16. Solution: (B).
Note that $(x - 1)$ is a factor of the polynomial $2x^3 - 3x^2 - 4x + 5$. After long division we have $(x - 1)(2x^2 - x - 5) = 0$. Now the other two solutions are obtained from the quadratic formula.

Problem 17. Solution: C.
By long division or the Remainder Theorem, the remainder is −1.

Problem 18. Solution: A.
The polynomial $f(x) = x^3 − 5x^2 + 7x + (m − 5)$ is divisible by $(x − 4)$ precisely when $f(4) = 0$ by the Factor Theorem. Thus $f(4) = 43 − 5 \cdot 42 + 7 \cdot 4 + m − 5 = 64 − 80 + 28 − 5 + m = 7 + m = 0$, which happens precisely when $m = −7$.

Problem 19. Solution: C.
Let the three roots be x_1, x_2, and x_3. By the Vieta's Theorem, $x_1 + x_2 + x_3 = 9$. Since x_1, x_2, and x_3 are distinct positive integers, the possible combinations of them are $(1, 2, 6)$, $(1, 3, 5)$, and $(2, 3, 4)$. Each combination will generate the different values of c and d. The answer is 3.

Problem 20. Solution: D.
Let $x = 0$, we get:
$f(0) = A(1)^3 + B(1)^2 + C(1) + D = 13$, or $A + B + C + D = 13$.

Problem 21. Solution: C
$a + b + c = − 2$.
$ab + bc + ca = − 4$.
$abc = 4$

$$\frac{1}{a}+\frac{1}{b}+\frac{1}{c}=-q \quad \Rightarrow \quad -(\frac{1}{a}+\frac{1}{b}+\frac{1}{c})=q \tag{1}$$

$$\frac{1}{a}\cdot\frac{1}{b}+\frac{1}{b}\cdot\frac{1}{c}+\frac{1}{c}\cdot\frac{1}{a}=r \tag{2}$$

$$\frac{1}{a}\cdot\frac{1}{b}\cdot\frac{1}{c}=-s \quad \Rightarrow \quad -\frac{1}{a}\cdot\frac{1}{b}\cdot\frac{1}{c}=s \tag{3}$$

$$q + r + s = -(\frac{1}{a}+\frac{1}{b}+\frac{1}{c}) + \frac{1}{ab}+\frac{1}{bc}+\frac{1}{ca} - \frac{1}{abc}$$
$$= -(\frac{bc+ab+ab}{abc}) + \frac{c+a+b}{abc} - \frac{1}{abc} = -(\frac{-4}{4}) + \frac{-2}{4} - \frac{1}{4} = 1 - \frac{1}{2} - \frac{1}{4} = \frac{1}{4}.$$

Problem 22. Solution: E.
By the Vieta's Theorem, the sum of the roots is $−(−2/1) = 2$.

Problem 23. Solution: C.

$P(x) = x^3 - 5x^2 + x - 5 = x^2(x-5) + x - 5 = (x-5)(x^2+1)$.
The only real root is $x = 5$. The answer is C.

Problem 24. Solution: A.
The given equation can be written as $2(x+1)(x+2)(x+4) - (x+1)(x+2)(x+3) = 0$
$\Rightarrow (x+1)(x+2)[2(x+4) - (x+3)] = 0 \Rightarrow (x+1)(x+2)(x+5) = 0$
The roots are $x = -1$, $x = -2$, and $x = -5$.

Problem 25. Solution: a.
Let $m = e^x$.
Then we have $m^5 + m^4 - 3m^3 - 17m^2 - 30m = 0$ \hfill (1)

By the Descartes' rule of sign changes, for a polynomial with real coefficients, the number of positive real roots equals the number of sign changes in the coefficients subtracted by 0 or a multiple of 2. Since m is always positive, the number real roots is 1. We see that $m = 3$ is a real root of (1).

Problem 26. Solution: 7.
Using long division, we get: $f(x) = \dfrac{3x^3 - 31x^2 + 62x + 56}{x^2 + x - 20} = 3x - 34 + \dfrac{156x - 624}{x^2 + x - 20}$.

$3x - 34 + \dfrac{156x - 624}{x^2 + x - 20} = 0 \Rightarrow 3x - 34 + \dfrac{156(x-4)}{(x-4)(x+5)} = 0 \Rightarrow 3x - 34 + \dfrac{156}{x+5} = 0$

$\Rightarrow (3x - 34)(x+5) = -156 \Rightarrow 3x^2 - 19x - 14 = 0 \Rightarrow (3x+2)(x-7) = 0$.
The integer solution is 7.

Problem 27. Solution: D.
$7200 = 3^2 \cdot 2^5 \cdot 5^2$.
The number of factors is $(2+1)(5+1)(2+1) = 54$.
Considering the negative values, we have $54 \cdot 2 = 108$.
By the Rational Root Theorem, the maximum possible number of distinct rational roots is 108.

1. DEFINITION

The symbol $\sqrt[n]{}$ is called a radical sign. The expression $\sqrt[n]{a}$ is a **radical**. The number a is called the radicand and n is a positive integer, called **the index** of the radical $\sqrt[n]{a}$. Note that the exponent $\frac{1}{n}$ and the radical sign $\sqrt[n]{}$ are both used to indicate the $\frac{1}{n}$ th root.

For example, $\sqrt[n]{x} = x^{\frac{1}{n}}$.

When $n = 2$, $\sqrt[2]{}$ is called **the square root**. It is customary to use the notation $\sqrt{}$ instead of $\sqrt[2]{}$ for the square root.

The radical symbol $\sqrt{}$ is used to represent the positive, or principal square root.

If x is a whole number, the principal square root of x is the nonnegative number y such that $y^2 = x$.

In real number system, we have $\sqrt{x} \geq 0$ and $x \geq 0$.

Example 1. The domain of the function $f(x) = \sqrt{5 - 2x}$ is:

A. $x \geq \frac{5}{2}$ B. $x \leq \frac{5}{2}$ C. $x \leq 0$ D. $x \geq 0$ E. none of these

Solution: B.

$\sqrt{5 - 2x} \geq 0 \quad \Rightarrow \quad 5 - 2x \geq 0 \quad \Rightarrow \quad x \leq \frac{5}{2}.$

Example 2. Find the solution set for $\sqrt{2x - 3} = -\sqrt{3x - 2}$.

A. $\{1\}$ B. $\{-1\}$ C. $\{1, -1\}$ D. $\{\ \}$ E. None of these

Solution: D.
We know that the radical symbol $\sqrt{}$ is used to represent the positive square root. $\sqrt{2x - 3} \geq 0$. Thus this equation has no real solutions.

Algebra II Through Competitions — Chapter 10 Radicals

Example 3. Find the maximum value of $f(x) = \sqrt{x^2 + 22x + 121} + \sqrt{x^2 - 26x + 169}$ over the interval $-12 \leq x \leq 12$.
A. 24 B. 26 C. 28 D. 30 E. 32

Solution: B.
Method 1:
$f(x) = \sqrt{x^2 + 22x + 121} + \sqrt{x^2 - 26x + 169} = \sqrt{(x+11)^2} + \sqrt{(x-13)^2}$.
Since $-12 \leq x \leq 12$, we have $f(x) = |x + 11| + 13 - x$.
When $x + 11 \geq 0$, we have $f(x) = x + 11 + 13 - x = 24$
When $x + 11 \leq 0$, we have $f(x) = -(x+11) + 13 - x = -x - 11 + 13 - x = 2 - 2x$.
The maximum value occurs at $x = -12$ and $f(-12) = 2 - 2 \cdot (-12) = 26$.

Method 2 (official solution):
Note that $f(x) = |x + 11| + |x - 13|$, whose maximum value over $[-12, 12]$ occurs at (-12); $f(-12) = 26$.

2. RADICAL IN SIMPLEST FORM

A radical is in its simplest form if
(a) The power of the radicand is less than the index.
(b). The radicand is not in a fraction form.
(c). No denominator contains a radical.

Example 4. Which of the following is equal $\sqrt[5]{729}$?
A. $2\sqrt[5]{3}$ B. $2\sqrt[5]{9}$ C. $3\sqrt[5]{3}$ D. $3\sqrt[5]{9}$ E. $9\sqrt[5]{3}$

Solution: C.
Method 1:
$\sqrt[5]{729} = \sqrt[5]{3^6} = \sqrt[5]{3^5 \cdot 3} = 3\sqrt[5]{3}$.
Method 2:
$\sqrt[5]{729} = (729)^{\frac{1}{5}} = (3^6)^{\frac{1}{5}} = 3 \cdot (3)^{\frac{1}{5}} = 3\sqrt[5]{3}$.

Example 5. A triangle has one side of length a and two sides of length b. The area of the triangle is

A. $\dfrac{a}{4}\sqrt{4b^2-a^2}$ B. $\dfrac{a}{2}\sqrt{4b^2-a^2}$ C. $ab\sqrt{\dfrac{3}{4}}$ D. $\dfrac{b}{4}\sqrt{4a^2-b^2}$

E. $\dfrac{b}{2}\sqrt{4a^2-b^2}$

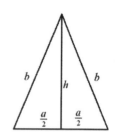

Solution: A.

The area is $\dfrac{1}{2}$ base × height = $\dfrac{1}{2} a \cdot \sqrt{b^2-(\dfrac{a}{2})^2} = \dfrac{a}{2}\sqrt{\dfrac{4b^2-a^2}{4}} = \dfrac{a}{4}\sqrt{4b^2-a^2}$.

3. RADICAL OPERATIONS

(1) $m\sqrt{a} + n\sqrt{a} = (m+n)\sqrt{a}, (a \geq 0)$

(2) $(\sqrt{a}+\sqrt{b})(\sqrt{a}-\sqrt{b}) = a-b, (a \geq 0, b \geq 0)$

(3) $x\sqrt[n]{a} \pm y\sqrt[n]{a} = (x \pm y)\sqrt[n]{a}, (a \geq 0$ if n is even)

(4) The Product Property: $\sqrt{ab} = \sqrt{a} \cdot \sqrt{b}$ and $\sqrt[n]{ab} = \sqrt[n]{a} \cdot \sqrt[n]{b}$, $(a \geq 0, b \geq 0$ if n is even).

(5) The Quotient Property $\sqrt{\dfrac{a}{b}} = \dfrac{\sqrt{a}}{\sqrt{b}}$ and $\sqrt[n]{\dfrac{a}{b}} = \dfrac{\sqrt[n]{a}}{\sqrt[n]{b}}$, $(a \geq 0, b > 0$ if n is even).

(6) $(\sqrt[n]{a})^m = \sqrt[n]{a^m}$, $(a \geq 0)$.

(7) $\sqrt[m]{\sqrt[n]{a}} = \sqrt[mn]{a}$, $(a \geq 0)$.

m and n are positive integers, $m, n \geq 2$.

Example 6. Simply: $\sqrt[x]{\sqrt{\sqrt[3]{25}}}$ where x is a natural number greater than 1.

a. 25^{3x} b. $25^{\frac{1}{3x}}$ c. $5^{\frac{1}{3x}}$ d. 5^{3x} e. none of these

Solution: C.

$\sqrt[x]{\sqrt{\sqrt[3]{25}}} = (\sqrt{\sqrt[3]{25}})^{\frac{1}{x}} = [(\sqrt[3]{25})^{\frac{1}{2}}]^{\frac{1}{x}} = [(25)^{\frac{1}{3}}]^{\frac{1}{2x}} = 25^{\frac{1}{6x}} = (5^2)^{\frac{1}{6x}} = 5^{\frac{1}{3x}}$.

Example 7. Which of the following is equal to $\sqrt[3]{108}$?

(A) $2\sqrt[3]{9}$ (B) $3\sqrt[3]{4}$ (C) $4\sqrt[3]{3}$ (D) $6\sqrt[3]{2}$ (E) $2\sqrt[3]{6}$

Solution: B.
$\sqrt[3]{108} = \sqrt[3]{36 \cdot 3} = \sqrt[3]{6^2 \cdot 3} = \sqrt[3]{3^2 \cdot 2^2 \cdot 3} = \sqrt[3]{3^3 \cdot 2^2} = 3\sqrt[3]{2^2} = 3\sqrt[3]{4}$.

Example 8. Find p if $\sqrt{\frac{x}{y}\sqrt{\frac{y^3}{x^3}\sqrt{\frac{x^5}{y^5}}}} = \left(\frac{x}{y}\right)^p$.

A. 5/8   B. 8/5   C. 1/8   D. 8/3   E. 3/8

Solution: E.
$\sqrt{\frac{x}{y}\sqrt{\frac{y^3}{x^3}\sqrt{\frac{x^5}{y^5}}}} = \sqrt{\frac{x}{y}\sqrt{\frac{y^3}{x^3} \cdot \frac{x^{5/2}}{y^{5/2}}}} = \sqrt{\frac{x}{y}\sqrt{\frac{y^{1/2}}{x^{1/2}}}} = \sqrt{\frac{x}{y} \cdot \frac{y^{1/4}}{x^{1/4}}} = \sqrt{\frac{x^{3/4}}{y^{3/4}}} = \frac{x^{3/8}}{y^{3/8}} = \left(\frac{x}{y}\right)^{3/8}$.

Example 9. Given that $x^2 + y^2 = 10$, $\sqrt[4]{xy} + \sqrt{xy} + 27 = 29$, $x > 0$ and $y > 0$. What is $x + y$?

(A) $2\sqrt{3}$ (B) $3\sqrt{3}$ (C) ½   (D) $\frac{\sqrt{3}}{2}$ (E) 8.

Solution: (A).
Let $\sqrt{xy} = m$.
$\sqrt[4]{xy} + \sqrt{xy} + 27 = 29$ $\Rightarrow$ $m^2 + m + 27 = 29$ $\Rightarrow$
$m^2 + m - 2 = 0$ $\Rightarrow$ $(m-1)(m+2) = 0$
Solving for m: $m = 1$ ($(m = -2$ ignored)
So we have $\sqrt{xy} = 1$ $\Rightarrow$ $xy = 1$ $\Rightarrow$ $2xy = 2$ (1)
Considering $x^2 + y^2 = 10$, we have $x^2 + 2xy + y^2 = 10 + 2$, or $(x+y)^2 = 12$.
From the answer key, we know that $x + y = \sqrt{12} = 2\sqrt{3}$.

4. INFINITELY NESTED RADICALS

Assume that $\sqrt{a + b\sqrt{a + b\sqrt{a + \cdots}}}$ is a real number when $a \geq 0$ and $b \geq 0$.
Let $L = \sqrt{a + b\sqrt{a + b\sqrt{a + \cdots}}}$.

Then $L = \sqrt{a+bL} \Rightarrow L^2 - bL - a = 0$.

By the quadratic formula, $L = \dfrac{b+\sqrt{b^2+4a}}{2}$.

Therefore $\sqrt{a+b\sqrt{a+b\sqrt{a+\cdots}}} = \dfrac{b+\sqrt{b^2+4a}}{2}$ \hfill (4.1)

Setting $a = 0$ in (4.1), we have $\sqrt{b\sqrt{b\sqrt{b\sqrt{\cdots}}}} = b$ \hfill (4.2)

Setting $a = b = 1$ in (4.1), $\sqrt{1+\sqrt{1+\sqrt{1+\cdots}}} = \dfrac{1+\sqrt{5}}{2}$, which is the golden ratio ϕ.

Generally, if $n > 0$, $\sqrt{n+\sqrt{n+\sqrt{n+\sqrt{n+\cdots}}}} = \dfrac{1}{2}(1+\sqrt{1+4n})$ \hfill (4.3)

$\sqrt{n-\sqrt{n-\sqrt{n-\sqrt{n-\cdots}}}} = \dfrac{1}{2}(-1+\sqrt{1+4n})$ \hfill (4.4)

Example 10. Which value best approximates $\sqrt{6+\sqrt{6+\sqrt{6+\sqrt{6+\sqrt{\cdots}}}}}$?

A. 2.9974   B. 3   C. π D. 6   E. infinity

Solution: B.
Method 1:
By formula (4.3), we have $\sqrt{6+\sqrt{6+\sqrt{6+\sqrt{6+\sqrt{\cdots}}}}} = \dfrac{1}{2}(1+\sqrt{1+4\times 6}) = 3$.

Method 2:
Setting $a = 6$ and $b = 1$ in (4.1), we have $x = \dfrac{1+\sqrt{1^2+4(6)}}{2} = \dfrac{1+5}{2} = 3$.

Method 3:
Let $x = \sqrt{6+\sqrt{6+\sqrt{6+\sqrt{6+\sqrt{\cdots}}}}}$.
$x^2 - 6 = x$ $\Rightarrow$ $(x-3)(x+2) = 0$.
$x = 3$ since $x > 0$.

Example 11. Determine the value of p where

$$\sqrt{p+\sqrt{p+\sqrt{p+\sqrt{p+\cdots}}}} = 7.$$

Solution: 42.
Method 1:

By formula (4.3), we have $7 = \frac{1}{2}(1+\sqrt{1+4p}) \Rightarrow 1+\sqrt{1+4p} = 14$

$\Rightarrow \sqrt{1+4p} = 13 \Rightarrow 1+4p = 169 \Rightarrow 4p = 168$

$\Rightarrow p = 42$.

Method 2:

Since the square roots are infinite, we can rewrite this expression as

$\sqrt{p+\sqrt{p+\sqrt{p+\sqrt{p+\cdots}}}} = 7$ as $S = \sqrt{p+S}$, where $S = \sqrt{p+\sqrt{p+\sqrt{p+\cdots}}}$, so

$S = \sqrt{p+S} = 7 \Rightarrow 7^2 = p+7 \Rightarrow p = 42$.

Example 12. If $\sqrt{x} = \sqrt{992+\sqrt{992+\sqrt{992+\sqrt{992+\cdots}}}}$, find the value of x.

Solution: 1024.
Square both sides: $x = 992+\sqrt{992+\sqrt{992+\sqrt{992+\cdots}}}$, or $x = 992+\sqrt{x}$.
Let $y = \sqrt{x} \geq 0$. We have $y^2 = 992+y \Rightarrow y^2-y-992 = 0 \Rightarrow (y-32)(y+31) = 0$.
Since $y \geq 0$, $y = 32$. Therefore $x = y^2 = 32^2 = 2^{10} = 1024$.

5. CONJUGATES

Sometimes, conjugates are used to simplify radical expressions.
$x+\sqrt{y}$ and $x-\sqrt{y}$ are conjugates.

Some other useful conjugates:

$p\sqrt{q} + r\sqrt{s}$ and $p\sqrt{q} - r\sqrt{s}$

$\sqrt[m]{a^n}$ and $\sqrt[m]{a^{m-n}}$

$a \pm \sqrt{b}$ and $a \mp \sqrt{b}$

$m\sqrt{a} \pm n\sqrt{b}$ and $m\sqrt{a} \mp n\sqrt{b}$

$\sqrt[3]{a} \pm \sqrt[3]{b}$ and $\sqrt[3]{a^2} \mp \sqrt[3]{ab} + \sqrt[3]{b^2}$.

Example 13. What value of x makes the following radical product equal to 13?

$$\sqrt{3\sqrt{x} - \sqrt{7x + \sqrt{4x-1}}} \cdot \sqrt{2x + \sqrt{4x-1}} \cdot \sqrt{3\sqrt{x} + \sqrt{7x + \sqrt{4x-1}}}.$$

A. 1   B. $3 + \sqrt{13}$ C. 7   D. $\sqrt{50}$ E. none of these.

Solution: C.

$$\left(\sqrt{3\sqrt{x} - \sqrt{7x + \sqrt{4x-1}}} \cdot \sqrt{3\sqrt{x} + \sqrt{7x + \sqrt{4x-1}}}\right) \cdot \sqrt{2x + \sqrt{4x-1}}$$

$$= \left(\sqrt{9x - (7x + \sqrt{4x-1})}\right) \cdot \sqrt{2x + \sqrt{4x-1}} = \sqrt{(2x)^2 - (4x-1)} = 13.$$

$2x - 1 = 13$ ⇒ $x = 7$.

Example 14. If a, b, and c are integers such that $(\sqrt[3]{4} + \sqrt[3]{2} - 2)(a\sqrt[3]{4} + b\sqrt[3]{2} + c) = 20$, find the value of $a + b - c$.

A. 10   B. 18   C. 6   D. $2\sqrt[3]{4}$ E. $\sqrt[3]{4}$

Solution: A.

If you expand the expression $(\sqrt[3]{4} + \sqrt[3]{2} - 2)(a\sqrt[3]{4} + b\sqrt[3]{2} + c) = 20$, you get $2^{\frac{2}{3}}(c + b - 2a) + 2^{\frac{1}{3}}(c + 2a - 2b) + 2(a + b - c) = 20$.

Since we know that a, b, and c are integers, the first two terms must be zero, and the final term, $2(a + b - c)$ must equal 20, so $a + b - c = 10$. Note: You can solve for a, b, and c, getting $(a, b, c) = (6, 8, 4)$, but there is no need to since the problem asks for $a + b - c$.

Algebra II Through Competitions   **Chapter 10 Radicals**

6. RATIONALIZING THE DENOMINATOR:

When we simplify a radical, we want to get rid of the radical sign in the denominator. The process of achieving this is called **rationalizing the denominator**.

$$\frac{\sqrt{a}}{\sqrt{b}} = \frac{\sqrt{ab}}{b}.$$

Example 15. Write the following without radicals in the denominator: $\dfrac{\sqrt{7}}{\sqrt{7}-\sqrt{2}}$.

A) $-\dfrac{\sqrt{2}}{2}$ B) $\dfrac{\sqrt{2}}{2}$ C) $\dfrac{7+\sqrt{2}}{5}$ D) $\dfrac{7-\sqrt{14}}{5}$ E) $\dfrac{7+\sqrt{14}}{5}$

Solution: E.

$$\frac{\sqrt{7}}{\sqrt{7}-\sqrt{2}} = \frac{\sqrt{7}(\sqrt{7}+\sqrt{2})}{(\sqrt{7}-\sqrt{2})(\sqrt{7}+\sqrt{2})} = \frac{7+\sqrt{14})}{7-2} = \frac{7+\sqrt{14}}{5}.$$

Example 16. Solve the equation $8^{\frac{1}{6}} + x^{\frac{1}{3}} = \dfrac{7}{3-\sqrt{2}}$.

A. 24   B. 27   C. 32   D. 64   E. none of A, B, C or D

Solution: B.

Note that $8^{\frac{1}{6}} = (2^3)^{\frac{1}{6}} = 2^{\frac{1}{2}}$. Rationalizing the denominator of the right side gives

$\dfrac{7}{3-\sqrt{2}} = \dfrac{7(3+\sqrt{2})}{(3-\sqrt{2})(3+\sqrt{2})} = \dfrac{21+7\sqrt{2}}{7} = 3+\sqrt{2}$. Thus the equation reduces to $x^{\frac{1}{3}} = 3$ or $x = 27$.

7. SOLVING RADICAL EQUATIONS

Example 17. Solve $\sqrt{2-x} + x = 1$. If there are multiple solutions, give the largest one.

A. 1   B. $\dfrac{1-\sqrt{5}}{2}$ C. $3-\sqrt{13}$ D. $\dfrac{1+\sqrt{5}}{2}$ E. No solution exists

Solution: B.

We note that $1 - x \leq 0$ or $x \leq 1$. Squaring both sides of the equation: $2 - x = (1 - x)^2 \Rightarrow$
$x^2 - x - 1 = 0 \Rightarrow x = \dfrac{1 - \sqrt{5}}{2}$.

Example 18. Solve the equation $\dfrac{\sqrt{x+1} + \sqrt{x-1}}{\sqrt{x+1} - \sqrt{x-1}} = 3$ for x.

A. 4/3   B. 5/3   C. 7/5   D. 9/5   E. none of A, B, C or D

Solution: B.
Cross multiplying and simplifying, we obtain $2\sqrt{x-1} = \sqrt{x+1}$.
Squaring both sides yields $4(x - 1) = x + 1$ or $3x = 5$. Thus, $x = 5/3$.

Example 19. The equation $\sqrt{(x+7)} + x = 13$ has
A. no roots   B. one root   C. two roots   D. three roots   E. none of A, B, C, or D.

Solution: B.
Subtract x from both sides and then square both sides to get $x + 7 = 169 - 26x + x^2$ which is equivalent to $x^2 - 27x + 162 = 0$. This can be factored to give the roots $x = 18$ and $x = 9$, but the former is extraneous, so there is just one root.

Example 20. Solve the radical equation $\sqrt{5x + 21} = x + 3$.

Solution:
$\sqrt{5x + 21} = x + 3$ $\Rightarrow$ $5x + 21 = x^2 + 6x + 9$ $\Rightarrow$ $x^2 + x - 12 = 0$.
$0 = (x + 4)(x - 3)$. $x = 3$, and $x = -4$ (extraneous).

Example 21. Solve the equation $\sqrt{x-4} + \sqrt{x+1} = 5$.
A. $x = 4$ B. $x = 14$ C. $x = 8$ D. $x = 0$ E. No solution

Solution: C.
We re-write the equation as $\sqrt{x-4} = 5 - \sqrt{x+1}$.
Square both sides: $x - 4 = (5 - \sqrt{x+1})^2$ $\Rightarrow$ $x - 4 = 25 + x + 1 - 10\sqrt{x+1}$
$\Rightarrow$ $\sqrt{x+1} = 3$.
We square both sides again: $x + 1 = 9 \Rightarrow$ $x = 8$.

Algebra II Through Competitions   **Chapter 10 Radicals**

PROBLEMS

Problem 1. The domain of the function $f(x) = \sqrt{7-x}$ over the real numbers is
A. $x > 7$ B. $x \geq 7$ C. $x < 7$ D. $x \leq 7$ E. $-\infty < x < \infty$

Problem 2. The range of the function $y = -3\sqrt{1-2x} + 1$ is:
A. $y \geq \dfrac{1}{2}$ B. $y \leq \dfrac{1}{2}$ C. $y \geq 1$ D. $y \leq 1$ E. none of these

Problem 3. Write the following without a radical in the denominator: $\dfrac{-3}{-\sqrt{2}+\sqrt{5}}$.
A) $-\sqrt{2}+\sqrt{5}$ B) $-\sqrt{2}-\sqrt{5}$ C) $\sqrt{2}+\sqrt{5}$ D) $\sqrt{2}-\sqrt{5}$ E) none of these

Problem 4. Write the following without a radical in the denominator: $\dfrac{\sqrt{5}}{\sqrt{5}+\sqrt{3}}$.
A) $\dfrac{5-\sqrt{15}}{8}$ B) $\dfrac{\sqrt{3}}{3}$ C) $\dfrac{5+\sqrt{3}}{3}$ D) $\dfrac{5-\sqrt{15}}{2}$ E) none of these

Problem 5. $(\sqrt{2}+\sqrt{3}-\sqrt{5}) \cdot (\sqrt{2}+\sqrt{3}+\sqrt{5})$ is equal to
A) $8+2\sqrt{6}$ B) $2\sqrt{6}$ C) $2\sqrt{10}+2\sqrt{15}+2\sqrt{6}$ D) $-2\sqrt{6}$ E) $2\sqrt{30}$

Problem 6. Which of the following is equal to $\sqrt[4]{128}$?
(A) $2\sqrt[4]{8}$ (B) $2\sqrt[4]{9}$ (C) $4\sqrt[4]{8}$ (D) $8\sqrt[4]{2}$ (E) $9\sqrt[4]{2}$

Problem 7. Simplify: $\sqrt{\dfrac{2^{x+4} - 2(2^{x+1})}{2(2^{x+3})}}$.
A. $\dfrac{3}{8}$ B. $\dfrac{\sqrt{3}}{4}$ C. 2^x D. $\dfrac{x\sqrt{3}}{4}$ E. $\dfrac{\sqrt{3}}{2}$.

Problem 8. Evaluate $(\sqrt[3]{\sqrt{6}})^{12}$.
A. 216   B. $\sqrt{6}$ C. 36   D. $\sqrt[4]{6}$ E. None of these.

Algebra II Through Competitions   **Chapter 10 Radicals**

Problem 9. Which value best approximates $\sqrt{3+\sqrt{3+\sqrt{3+\sqrt{3+\cdots}}}}$?

A. 1.499987   B. $\dfrac{1}{2}+\dfrac{\sqrt{13}}{2}$ C. 3   D. infinity   E. none of these

Problem 10. The expression $\sqrt{5+\sqrt{5+\sqrt{5+\sqrt{5+\cdots}}}}$ reduces to which of the following number?

A. $\dfrac{1+\sqrt{21}}{2}$ B. $\dfrac{1+2\sqrt{5}}{2}$ C. $\dfrac{5+\sqrt{5}}{2}$ D. $\dfrac{5-\sqrt{5}}{2}$ E. 3.

Problem 11. (2009 Illinois Algebra I) The value of

$x=\sqrt{2010-\sqrt{2010-\sqrt{2010-\sqrt{2010-\cdots}}}}$ can be expressed $\dfrac{-1+\sqrt{k}}{w}$ where k and w are positive integers. Find the smallest possible value of $(k+w)$.

Problem 12. An isosceles right triangle region of area 36 is cut from a corner of a rectangular region with sides of length $6\sqrt{2}$ and $6(\sqrt{2}+1)$. What is the perimeter of the resulting trapezoid?

A. 36   B. $18\sqrt{2}+18$ C. 30   D. $12\sqrt{2}+24$ E. $24\sqrt{2}+12$

Problerm 13. A 25 foot tall ladder is placed along the vertical wall of a house. The foot of the ladder is 20 feet from the bottom of the house. If the top of the ladder slips 8 feet, then the foot of the ladder will slide how many feet?
A. 3 ft.   B. 5 ft.   C. 8 ft.   D. 4 ft.   E. 7 ft.

Problem 14. If the sum of the squares of the lengths of all the sides of a rectangle is 100, then the length of a diagonal of the rectangle is
A. $2\sqrt{5}$.   B. $2\sqrt{13}$.   C. $4\sqrt{3}$.   D. $5\sqrt{2}$.   E. 10.

Problem 15. Solve for x: $\sqrt{1+\sqrt{3-\sqrt{1+\sqrt{2+\sqrt{x}}}}}=1$.

A. 3844   B. 62   C. 62   D. 64   E. 4096

Algebra II Through Competitions   **Chapter 10 Radicals**

Problem 16. How many real number solutions does the equation $5\sqrt{x} = 6 - x$ have?
A. 0   B. 1   C. 2   D. 3   E. None of these

Problem 17. What is the sum of the digits of the integer solution to
$\sqrt{14 + \sqrt{27 - \sqrt{x-1}}} = 4$?
A. 5   B. 6   C. 8   D. 9   E. 11.

Problem 18. Let z denote the real number solution to
$\sqrt{3 + \sqrt{x-1}} = 5$. What is the sum of the digits of z?
A. 13   B. 14   C. 15   D. 16   E. 17.

Problem 19. The number of real solutions of the equation $(\sqrt{3x-2} + \sqrt{2x-3}) = 1$ is:
A. 0   B. 1   C. 2   D. 3   E. 4

Problem 20. Suppose that $\sqrt{x+1} = 1 - x$. Which of the following statements is correct?
A. There are no solutions.
B. There are two solutions. The larger solution is greater than 2.
C. There are two solutions. The larger solution is less than or equal to 2.
D. There is only one solution. This solution is greater than 2.
E. There is only one solution. This solution is less than 2.

Problem 21. What is the sum of the solutions of $\sqrt[3]{x^2} + \sqrt[3]{x} = 6$?
A. −19   B. −18   C. 6   D. 8   E. 18

Problem 22. The sum of the solutions to the equation $x - 2 = \sqrt{2x-1}$ is:
A. 4   B. 5   C. 6   D. 7   E. none of these

Problem 23. The solution to the equation $\sqrt{3x+4} = 1 + \sqrt{x+5}$ is:
A. 1   B. −1   C. −1 and 4   D. 4   E. 1 or −4.

Problem 24. How many real solutions does the following equation have?

$\sqrt{1+x+\sqrt{x}} = \sqrt{x+\sqrt{x+7}}$

A. 4 B. 3 C. 2 D. 1 E. 0

Problem 25. What is the sum of all the real solutions to the equation

$\sqrt{4x+\sqrt{17x^2+2}} = x+2$?

A. 0 B. $\sqrt{2}$ C. $\sqrt{7}$ D. $\sqrt{2} + \sqrt{7}$ E. none of these

Problem 26. Find the minimum value of $\sqrt{x^2+y^2}$ if $5x+12y = 60$.

A. 60/13. B. 13/12. C. 1. D. 0. E. None of the above.

Problem 27. Suppose N is a positive integer that is a perfect cube. Which of the following represents the next positive integer that is a perfect cube?

A. $N^3 + 3\sqrt[3]{N} + 1$ B. $N + 3\sqrt[3]{N^2} + 3\sqrt[3]{N} + 1$ C. $N^3 + 3N^2 + 3N + 1$
D. $N^3 + N^2 + N + 1$ E. N^3

Algebra II Through Competitions   **Chapter 10 Radicals**

SOLUTIONS

Problem 1. Solution: D.
$\sqrt{7-x} \geq 0 \implies 7-x \geq 0 \implies x \leq 7$.

Problem 2. Solution: D.
Since $\sqrt{1-2x} \geq 0$, $-3\sqrt{1-2x}$ is negative.
The maximum value of y is $y_{max} = -3 \times 0 + 1 = 1$. Thus $y \leq 1$.

Problem 3. Solution: B.
$$\frac{-3}{-\sqrt{2}+\sqrt{5}} = \frac{-3}{\sqrt{5}-\sqrt{2}} = \frac{-3(\sqrt{5}+\sqrt{2})}{(\sqrt{5}-\sqrt{2})\sqrt{5}+\sqrt{2}} = \frac{-3(\sqrt{5}+\sqrt{2})}{3} = -\sqrt{2}-\sqrt{5}.$$

Problem 4 Solution: D.
$$\frac{\sqrt{5}}{\sqrt{5}+\sqrt{3}} = \frac{\sqrt{5}(\sqrt{5}-\sqrt{3})}{(\sqrt{5}+\sqrt{3})(\sqrt{5}-\sqrt{3})} = \frac{5-\sqrt{15}}{2}.$$

Problem 5. Solution: B
$(\sqrt{2}+\sqrt{3}-\sqrt{5}) \cdot (\sqrt{2}+\sqrt{3}+\sqrt{5}) = (\sqrt{2}+\sqrt{3})^2 - (\sqrt{5})^2 = 2+3+2\sqrt{6}-5 = 2\sqrt{6}$.

Problem 6. Solution: A.
$\sqrt[4]{128} = \sqrt[4]{2^4 \times 2^3} = 2\sqrt[4]{2^3} = 2\sqrt[4]{8}$.

Problem 7. Solution: E.
$$\sqrt{\frac{2^{x+4}-2(2^{x+1})}{2(2^{x+3})}} = \sqrt{\frac{2^{x+4}-2^{x+2}}{2^{x+4}}} = \sqrt{\frac{2^{x+4}(1-2^{-2})}{2^{x+4}}} = \sqrt{1-\frac{1}{4}} = \sqrt{\frac{3}{4}} = \frac{\sqrt{3}}{2}.$$

Problem 8. Solution: C.
$$(\sqrt[3]{\sqrt{6}})^{12} = \left(\sqrt[3]{6^{\frac{1}{2}}}\right)^{12} = \left(6^{\frac{1}{6}}\right)^{12} = 6^2 = 36.$$

Problem 9. Solution: B.

Let $\sqrt{3+\sqrt{3+\sqrt{3+\sqrt{3+\cdots}}}}$ Because the nested root repeats infinitely,

$x^2 - 3 = x \quad \Rightarrow \quad x = \dfrac{1 \pm \sqrt{1^2 - 4 \cdot 1 \cdot (-3)}}{2} \quad \Rightarrow \quad x = \dfrac{1+\sqrt{13}}{2}$ because $x > 0$

Problem 10. Solution: A.

Let $x = \sqrt{5+\sqrt{5+\sqrt{5+\sqrt{5+\cdots}}}}$ then $x = \sqrt{5+x} \Rightarrow x^2 - x - 5 = 0 \Rightarrow x = \dfrac{1+\sqrt{21}}{2}$.

Problem 11. Solution: 8043.

By squaring both sides of $x = \sqrt{2010 - \sqrt{2010 - \sqrt{2010 - \sqrt{2010 - \cdots}}}}$, we get

$x^2 = 2010 - \sqrt{2010 - \sqrt{2010 - \sqrt{2010 - \cdots}}} \quad \Rightarrow \quad x^2 = 2010 - x \Rightarrow x^2 + x - 2010 = 0$.

Since x is positive, $x = \dfrac{-1+\sqrt{1^2 - 4 \times (-2010)}}{2} = \dfrac{-1+\sqrt{8041}}{2}$.

$k = 8041$ and $w = 2$. $k + w = 8043$.

Problem 12. Solution: E.

$x = 6(\sqrt{2}+1) - 6\sqrt{2} = 6$
$y = 6\sqrt{2} \cdot \sqrt{2} = 12$
Perimeter $= 6 + 6\sqrt{2} + 6(\sqrt{2}+1) + 12 = 24 + 12\sqrt{2}$

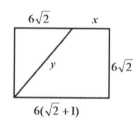

Problem 13. Solution: D.
The foot of the 25-foot ladder is 20 feet from the wall and the top of the ladder is 15 feet from the ground. (This is a 3-4-5 right triangle.) When the top slips down 8 feet, the base will be $\sqrt{25^2 - 7^2} = \sqrt{625 - 49} = \sqrt{576} = 24$ feet from the wall, so it has moved 4 feet.

Problem 14. Solution: D.
Let x and y be the dimensions of the rectangle. Then $2x^2 + 2y^2 = 100 \Rightarrow x^2 + y^2 = 50$
$\Rightarrow \sqrt{x^2 + x^2} = \sqrt{50} = 5\sqrt{2}$ and the length of the diagonal is $\sqrt{x^2 + y^2}$.

Problem 15. Solution: A.

Square both sides and get $1+\sqrt{3-\sqrt{1+\sqrt{2+\sqrt{x}}}}=1$.

So $\sqrt{3-\sqrt{1+\sqrt{2+\sqrt{x}}}}=0$.

So again square both sides and get $3-\sqrt{1+\sqrt{2+\sqrt{x}}}=0$.

So $\sqrt{1+\sqrt{2+\sqrt{x}}}=3$ so square both side and get $1+\sqrt{2+\sqrt{x}}=9$ then $\sqrt{2+\sqrt{x}}=8$.

So again square both sides and have $2+\sqrt{x}=64 \Rightarrow \sqrt{x}=62$.

So square both sides on last time to get $x = 3844$.

Problem 16. Solution: B.
Squaring both sides and simplifying gives $x^2 - 37x + 36 = (x - 36)(x - 1) = 0$. So there are two solutions, $x = 36$ and $x = 1$. However, checking for extraneous roots, only $x = 1$ satisfies the original equation.

Problem 17. Solution: C.

Square both sides to get $14+\sqrt{27-\sqrt{x-1}}=16$, then massage it, square again, and solve $27-\sqrt{x-1}=4$ to get $x-1 = 529$. Thus $x = 530$ and the sum of the digits is $5 + 3 + 0 = 8$.

Problem 18. Solution: E.
Square both sides twice to get $z = (5^2 - 3)^2 + 1 = 485$, so the sum of the digits is $4 + 8 + 5 = 17$.

Problem 19. Solution: A.

Method 1:

Re-write the given equation as: $\sqrt{3x-2}=1-\sqrt{2x-3}$. Square both sides:

$3x-2 = (1-\sqrt{2x-3})^2 \quad \Rightarrow \quad 3x-2 = 1+2x-3-2\sqrt{2x-3} \Rightarrow x=-2\sqrt{2x-3}$

Square both sides again: $x^2 = 4(2x-3) \Rightarrow x^2 - 8x + 12 = 0 \Rightarrow (x-6)(x-2) = 0$.
$x = 2, 6$.
Neither solution works when checked in the original equation. Thus, there are 0 real solutions.

Method 2 (official solution):

This requires rearranging sides and squaring. $(\sqrt{3x-2}+\sqrt{2x-3})^2 = (1)^2$

$2\sqrt{6x^2 - 13x + 6} = 6 - 5x$

$4(6x^2 - 13x + 6) = 36 - 60x + 25x^2 \Rightarrow x^2 - 8x + 12 = 0 \Rightarrow (x-6)(x-2) = 0.$

$x = 2, 6.$

Neither solution works when checked in the original equation. Thus, there are 0 real solutions.

Problem 20. Solution: E.

Square both sides of the equation and reduce to get $x^2 - 3x = 0$. This has two solutions, $x = 3$ and $x = 0$. But $x = 3$ is extraneous for the given equation. Thus there is only one solution, and it is less than 2.

Problem 21. Solution: A.

$\sqrt[3]{x^2} + \sqrt[3]{x} = 6$ can be written as $(\sqrt[3]{x})^2 + \sqrt[3]{x} - 6 = 0$ \hfill (1)

Let $y = \sqrt[3]{x}$. (1) becomes: $y^2 + y - 6 = 0 \Rightarrow (y-2)(y+3) = 0$.

So we have $y = 2 \Rightarrow \sqrt[3]{x} = 2 \Rightarrow x = 2^3 = 8$

or $y = -3 \Rightarrow \sqrt[3]{x} = -3 \Rightarrow x = (-3)^3 = -27$.

The sum of the roots is $8 - 27 = -19$.

Problem 22. Solution: B.

Square both sides of the given equation: $(x-2)^2 = 2x - 1 \Rightarrow x^2 - 4x + 4 - 2x + 1 = 0$

$\Rightarrow x^2 - 6x + 5 = 0 \Rightarrow (x-1)(x-5) = 0$

$x = 5$ and $x = 1$ (extraneous).

The equation has only one root. The answer is B.

Problem 23. Solution: D.

Square both sides of the given equation: $3x + 4 = (1 + \sqrt{x+5})^2 \Rightarrow$

$3x + 4 = 1 + x + 5 + 2\sqrt{x+5} \Rightarrow 2x - 2 = 2\sqrt{x+5} \Rightarrow x - 1 = \sqrt{x+5}$.

Square both sides again: $(x-1)^2 = x + 5 \Rightarrow x^2 - 2x + 1 = x + 5 \Rightarrow$

$x^2 - 3x + 4 = 0 \Rightarrow (x-4)(x+1) = 0$

$x = 4$ or $x = -1$ (extraneous).

Problem 24. Solution: D.

$\sqrt{1 + x + \sqrt{x}} = \sqrt{x + \sqrt{x+7}} \Rightarrow 1 + x + \sqrt{x} = x + \sqrt{x+7} \Rightarrow$

$1 + \sqrt{x} = \sqrt{x+7} \Rightarrow (1+\sqrt{x})^2 = (\sqrt{x+7})^2 \Rightarrow$

$1 + 2\sqrt{x} + x = x + 7 \Rightarrow 2\sqrt{x} = 6 \Rightarrow \sqrt{x} = 3 \Rightarrow x = 9$

Problem 25. Solution: C.

To solve this, we first must square both sides to remove the first radical, yielding $4x + \sqrt{17x^2 + 2} = (x+2)^2 = x^2 + 4x + 4 \Rightarrow \sqrt{17x^2 + 2} = x^2 + 4$. No we need to square again to get rid of the remaining radical. So $17x^2 + 2 = (x^2 + 4)^2 \Leftrightarrow 17x^2 + 2 = x^4 + 8x + 16 \Leftrightarrow x^4 - 9x^2 + 14 = 0$. We can solve this by factoring, so $x^4 - 9x^2 + 14 = 0 \Leftrightarrow (x^2 - 2)(x^2 - 7) = 0 \Rightarrow x = \pm\sqrt{2}, \pm\sqrt{7}$. But when you square both sides of an equation, you can introduce extraneous roots. Clearly $x = -\sqrt{7}$ cannot be a solution since the right side of the original equation would be negative, so we should check $-\sqrt{2}$ as well. Substituting this into the original equation yields

$$\sqrt{-4\sqrt{2} + \sqrt{17(-\sqrt{2})^2 + 2}} = \sqrt{-4\sqrt{2} + \sqrt{36}} = \sqrt{-4\sqrt{2} + 6} = \sqrt{(-\sqrt{2} + 2)^2} = -\sqrt{2} + 2,$$

which is fine. But what about the other "solutions"? Again trying each of these yields

$$\sqrt{4\sqrt{2} + \sqrt{17(2) + 2}} = \sqrt{4\sqrt{7} + \sqrt{121}} = \sqrt{4\sqrt{7} + 11} = \sqrt{(\sqrt{7} + 2)^2} = \sqrt{7} + 2, \text{ so this one}$$

checks. Similarly,

$$\sqrt{4\sqrt{7} + \sqrt{17(7) + 2}} = \sqrt{4\sqrt{7} + \sqrt{121}} = \sqrt{4\sqrt{7} + 11} = \sqrt{(\sqrt{7} + 2)^2} = \sqrt{7} + 2, \text{ so this one}$$

checks as well, so the sum of the solutions is $-\sqrt{2} + \sqrt{2} + \sqrt{7} = \sqrt{7}$.

Problem 26. Solution: A.

$5x + 12y = 60 \Rightarrow y = 5 - \frac{5}{12}x$, so $\sqrt{x^2 + y^2} = \sqrt{x^2 + (5 - \frac{5}{12}x)^2} = \sqrt{\frac{169}{144}x^2 + \frac{25}{6}x + 25}$.

$\sqrt{\frac{169}{144}x^2 + \frac{25}{6}x + 25} = \sqrt{\left(\frac{13}{12}x + \frac{25}{13}\right)^2 + \left(25 - (\frac{25}{13})^2\right)}$

$= \sqrt{\left(\frac{13}{12}x + \frac{25}{13}\right)^2 + \frac{25 \cdot 169 - 25^2}{169}} = \sqrt{\left(\frac{13}{12}x + \frac{25}{13}\right)^2 + \left(\frac{25 \cdot 144}{169}\right)}$ When the term involving the variable x is the expression is zero (as small as you can make it), the remaining term is $\frac{60}{13}$.

Problem 27. Solution: B.

If $N = x^3$, then $(x+1)^3 = x^3 + 3x^2 + 3x + 1 = N + 3\sqrt[3]{N^2} + \sqrt[3]{N} + 1$.

Chapter 11 Exponential Functions

1. EXPRESSIONS:

$$\underbrace{a \times a}_{2 \text{ a's}} = a^2 \qquad \underbrace{a \times a \times a}_{3 \text{ a's}} = a^3 \qquad \underbrace{a \times a \times a \times a}_{4 \text{ a's}} = a^4 \qquad \underbrace{a \times a \times a \times a \times a \ldots a}_{n \text{ times}} = a^n$$

a is an integer, decimal, or fraction and is called **the base**. n is any counting number and is called **the exponent**.

2. PROPERTIES OF EXPONENTS:

Property 1: $a^0 = 1$
Note: $(0)^0$ is not defined and $(0)^0 \neq 1$.

Property 2: $a^1 = a$

Property 3: $a^m \times a^n = a^{m+n} \qquad \Leftrightarrow \qquad a^{m+n} = a^m \times a^n$ (Power rule)
Note: $a^m + a^n \neq a^{m+n}$

Example 1. Let $f(x) = x^x$. What is $f(a+b)$?
A. $(a+b)^a (a+b)^b$ B. $a^a + b^b$ C. $a^a + b$ D. $(a+b)^x$ E. $a^a \times b^b$

Solution: A.
$f(x) = x^x \implies f(a+b) = (a+b)^{(a+b)} = (a+b)^a (a+b)^b$.

Example 2. If $3^x = 5$, then $3^{2x+3} =$
A. 37.   B. 75.   C. 270.   D. 325.   E. 675.

Solution: E.
$3^x = 5 \implies 3^{2x} = 5^2 = 25$, so $3^{2x+3} = 3^{2x} \times 3^3 = 25 \times 27 = 675$.

Example 3. If S and T are sets such that S has 2 more elements than T and 96 more subsets than T, how many elements are in S?
A. 3   B. 5   C. 7   D. 9   E. 11

Solution: C.
Let the number of elements in S and T be m and n respectively. We know that the number of subsets of a set with m elements is 2^m, so we have $m - n = 2$ and $2^m - 2^n = 96$. Let $m = n + 2$ so that $2^m - 2^n = 2^{n+2} - 2^n = 2^n(2^2 - 1) = 96 \implies 2^n \cdot 3 = 96 \implies 2^n = 32 \implies n = 5 \implies m = 7$.

Algebra II Through Competitions — Chapter 11 Exponential Functions

Property 4: $\dfrac{a^m}{a^n} = a^{m-n}$ (Quotient rule).

Property 5: $(a^m)^n = a^{mn}$ $\qquad\qquad (ab)^n = a^n b^n$ (Power rule)

Example 4. The numbers x and y satisfy $2^x = 15$ and $15^y = 32$. What is the value of xy?
A. 3   B. 4   C. 5   D. 6   E. none of A, B, C or D

Solution: C.
Note that $(2^x)^y = 15^y = 32$ so $2^{xy} = 2^5$ and $xy = 5$.

Example 5. Let a, b, $c \geq 2$ be natural numbers and $a^{(b^c)} = (a^b)^c$. Which one(s) of a, b, c can have arbitrary values?
A. a B. b C. c D. both a and b E. all three

Solution: A.
The given equation is equivalent to $b^c = bc$ with a arbitrary.
Thus $b^{c-1} = c$. With natural b, $c \geq 2$, one solution is $b = c = 2$, but $a \geq 2$ is arbitrary.

Example 6. $4^{1/4} \times 256^{1/4} =$
A. $4\sqrt{2}$ B. $5\sqrt{2}$ C. 8   D. 10   E. none of these

Solution: A.
$4^{1/4} \times 256^{1/4} = (2^2)^{1/4} \times (2^8)^{1/4} = 2^{1/2} \times 2^2 = 4\sqrt{2}$.

Property 6: $\dfrac{a^m}{b^m} = \left(\dfrac{a}{b}\right)^m \quad \Leftrightarrow \quad \left(\dfrac{a}{b}\right)^m = \dfrac{a^m}{b^m}$ (Power rule)

Property 7: $a^{-n} = \dfrac{1}{a^n}$ (Negative exponents)

Property 8: $a^{\frac{m}{n}} = \left(a^{\frac{1}{n}}\right)^m$ (Rational exponents)

Example 7. If $10^{2y} = 25$, then 10^{-y} equals
(A) $-1/5$ (B) $1/625$ (C) $1/50$ (D) $1/25$ (E) $1/5$.
Solution: E.

$10^{2y} = 25 \Rightarrow (10^y)^2 = 5^2 \Rightarrow 10^y = 5 \Rightarrow (10^y)^{-1} = (5)^{-1}$
$\Rightarrow 10^{-y} = 1/5.$

Example 8. What non-zero real value for x satisfies $(7x)^{14} = (14x)^7$?
A. 1/7   B. 2/7   C. 1   D. 7   E. 14

Solution: B.
Taking the seventh root of both sides, we get $(7x)^2 = 14x$.
Simplifying gives $49x^2 = 14x$, which then simplifies to $7x = 2$. Thus $x = 2/7$.

Example 9. Which of the following is equivalent to $\dfrac{x^5 y^2 z^8}{(xy)^{-3}}$?

A. $\dfrac{x^2 z^8}{y}$ B. $x^{12} y^8 z^8$ C. $\dfrac{x^{-4} y z^8}{3}$ D. $x^8 y^5 z^8$ E. $x^{-2} y z^8$

Solution: D.
$\dfrac{x^5 y^2 z^8}{(xy)^{-3}} = \dfrac{x^5 y^2 z^8}{x^{-3} y^{-3}} = x^{5+3} y^{2+3} z^8 = x^8 y^5 z^8.$

3. SOLVING EXPONENT EQUATIONS

(1). $a^m = a^n$ if and only if $m = n$.
(2). $a^m = b^n$ if and only if $a = b$ and $m = n$.
(3). $(ax^2 + bx + c)^{dx^2 + ex + f} = 1$.
   Case 1: Solving $ax^2 + bx + c = 1$ (Ignoring $dx^2 + ex + f$).
   Case 2: Solving $dx^2 + ex + f = 0$ with $ax^2 + bx + c \neq 0$.
   Case 3: $ax^2 + bx + c = -1$ with even $dx^2 + ex + f$.

Example 10. Solve the equation. $27^{2t-1} = 81^{t+2}$
A. 3   B. –3   C. –1/2.   D. –2   E. 11/2.

Solution: E.
$27^{2t-1} = 81^{t+2} \Rightarrow (3)^{3(2t-1)} = (3)^{4(t+2)} \Rightarrow 3(2t-1) = 4(t+2) \Rightarrow 6t - 3 = 4t + 8 \Rightarrow 2t = 11 \Rightarrow t = 11/2.$

Example 11. Suppose $a, b < 1$. If $a^x = b^y$ and $a^y = b^x$, then
A. $a = b$ B. $a > b$ C. $a < b$ D. $a \neq b$ E. None of these

Solution: A.
$a^x = b^y$ (1)
$a^y = b^x$ (2)
(1) × (2): $a^{x+y} = b^{y+x}$
Thus we have $a = b$.

Example 12. Solve: $\frac{1}{25} = 125^{3x-4}$

A. $\frac{2}{9}$ B. $\frac{14}{9}$ C. $\frac{2}{3}$ D. $\frac{10}{9}$ E. $\frac{9}{10}$

Solution: D.
$5^{-2} = 5^{3(3x-4)}$ ⇒ $-2 = 9x - 12$ ⇒ $x = 10/9$.

Example 13. If $4^x - 4^{x-1} = 24$, then $(2x)^x$ equals

A. $5\sqrt{5}$ B. $\sqrt{5}$ C. $25\sqrt{5}$ D. 125   E. 25

Solution: C.
$4^x - 4^{x-1} = 24$ ⇒ $4^x - \frac{4^x}{4} = 24$ ⇒ $\frac{4 \times 4^x}{4} - \frac{4^x}{4} = 24$ ⇒ $\frac{3 \times 4^x}{4} = 24$

⇒ $3 \times 4^x = 24 \times 4$ ⇒ $4^x = 8 \times 4$ ⇒ $2^{2x} = 2^5$.

Solving we get $x = 5/2$.
So $2x = 5$ $(2x)^x = (5)^{\frac{5}{2}} = 25\sqrt{5}$.

Example 14. How many real solutions are there to the equation $(5^{6x+3})(25^{3x+6}) = 125^{4x+5}$?

A. 0   B. exactly 1   C. exactly 2   D. exactly 3   E. none of these

Solution: E
$(5^{6x+3})(25^{3x+6}) = 125^{4x+5}$ ⇒ $(5^{6x+3})(5^{2(3x+6)}) = 5^{3(4x+5)}$ ⇒ $5^{6x+3 + 2(3x+6)}$
$= 5^{12x+15}$ ⇒ $6x + 3 + 2(3x + 6) = 12x + 15$.
Simplifying we get $12x + 15 = 12x + 15$ which is always true no matter what value for x.
So the answer is E.

Example 15. Solve $9^x - 3^x = 12$.

A. $\log_4 3$ B. $\log_3 12$ C. $\log_3 4$ D. $11 - 2$ E. $\log_6 12$

Solution: C.
$9^x - 3^x = 12 \Rightarrow 3^{2x} - 3^x = 12$ (1)
Let $y = 3^x$.
(1) becomes: $y^2 - y = 12 \Rightarrow (y-4)(y+3) = 0$.
$y = 4$ or $y = -3$ (ignored)
That is, $3^x = 4 \Rightarrow x = \log_3 4$.

Example 16. How many integers x satisfy the equation $(x^2 - x - 1)^{x+2} = 1$?
A. 2.   B. 3.   C. 4.   D. 5.   E. None of these.

Solution: C.
Case 1: $x^2 - x - 1 = 1$.
If $x^2 - x - 1 = 1$, then $x^2 - x - 2 = 0 \Rightarrow x = 2$ or $x = -1$.
Case 2: $x + 2 = 0$ and $x^2 - x - 1 \neq 0$.
When $x = -2$, $(x^2 - x - 1)^{x+2} = 1 \neq 0$. So $x = -2$ is a solution.
Case 3: $x^2 - x - 1 = -1$ and $x + 2$ is even.
$x^2 - x - 1 = -1 \Rightarrow x^2 - x = 0 \Rightarrow x = 0$ or $x = 1$. If $x = 0$, then $x + 2$ is even. If $x = 1$, then $x + 2$ is odd and $(-1)^3 \neq 1$. So case 3 gives us one solution: $x = 0$.
Therefore the solutions are $x = 2$, $x = -1$ (from case 1), $x = -2$ (from case 2), and $x = 0$ (from case 3).

Example 17. One of the values for x that satisfies the equation $4^{(x^2 - 3x)} = 8^{\frac{1}{2}} \cdot 2^{\frac{3}{2}}$ can be expressed in reduces simplest form $\dfrac{k + \sqrt{w}}{f}$ where k, w, and f are positive integers. Find the value of $(k + w + f)$.

Solution: 20.
$4^{(x^2 - 3x)} = 8^{\frac{1}{2}} \cdot 2^{\frac{3}{2}} \Rightarrow 2^{2(x^2 - 3x)} = 2^{\frac{3}{2}} \cdot 2^{\frac{3}{2}} \Rightarrow 2^{2(x^2 - 3x)} = 2^3$.
Thus we get $2(x^2 - 3x) = 3 \Rightarrow 2x^2 - 6x - 3 = 0$
$x_{1,2} = \dfrac{6 \pm \sqrt{6^2 - 4 \times 2 \times (-3)}}{2 \times 2} = \dfrac{6 \pm \sqrt{36 + 24}}{4} = \dfrac{6 \pm 2\sqrt{15}}{4} = \dfrac{3 \pm \sqrt{15}}{2}$.
$k + w + f = 3 + 15 + 2 = 20$.

Algebra II Through Competitions — **Chapter 11 Exponential Functions**

Example 18. If $x \neq 0$ and $\dfrac{x^k}{x^{(3(k+2))}} = x^2$, find the value of k.

Solution: –4.

$$\dfrac{x^k}{x^{(3(k+2))}} = x^2 \Rightarrow x^{k-3k-6} = x^2 \Rightarrow k-3k-6=2 \Rightarrow k=-4.$$

Example 19. Find x in $\dfrac{b^{2y}\sqrt{a^3}}{a^{4x}b^{1/3}} = a^{7/10}b^{1/15}$.

A. 2/5   B. 1/10   C. 1/5   D. 5   E. None of these

Solution: C.

$$\dfrac{b^{2y}\sqrt{a^3}}{a^{4x}b^{1/3}} = a^{7/10}b^{1/15} \Rightarrow \dfrac{b^{2y}a^{\frac{3}{2}-4x}}{b^{1/3}} = a^{7/10}b^{1/15}$$

Since we only need to find x, we focus on the terms containing x. So we can have

$$a^{\frac{3}{2}-4x} = a^{7/10} \Rightarrow \dfrac{3}{2}-4x = \dfrac{7}{10} \Rightarrow 4x = \dfrac{3}{2}-\dfrac{7}{10} = \dfrac{8}{10} = \dfrac{4}{5} \Rightarrow x = \dfrac{1}{5}.$$

Example 20. Solve the equation for x: $y = 2e^{2x} - 1$.

A. $\dfrac{1}{4}\ln\dfrac{y}{2}$ B. $\dfrac{1}{2}\ln\dfrac{y}{2}$ C. $\ln\dfrac{y}{2}$ D. $\dfrac{1}{2}\ln\dfrac{y+1}{2}$ E. $\ln\dfrac{y+1}{2}$

Solution: D.
We re-write $y = 2e^{2x} - 1$ as $(y+1)/2 = e^{2x}$.
Take the natural logarithm of both sides of the equation:

$$2x = \ln\left(\dfrac{y+1}{2}\right) \Rightarrow x = \dfrac{1}{2}\ln\left(\dfrac{y+1}{2}\right).$$

PROBLEMS

Problem 1. Let $f(x) = 2^x$. For which exact value of x is $f(x-2) = f(x) - 2$?
A. 8/3   B. 1.415   C. 2   D. $\log_2 3$ E. $3 - \log_2 3$

Problem 2. Find the intersection of $y = 2^{x+2}$ and $y = 4^{3x-4}$.
A. (0, 4)   B. (0.5, 32)   C. (1, 8)   D. (2, 16)   E. (3, 32)

Problem 3. If $1 \times 10^a + 2 \times 10^b + 3 \times 10^c + 4 \times 10^d = 24{,}130$ and $a \neq b \neq c \neq d$, then what does $a/2 + b/4 + c/8 + d/16$ equal?
A. $2\frac{5}{16}$ B. $\frac{5}{16}$ C. $2\frac{3}{16}$ D. $1\frac{7}{16}$ E. $2\frac{3}{8}$.

Problem 4. Let $2^a + 2^b + 2^c = 42$ where $a \neq b \neq c$ and $a, b, c \in \mathbf{N}$. Determine $1/a + 1/b + 1/c$.
A. 23/15   B. 8/5   C. 47/60   D. 42/23   E. 19/15

Problem 5. Solve the equation for x in terms of y: $y = 3e^{2x-1} + 5$, given that $y > 5$.
A. $\frac{1}{3}e^{2y+1} - 5$ B. $\ln(\frac{y-5}{3}) + 1$ C. $\frac{1}{2}\left(\ln\left(\frac{y-5}{3}\right) + 1\right)$ D. $3\ln(2y-1) + 5$
E. none of these

Problem 6. Let $g(x) = \dfrac{a^x - a^{-x}}{a^x + a^{-x}}$, where $a > 0$ and $a \neq 1$. If $g(p) = \dfrac{1}{3}$, then $g(4p)$ is equal to
A. $\dfrac{15}{17}$ B. $\dfrac{7}{8}$ C. $\dfrac{13}{15}$ D. $\dfrac{6}{7}$ E. $\dfrac{11}{13}$

Problem 7. Solve for b: $e^{2b} + 5e^b = 14$.
A. 0   B. $\ln 3 - \ln 2$ C. $\ln 3$ D. 1.2345   E. $\ln 2$

Problem 8. Which answer below is the x-coordinate of an intersection between: $y = e^{bx}$ and $y = 2(x+1)^2$ if $b = \ln 2$?
A. 0.5 – 1   B. 6.120394   C. 7   D. $5 + \ln 7$ E. curves don't intersect.

Algebra II Through Competitions **Chapter 11 Exponential Functions**

Problem 9. Write x, y and z in order from largest to smallest if $x = 2^{100}$, $y = 3^{75}$ and $z = 5^{50}$.
A. $x > y > z$ B. $x > z > y$ C. $y > x > z$ D. $y > z > x$ E. $z > x > y$.

Problem 10. Solve the equation, $12 + 64^x = 8^{x+1}$.
A. $x = \dfrac{1}{3}$ only   B. $x = \dfrac{1}{3}$ or $\dfrac{\log_2 6}{8}$ C. $x = \dfrac{1}{3}$ or $\dfrac{\log_2 6}{3}$ D. $x = \log_2 12$ only
E. $\dfrac{\sqrt{3}}{3}$ or $\dfrac{1}{3}$.

Problem 11. Solve for x: $8^{3x+1} - 8^{3x} = 448$.
A. 2/3   B. 3/2   C. 2   D. 2.93578   E. none of the above

Problem 12. If $3^{x+2} = 2^{2x-1}$, then x is:

a. $\dfrac{\log 18}{\log \frac{4}{3}}$ b. $\dfrac{\log 12}{\log \frac{4}{3}}$ c. $\dfrac{\log 27}{\log 2}$ d. $\dfrac{\log 3}{\log 2}$ e. $\dfrac{\log \frac{4}{3}}{\log 18}$

Problem 13. Solve $\dfrac{e^x - e^{-x}}{2} = 2$ for x, where $x > 0$.
A. $\ln(2 + \sqrt{5})$ B. $\ln(2 \pm \sqrt{5})$ C. $\ln 2 + \ln\sqrt{5}$ D. 4   E. none of the above

Problem 14. Solve for x: $3^{2x} + 3^x - 20 = 0$.
A. $\dfrac{\ln 3}{\ln 4}$ B. $\dfrac{\ln 3}{\ln 20}$ C. $\dfrac{\ln 20}{\ln 3}$ D. $\dfrac{\ln 4}{\ln 3}$ E. none of these

Problem 15. The sum of the solutions to the equation $12 - 8 \cdot 5^x + 25^x = 0$ is:
A. $\log_5 12$ B. $\log_{25} 6$ C. 6   D. 17   E. none of these

Problem 16. If $2^x \cdot 4^x \cdot 8^x = \sqrt{2}$, then what is x?
A. 1/4   B. 1/6   C. 1/8   D. 1/10   E. 1/12

Problem 17. If $a = 2b^2$, $b = 4c^3$, and $c = 8d^4$, then which of the following must be true?
A. $a = 2^{23} d^{36}$ B. $a = 2^{23} d^{24}$ C. $a = 2^{11} d^{36}$ D. $a = 2^{11} d^{24}$ E. $a = 2^{11} d^{12}$

Problem 18. The absolute value of the difference of the solutions to the equation $3^{2x} -10 \times 3^x + 9 = 0$ is
A. 1   B. 2   C. 3   D. 4   E. none of these

Problem 19. The sum of the solutions for the equation $8^{6x^2+4x} = 4^{9x^2-9x+6}$ is:
A. 1   B. -2 C. 0   D. 2/5   E. 16/15.

Problem 20. Solve for x: $9^{x+1} + 9^{x+2} + 9^{x+3} + 9^{x+4} + 9^{x+5} = 22143$
A. $-1/2$ B. $1/2$ C. $-1/3$ D. $1/3$ E. none of these

Problem 21. If $\dfrac{9+3^{2x}}{10} = 3^x$, then the value of $x^2 + x + 1$ is
A. 0 or 2   B. 0 only   C. 1 or 7   D. 7 only   E. 1 only

Problem 22. Solve the equation. $3p^{3/2} = 24$.
A. 4   B. $16\sqrt{2}$ C. $4\sqrt[3]{3}/3$ D. 16   E. 16/3.

Problem 23. How many real number solutions are there to the equation $(x^2 - 4x + 2)^{x^2-1} = 1$?
A. 2   B. 3   C. 4   D. 5   E. 6

Problem 24. The number of distinct real values x which satisfy the equation $(x^2 - 5x + 5)^{x^2 - 9x + 20} = 1$ is
A. 0   B. 2   C. 3   D. 5   E. 6

SOLUTIONS

Problem 1. Solution: E.
$2^{x-2} = 2^x - 2$. Let $y = 2^{x-2}$. $y = 4y - 2$
$y = 2^{x-2} = 2/3$. $x = 2 + \log_2(2/3) = 2 + \log_2 2 - \log_2 3 = 3 - \log_2 3$.

Problem 2. Solution: D.
$2^{x+2} = 4^{3x-4} \Rightarrow 2^{x+2} = (2^2)^{3x-4} \Rightarrow 2^{x+2} = 2^{2(3x-4)} \Rightarrow x + 2 = 6x - 8$
$\Rightarrow x = 2, y = 16$.

Problem 3. Solution: A.
Rearrange the terms so that $24{,}130 = 2 \times 10^b + 4 \times 10^d + 1 \times 10^a + 3 \times 10^c$. Now $b = 4$, $d = 3$, $a = 2$, and $c = 1$, making $a/2 + b/4 + c/8 + d/16 = 2/2 + 4/4 + 1/8 + 3/16 = 2\dfrac{5}{16}$.

Problem 4. Solution: A.
$42 = 32 + 8 + 2 = 2^5 + 2^3 + 2^1 \Rightarrow 1/5 + 1/3 + 1/1 = 23/15$.

Problem 5. Solution: C.
We re-write $y = 3e^{2x-1} + 5$ as $(y-5)/3 = e^{2x-1}$.
Take the natural logarithm of both sides of the equation:
$2x - 1 = \ln\left(\dfrac{y-5}{3}\right) \Rightarrow x = \dfrac{1}{2}\left(\ln\left(\dfrac{y-5}{3}\right) + 1\right)$.

Problem 6. Solution: A.
$g(x) = \dfrac{a^x - a^{-x}}{a^x + a^{-x}} = \dfrac{a^x - \dfrac{1}{a^x}}{a^x + \dfrac{1}{a^x}} = \dfrac{a^{2x} - 1}{a^{2x} + 1}$.

$g(p) = \dfrac{a^{2p} - 1}{a^{2p} + 1} = \dfrac{1}{3} \Rightarrow 3(a^{2p} - 1) = a^{2p} + 1 \Rightarrow 2a^{2p} = 4 \Rightarrow a^{2p} = 2$
$\Rightarrow (a^{2p})^4 = 2^4 \Rightarrow a^{8p} = 16$.

Thus $g(4p) = \dfrac{a^{8p} - 1}{a^{8p} + 1} = \dfrac{16 - 1}{16 + 1} = \dfrac{15}{17}$.

Problem 7. Solution: E.
$(e^b)^2 + 5(e^b) = 14$. Let $x = e^b$.
$x^2 + 5b - 14 = 0 \Rightarrow (x + 7)(x - 2) = 0$
$x = 2$ beause $e^b > 0$ for all b. $x = e^b = 2 \Rightarrow b = \ln 2$.

Problem 8. Solution: C.
Assume that they intersect. We have $e^{bx} = e^{(\ln 2)x} = 2^x \Rightarrow 2^x = 2(x+1)^2 \Rightarrow$
$$2^{x+1} = 2^2(x+1)^2 \qquad (1)$$
Let $x + 1 = m$. (1) becomes $2^m = 2^2 m^2 \qquad (2)$
We see that (2) is true when $m = 8$: $2^8 = 2^2 \times 8^2 = 2^2 \times 2^6 = 2^8$.
Thus $x = 7$.

Problem 9. Solution: D.

$x = 2^{100}$ $\qquad X = (2^{100})^{\frac{1}{25}} = 2^4 = 16.$

$y = 3^{75}$ $\qquad Y = (3^{75})^{\frac{1}{25}} = 3^3 = 27.$

$z = 5^{50}$ $\qquad Z = (5^{50})^{\frac{1}{25}} = 5^2 = 25.$

Since $Y > Z > X$, we conclude that $y > z > x$.

Problem 10. Solution: C.
$12 + 64^x = 8^{x+1} \Rightarrow 12 + (8^x)^2 = 8^x \cdot 8 \Leftrightarrow w^2 - 8w + 12 = 0$, where $w = 8^x$, so
$(w - 6)(w - 2) = 0 \Rightarrow w = 6, 2 \therefore 8^x = 2^{3x} = 6, 2 \Rightarrow 3x = \log_2 6 \Rightarrow x = \dfrac{\log_2 6}{3}$, or
$2^{3x} = 2 \Rightarrow x = \dfrac{1}{3}$.

Problem 11. Solution: A.
Factor out 8^{3x} to get $8^{3x} \cdot (8 - 1) = 448$ so $8^{3x} \cdot 7 = 448$ and $8^{3x} = 64$. Thus $8^{3x} = 8^2$ so $3x = 2$ and $x = 2/3$.

Problem 12. Solution: A.
$\log 3^{x+2} = \log 2^{2x-1} \Rightarrow (x+2)\log 3 = (2x-1)\log 2 \Rightarrow x = \dfrac{\log 2 + 2\log 3}{2\log 2 - \log 3} = \dfrac{\log 18}{\log \dfrac{4}{3}}.$

Problem 13. Solution: A.
$\dfrac{e^x - e^{-x}}{2} = 2 \Leftrightarrow e^x - e^{-x} = 4$. Now let $w = e^x$ and rewrite the expression as $w + w^{-1} = 4 \Leftrightarrow$
$w^2 - 4w - 1 = 0$, and solve for w: $w = \dfrac{4 \pm \sqrt{20}}{2} = 2 \pm \sqrt{5}$. Since $w = e^x$, it must be positive, so $e^x = 2 + \sqrt{5}$, making $x = \ln(2 + \sqrt{5})$.

Problem 14. Solution: D.
Let $y = 3^x$.
$3^{2x} + 3^x - 20 = 0 \Rightarrow y^2 + y - 20 = 0 \Rightarrow (y-4)(y+5) = 0$.
So $y = 4$ or $y = -5$ (ignored since it is negative).
Then $3^x = 4 \Rightarrow x \ln 3 = \ln 4 \Rightarrow x = \dfrac{\ln 4}{\ln 3}$.

Problem 15. Solution: A.
$12 - 8 \cdot 5^x + 25^x = 0 \Rightarrow 12 - 8 \cdot 5^x + 5^{2x} = 0$.
Let $y = 5^x$. $12 - 8y + y^2 = 0 \Rightarrow y^2 - 8y + 12 = 0 \Rightarrow (y-2)(y-6) = 0$.
So $y = 2$ or $y = 6$ (ignored since it is negative).
Then $5^x = 2 \Rightarrow x \log 5 = \log 2 \Rightarrow x = \dfrac{\log 2}{\log 5} = \log_5 2$.

$5^x = 6 \Rightarrow x \log 5 = \log 6 \Rightarrow x = \dfrac{\log 6}{\log 5} = \log_5 6$.

The sum is $\log_5 2 + \log_5 6 = \log_5 12$.

Problem 16. Solution: E.
$2^x \cdot 4^x \cdot 8^x = \sqrt{2} \Rightarrow 2^x \cdot 2^{2x} \cdot 2^{3x} = \sqrt{2} \Rightarrow 2^{x+2x+3x} = 2^{\frac{1}{2}}$.
Thus we have $x + 2x + 3x = 1/2 \Rightarrow 6x = 1/2 \Rightarrow x = 1/12$.

Problem 17. Solution: B.
$a = 2b^2 = 2(4c^3)^2 = 2^5 c^6 = 2^5(8d^4)^6 = 2^5(2^3 d^4)^6 = 2^5(2^{18} d^{24}) = 2^{23} d^{24}$.

Problem 18. Solution: B.
Let $y = 3^x$.
$3^{2x} - 10 \times 3^x + 9 = 0 \Rightarrow y^2 - 10y + 9 = 0 \Rightarrow (y-1)(y-9) = 0$.
So $y = 1$ or $y = 9$.
Then $3^x = 1 \Rightarrow x = 0$.
$3^x = 9 \Rightarrow x = 2$.

Problem 19. Solution: D.
$8^{6x^2+4x} = 4^{9x^2-9x+6} \Rightarrow 2^{3(6x^2+4x)} = 2^{2(9x^2-9x+6)}$
Now with the bases the same we know that the exponents must be the same, so $3(6x^2 + 4x) = 2(9x^2 - 9x + 6)$. This simplifies to $30x = 12 \Rightarrow x = 2/5$.

Problem 20. Solution: A.
$9^{x+1} + 9^{x+2} + 9^{x+3} + 9^{x+4} + 9^{x+5} = 22143 \Rightarrow 9^x(9 + 9^2 + 9^3 + 9^4 + 9^5) = 22143$

$\Rightarrow \quad 9^x(66429) = 22143 \quad \Rightarrow \quad 9^x = 1/3 \quad \Rightarrow \quad (3)^{2x} = 3^{-1} \quad \Rightarrow$
$2x = -1 \quad\quad\quad\quad\quad \Rightarrow \quad x = -1/2$.

Problem 21. Solution: C.
$\dfrac{9+3^{2x}}{10} = 3^x \Rightarrow 3^2 + 3^{2x} = 10 \cdot 3^x \Rightarrow 3^{2x} - 10 \cdot 3^x + 9 = 0 \Rightarrow$
$(3^x - 9)(3^x - 1) = 0 \Rightarrow 3^x - 9 = 0$ or $3^x - 1 = 0 \Rightarrow 3^x = 9$ or $3^x = 1 \Rightarrow x = 2$ or $x = 1$
If $x = 2$, $x^2 + x + 1 = 7$ and if $x = 0$, $x^2 + x + 1 = 0$, so there are 2 solutions, 1 or 7.

Problem 22. Solution: A.
$3p^{3/2} = 24 \quad \Rightarrow \quad p^{3/2} = 8 \Rightarrow \quad p^{3/2} = 2^3 \Rightarrow \quad p^{3/2} = (\sqrt{4})^3 = (4)^{3/2}$. Thus $p = 4$.

Problem 23. Solution: D.
Case 1: $x^2 - 4x + 2 = 1$
Case 2: $x^2 - 1 = 0$ with $x^2 - 4x + 2 \neq 0$
Case 3: $x^2 - 4x + 2 = -1$ with even $x^2 - 1$.

For case 1, we have $x^2 - 4x + 1 = 0 \quad \Rightarrow \quad x_{1,2} = \dfrac{4 \pm \sqrt{4^2 - 4 \cdot 1 \cdot 2}}{2} = \dfrac{4 \pm 2\sqrt{2}}{2} = 2 \pm \sqrt{2}$.
For case 2, we have $x^2 = 1 \quad \Rightarrow \quad x_{3,4} = \pm 1$. We check that $x^2 - 4x + 2 \neq 0$.
For case 3, we have $x^2 - 4x + 3 = 0 \quad \Rightarrow \quad (x-1)(x-3) = 0$. $x = 1$ or $x = 3$. We check that $x^2 - 1$ is even.
Therefore, there are 5 solutions ($x_{1,2} = 2 \pm \sqrt{2}$, $x_{3,4} = \pm 1$, and or $x_5 = 3$).

Problem 24. Solution (official): D.
$\left(x^2 - 5x + 5\right)^{x^2 - 9x + 20} = 1 \Rightarrow x^2 - 5x + 5 = 1$ or $x^2 - 9x + 20 = 0$. Solving the first we get
$x^2 - 5x + 5 = 1 \Leftrightarrow x^2 - 5x + 4 = 0 \Leftrightarrow (x - 4)(x - 1) = 0$, so $x = 4$ or 1. Solving the second
we get $x^2 - 9x + 20 = 0 \Leftrightarrow (x - 4)(x - 5) = 0$, so $x = 4$ or 5. Thus the solutions are $x = 1, 4,$ or 5.
Note 1: this is the last problem in the test.
Note 2: This official solution is insufficient.
There is one more case the official solution missed:
When $x^2 - 5x + 5 = -1$ and $x^2 - 9x + 20$ is even.
$x^2 - 5x + 5 = -1 \quad \Rightarrow \quad x^2 - 5x + 6 = 0 \quad \Rightarrow (x - 2)(x - 3) = 0$
$\Rightarrow x = 2$ or $x = 3$. We checked and both are the solutions. So we have five solutions.

Algebra II Through Competitions — Chapter 12 Logarithmic Functions

1. DEFINITION

If $a \neq 1$ and x are positive real numbers then y is the logarithm to the base a of x written

$$y = \log_a x \qquad (1.1)$$

if and only if $\qquad x = a^y \qquad (1.2)$

A quick way to convert the logarithm to the exponent:

$$\log_a(x) = y \;\Rightarrow\; a^y = x$$

Notes:

(1) The logarithm base 10 is called the **common logarithm**. $\log x$ always refers to log base 10, i.e., $\log x = \log_{10} x$.

(2) The logarithm base e is called the **natural logarithm**. $\ln x = \log_e x$.

(3) Zero and negative numbers have no logarithm expressions. You will not be able to see any expressions like $\log_a 0$, $\log_a(-3)$, or $\log_{-10} x$.

(4) The graphs of logarithmic functions $f(x) = \log_a x$ all have x-intercept 1, and are increasing when $a > 1$ and decreasing when $a < 1$.

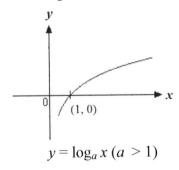
$y = \log_a x \;(a > 1)$

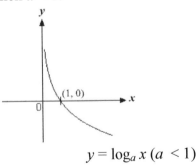
$y = \log_a x \;(a < 1)$

Example 1. If $\log 3 = A$ and $\log 7 = B$, then, in terms of A and B, $\log_7 9 =$

A. $2A/B$ B. $2A - B$ C. $2B/A$ D. $A/(2B)$ E. $B/(2A)$.

Solution: A.
$\log 3 = A \qquad \Rightarrow \qquad 10^A = 3 \qquad (1)$

$\log 7 = B \quad \Rightarrow \quad 10^B = 7$ (2)
$\log_7 9 = X \quad \Rightarrow \quad 7^X = 9$ (3)

Substituting (1) and (3) into (3): $(10^B)^X = 10^A \cdot 10^A = 10^{2A}$ (4)
Take logs on both sides of (4): $\log(10^B)^X = \log 10^{2A} \quad \Rightarrow \quad X \log(10^B) = \log 10^{2A}$
$\Rightarrow \quad X B \log(10^B) = 2A \quad \Rightarrow \quad X = 2A/B.$

2. PROPERTIES OF LOGARITHMS:

If $a \neq 1$, $b \neq 1$, x and y are positive real numbers r is any real number then we have

$\log_a 1 = 0$ (2.1)

$\log_a a = 1$ (2.2)

$a^{\log_a x} = x$ (2.3)

$\log_a a^r = r$ (2.4)

$\log_a b \log_b c = \log_a c$ (2.5)

Example 2. $\log_5 625^{10} =$
A. 12.5 B. 40 C. 50 D. 125 E. none of these

Solution: B.
$\log_5 625^{10} = \log_5 (5)^{40} = 40 \log_5 (5) = 40.$

Example 3. The equation $x^{\log x^4} = \dfrac{x^8}{1000}$ has two positive solutions. What is the ratio of the larger to the smaller?
A. 10 B. $10\sqrt{10}$ C. 100 D. $100\sqrt{10}$ E. 1000

Solution: A.
$x^{\log x^4} = \dfrac{x^8}{1000} \quad \Rightarrow \quad \log(x^{\log x^4}) = \log(\dfrac{x^8}{1000}) \quad \Rightarrow \quad \log(x^{4\log x}) = \log x^8 - \log 1000$
$\Rightarrow \quad 4(\log x)(\log x) = 8 \log x - 3 \quad \Rightarrow \quad 4(\log x)^2 = 8 \log x - 3$ (1)
Let $\log x = y$. (1) becomes: $4y^2 = 8y - 3 \quad \Rightarrow \quad 4y^2 - 8y + 3 = 0 \quad \Rightarrow$
$(2y - 1)(2y - 3) = 0.$

We get $2y - 1 = 0 \Rightarrow y = \dfrac{1}{2}$. $\log x = \dfrac{1}{2} \Rightarrow x = 10^{\frac{1}{2}}$.

$2y - 3 = 0 \Rightarrow y = \dfrac{3}{2}$. $\log x = \dfrac{3}{2} \Rightarrow x = 10^{\frac{3}{2}}$.

The ratio of the larger to the smaller is $\dfrac{10^{\frac{3}{2}}}{10^{\frac{1}{2}}} = 10^{\frac{3}{2} - \frac{1}{2}} = 10$.

3. THE LAWS OF LOGARITHMS:

Law 1: the Product Identity

$$\log_a xy = \log_a x + \log_a y \qquad (3.1)$$

Law 2: the Quotient Identity

$$\log_a \left(\dfrac{x}{y}\right) = \log_a x - \log_a y \qquad (3.2)$$

When we use the law of logarithm, we need to be careful about the restrictions.

When we apply **the Law 1 the Product Identity** $\log_a xy = \log_a x + \log_a y$ or **Law 2: the Quotient Identity** $\log_a \left(\dfrac{x}{y}\right) = \log_a x - \log_a y$, we need to know that in the expressions $\log_a xy$ and $\log_a \left(\dfrac{x}{y}\right)$, both x and y can be negative. However, in the expressions $\log_a x + \log_a y$ or $\log_a x - \log_a y$, none of them can be negative.

Example 4. Which of the following expressions is equivalent to $\log_4 (8 \cdot 2^x)$?
A. $64x$ B. $2x + 6$ C. $(x + 3)/2$ D. $2x$ E. $16x$

Solution: C.

$$\log_4 (8 \cdot 2^x) = \log_4 8 + \log_4 2^x = \dfrac{3}{2} + \dfrac{\log_2 2^x}{\log_2 4} = \dfrac{3}{2} + \dfrac{x}{2}.$$

Example 5. (If $\log_2 (a^3b) = x$ and $\log_2 \left(\dfrac{3a}{b}\right) = y$, what is the value of $\log_2 a$?

A. $\dfrac{x+y}{4} - \log_2 \sqrt[4]{3}$ B. $\dfrac{x+y}{4}$ C. $\log_2 \sqrt[4]{3}$ D. $\dfrac{x-y}{4}$ E. $\dfrac{x-y}{4} - \log_2 \sqrt[4]{3}$

Solution: A.
$\log_2 (a^3b) = x \Leftrightarrow 3\log_2 a + \log_2 b = x$.
$\log_2 \left(\dfrac{3a}{b}\right) = y \Leftrightarrow \log_2 3 + \log_2 a - \log_2 b = y$. Adding these last two expressions give us

$4\log_2 a + \log_2 3 = x + y$. Solving for $\log_2 a$ yields $\log_2 a = \dfrac{x+y}{4} - \dfrac{1}{4}\log_2 3 = \dfrac{x+y}{4} - \log_2 \sqrt[4]{3}$.

Law 3: the Power Identity

Formula 1: $\log_a x^r = r \log_a x$ \hfill (3.3)

Formula 2: $\log_{a^m} b^n = \dfrac{n}{m} \log_a b$ $(a, b > 0, a \neq 1)$ \hfill (3.4)

When we apply **Law 3: the Power Identity** $\log_a x^r = r \log_a x$, we need to know that in the expressions $\log_a x^r$, x can be negative if r is even. However, x must be positive in the expression $r \log_a x$.

Example 6. Which of the following is equivalent to $-\log_2 \left(x - \sqrt{x^2 - 1}\right)$?

A. $\dfrac{1}{2}\log_2 \left(\dfrac{x^2-1}{x}\right)$ B. $\left(\log_2 x - \dfrac{1}{2}\log_2(x^2-1)\right)^{-1}$ C. $\log_2 \left(x + \sqrt{x^2-1}\right)$

D. $\dfrac{1}{2}\log_2(x^2-1) - \log_2 x$ E. none of these.

Solution: C.
$-\log_2 \left(x - \sqrt{x^2-1}\right) = \log_2 \left(x - \sqrt{x^2-1}\right)^{-1} = \log_2 \left(\dfrac{1}{x - \sqrt{x^2-1}}\right) =$

$$\log_2\left(\frac{1}{x-\sqrt{x^2-1}} \cdot \frac{x+\sqrt{x^2-1}}{x+\sqrt{x^2-1}}\right) = \log_2\left(\frac{x+\sqrt{x^2-1}}{x^2-(x^2-1)}\right) = \log_2\left(x+\sqrt{x^2-1}\right)$$

Example 7. If $\ln\sqrt[a]{e^5} + \dfrac{2\ln\sqrt[b]{e^4}}{a} - \dfrac{5\ln\sqrt[a]{e^3}}{3} = 16$ where a and b are natural numbers greater than 1, then the value of ab is:

A. $\dfrac{1}{2}$ B. $\dfrac{1}{4}$ C. 4   D. 6   E. can not be determined

Solution: A.

$$\ln\sqrt[a]{e^5} + \frac{2\ln\sqrt[b]{e^4}}{a} - \frac{5\ln\sqrt[a]{e^3}}{3} = 16 \Rightarrow \ln e^{\frac{5}{a}} + \frac{2}{a}\ln e^{\frac{4}{b}} - \frac{5}{3}\ln e^{\frac{3}{a}} = 16$$

$$\Rightarrow \frac{5}{a} + \frac{2}{a}\cdot\frac{4}{b} - \frac{5}{3}\cdot\frac{3}{a} = 16 \Rightarrow 5b + 9 - 5b = 16ab \quad\Rightarrow\quad ab = \frac{1}{2}.$$

Example 8. Which expression below is equivalent to $-3\log_8 4$?

A. $\dfrac{\sqrt[5]{32^4}}{32}$ B. $\ln\dfrac{1}{e^2}$ C. $\sqrt{-4}$ D. $\log 1 - \log 0.01$ E. $\log_5\sqrt{5} - \log_5 5\sqrt{5}$.

Solution: B.

By (3.4), $-3\log_8 4 = -3\cdot\dfrac{2}{3}\log_2 2 = -2$

Note that $\ln\dfrac{1}{e^2} = \ln 1 - \ln e^2 = 0 - 2 = -2$.

Also note that $\log 1 - \log 0.01 = 0 - (-2) = 2$.

4. CHANGE – OF – BASE THEOREM:

Formula 1: $\log_a x = \dfrac{\log_b x}{\log_b a}$ (4.1)

Formula 2: $\log_a N = \log_{a^m} N^m, \quad m \neq 0$ (4.2)

Formula 3: $\log_a N = \dfrac{1}{\log_N a}, \quad N \neq 1$ (4.3)

Formula 4: $\log_a N = -\log_a \dfrac{1}{N}$ (4.4)

Algebra II Through Competitions **Chapter 12 Logarithmic Functions**

Formula 5: $\log_a N = -\log_{\frac{1}{a}} N$ (4.5)

Formula 6: $\log_a N = m \log_{a^m} N$, $m \neq 0$ (4.6)

Formula 7: $\dfrac{\log_a M}{\log_a N} = \dfrac{\log_b M}{\log_b N}$ ($a > 0$, $a \neq 1$, $b > 0$, $b \neq 1$, $M, N > 0$, $N \neq 1$) (4.7)

Formula 8: $\log_a \sqrt[n]{N} = \dfrac{1}{n} \log_a N$, N is integer greater than 1. (4.8)

Example 9. Suppose that $\log_b 3 = 1.0986$ and $\log_b 5 = 1.6094$. Then the value of $\log_b 75$, accurate to four decimal places, is
A. 2.2422   B. 2.8456   C. 3.5362   D. 3.8066   E. 4.3174

Solution: E.
Using properties of logs, we have
$\log_b 75 = \log_b (3 \cdot 5^2) = \log_b 3 + 2\log_b 5 = 1.0986 + 2(1.6094) = 4.3174$

Example 10. If $\log_5 \sqrt{2} = x$, then $\log_{\sqrt{2}} 5$ equals:
A. $-x$ B. x^2 C. $-\sqrt{x}$ D. $1/x$ E. $\sqrt{x}$

Solution: D.
Method 1:
By (4.3), $\log_5 \sqrt{2} = \dfrac{1}{\log_{\sqrt{2}} 5} = \dfrac{1}{x}$.

Method 2:
$\log_5 \sqrt{2} = \dfrac{\log \sqrt{2}}{\log 5} = x \Rightarrow \dfrac{\log 5}{\log \sqrt{2}} = \log_{\sqrt{2}} 5 = \dfrac{1}{x}$

Example 11. Simplify $\log_2 (7a) + \log_4 (7a^2) + \log_8 (7a^3) + \log_{64} (7a^6)$.
A. $4 \cdot \log_2 (7a^3)$ B. $2 \cdot \log_2 (7a^2)$ C. $4 \cdot \log_2 (7a^2)$ D. $6 \cdot \log_2 (7a^3)$ E. $6 \cdot \log_2 (7a^6)$
Solution: B.

$\log_2 7 + \log_2 a + (\log_2 7 + 2 \cdot \log_2 a)/2 + (\log_2 7 + 3 \cdot \log_2 a)/3 + (\log_2 7 + 6 \cdot \log_2 a)/6 = 2 \cdot \log_2 7 + 4 \cdot \log_2 a = 2 \cdot \log_2 (7a^2)$.

Example 12. Which of the following expressions is equivalent to $\log_8(32x) - \log_4(8x) + \log_2(x)$?

A. $\log_2(3x)$ B. $\frac{1}{6}\log_2(2x^5)$ C. $\frac{1}{3}\log_2(6x^2)$ D. $\log_2(0.5x^6)$ E. $\frac{1}{2}\log_2(4x)$

Solution: B.
$$\log_8(32x) - \log_4(8x) + \log_2(x) = \frac{\log_2(32x)}{\log_2 8} - \frac{\log_2(8x)}{\log_2 4} + \log_2(x)$$
$$= \frac{\log_2(32x)}{3} - \frac{\log_2(8x)}{2} + \log_2(x) = \frac{1}{6}(2\log_2(32x) - 3\log_2(8x) + 6\log_2 x)$$
$$= \frac{1}{6}\left(\log_2\left(\frac{(32x)^2 \cdot x^6}{(8x)^3}\right)\right) = \frac{1}{6}\left(\log_2\left(\frac{2^{10}x^8}{2^9 x^3}\right)\right) = \frac{1}{6}(\log_2 2x^5).$$

5. SOLVING LOGARITHM EQUATIONS

If x, y and $a \neq 1$ are positive real numbers we have that $x = y$ if and only if $\log_a x = \log_a y$. Similarly $x^a = y^a$ if and only if $x = y$.

Example 13. Find the solution set of $\log(x+1) + 2\log(x-1) = 0$.

A. $\left\{\frac{1+\sqrt{5}}{2}\right\}$ B. $\left\{0, \frac{1\pm\sqrt{5}}{2}\right\}$ C. $\left\{\frac{\sqrt{5}-1}{2}\right\}$ D. $\left\{\frac{2}{3}\right\}$ E. $\left\{\frac{1\pm\sqrt{-3}}{2}\right\}$

Solution: A.
If $\log(x+1) + 2\log(x-1) = 0 \Rightarrow \log((x+1)(x-1)^2) = 0 \Rightarrow (x+1)(x-1)^2 = 1$, so we have $(x+1)(x-1)^2 = x^3 - x^2 - x + 1 = 1 \Leftrightarrow x^3 - x^2 - x = 0$. Now factoring yields $x(x^2 - x - 1) = 0 \Rightarrow x = 0$, $x = \frac{1\pm\sqrt{5}}{2}$, but only values greater than -1 can be used in the log expression.

Example 14. Solve for a: $\log_7(5a) - \log_7(a-4) = 1$.
A. 0.75   B. 6   C. 9.75   D. 14   E. 24

Solution: D.
$\log_7(5a) - \log_7(a-4) = 1 \Rightarrow \log_7 5a/(a-4) = 1 \Rightarrow 5a/(a-4) = 7 \Rightarrow a = 14$.

Example 15. The sum of the solutions to equation $12 - 8\log_2 x + (\log_2 x)^2 = 0$ is:

A) 68   B) 256   C) 4   D) 64   E) none of these

Solution: A
Let $\log_2 x = y$. The given equation becomes: $y^2 - 8y + 12 = 0 \Rightarrow (y-2)(y-6) = 0$.
$y - 2 = 0 \Rightarrow y = 2$. So $\log_2 x = 2 \Rightarrow x = 4$.
$y - 6 = 0 \Rightarrow y = 6$. So $\log_2 x = 6 \Rightarrow x = 64$.
The sum of the two solutions is $64 + 4 = 68$.

6. THE NUMBER OF DIGITS

Theorem: The number of digits in base b of a positive integer k is

$\lfloor \log_b k \rfloor + 1 = \lceil \log_b (k+1) \rceil$, with the right side of the equation also holding true for $k = 0$.

Example 16. How many digits are required to represent the number 2007^{2007} in decimal form?
A. 6628   B. 6629   C. 6630   D. 6631   E. 7321

Solution: B.
The one digit numbers, 1 through 9, all have common logarithm value in the interval [0; 1). Two digit numbers have common logarithm value between 1 and 2. In general, for a positive integer N, N has $k+1$ digits if the common logarithm of N satisfies $k \cdot \log N < k + 1$). Since $\log(2007^{2007}) = 2007 \cdot \log(2007) = 6628.2125...$, it follows that 2007^{2007} has 6629 digits.

Algebra II Through Competitions — Chapter 12 Logarithmic Functions

PROBLEMS

Problem 1. If $3^{x+2} = 2^{2x-1}$, then x is:

A. $\dfrac{\log 18}{\log \dfrac{4}{3}}$ B. $\dfrac{\log 12}{\log \dfrac{4}{3}}$ C. $\dfrac{\log 27}{\log 2}$ D. $\dfrac{\log 3}{\log 2}$ E. $\dfrac{\log \dfrac{4}{3}}{\log 18}$

Problem 2. If $2^x = 3$, then $3^{\frac{1}{x}}$ is equal to

A. 3.   B. 2.   C. 1/2   D. 1/3.   E. 7

Problem 3. If $\log_7 3 = a$ and $\log_7 4 = b$, find x in terms of a and b if $9^x = 28$.

A. $x = \dfrac{1+b}{2a}$ B. $x = \dfrac{7b}{a^2}$ C. $x = \dfrac{b-a}{2}$ D. $x = \dfrac{2a}{b+1}$ E. none of the above

Problem 4. (2012 Indiana Algebra II) Solve $e^{2x-5} = 71$.

A. $\dfrac{\ln 71 - 5}{2}$ B. $\dfrac{\ln 71}{2x - 5}$ C. $\ln 71 + 5$ D. $\dfrac{\ln 71 + 5}{2}$ E. none of these

Problem 5. Let $\log_7(a \cdot b) + \log_7(b \cdot c) + \log_7(a \cdot c) = 10$ for positive a, b, and c. What is the value of $a \cdot b \cdot c$?

A. 49   B. 7   C. 16807   D) $\sqrt{343}$ E) none of these

Problem 6. If $\log_b z = \dfrac{1}{3} \log_b x + \log_b y$, write z in terms of x and y.

A. $y\sqrt[3]{x}$ B. $(x+y)^{\frac{1}{3}}$ C. $(xy)^{\frac{1}{3}}$ D. $\dfrac{x}{3} + y$ E. none of these

Problem 7. $3 \log_5 25 + \log_{25} 625^{10} =$

A. 56   B. 46   C. 36   D. 26   E. none of these

Problem 8. If $\log_3 x + \log_3(x^3) = 8$ then $x =$

A. $\sqrt{3}$ B. 3   C. 9   D. 81   E. None of these

Problem 9. If $\log (x^2 -1) - \log (x -1) = 2$, solve for x.
A. 80   B. 99   C. $2^{10}+1$ D. 101   E. None of these

Problem 10. The largest solution of the equation $2\log_{10} x = \log_{10}(3x - 20) + 1$ is
A. 10   B. 20/3   C. 23/3   D. 20   E. 100

Problem 11. State the inverse function of $f(x) = \log_b x$.
A. $f^{-1}(x) = e^x$ B. $f^{-1}(x) = x^b$ C. $f^{-1}(x) = \log_x b$
D. $f^{-1}(x) = b^x$ E. none of these

Problem 12. The equation $\log_b \sqrt{x} = \sqrt{\log_b x}$ has two solutions, the sum of which is 10. What is b?
A. $\sqrt{2}$ B. $\sqrt{3}$ C. 2   D. $\sqrt{5}$ E. $\sqrt{6}$

Problem 13. Solve $\dfrac{e^x - e^{-x}}{2} = 2$ for x, where $x > 0$.
A. $\ln(2+\sqrt{5})$ B. $\ln(2\pm\sqrt{5})$ C. $\ln 2 + \ln \sqrt{5}$ D. 4   E. none of the above

Problem 14. There exist positive integers A, B, and C, with no common factor greater than 1, such that
$A\log_{200} 5 + B\log_{200} 2 = C$. What is $A + B + C$?
A. 6.   B. 7.   C. 8.   D. 9.   E. 10.

Problem 15. If $\log\left(3\sqrt{3\sqrt{3\sqrt{3\sqrt{3}}}}\right) = A\log 3$, what is A?

A. $\dfrac{31}{16}$ B. $\dfrac{31}{32}$ C. $\dfrac{16}{15}$ D. $\dfrac{32}{17}$ E. $\dfrac{13}{32}$

Problem 16. Find the value of $\displaystyle\prod_{x=5}^{15624} \log_x (x+1)$.

Algebra II Through Competitions   **Chapter 12 Logarithmic Functions**

Problem 17. Suppose the number a satisfies $a\log a + \log\log a - \log(\log\log 2 - \log\log a) = 0$. What is the value of $a^{(a^{(a^a)})}$?

A. 1   B. 2   C. 4   D. 8   E. 16

Problem 18. If $\log 3 = A$ and $\log 7 = B$, then, in terms of A and B, $\log_7 9 =$

A. $2A/B$ B. $2A - B$ C. $2B/A$ D. $(A/B)2$ E. $B/(2A)$.

Problem 19. Solve for x: $(\log(x))^3 = \log(x)^{16}$

I. 10^4 II. 10^0 III. $10^{16/3}$ IV. 13   V. 1/10,000.

A. I only   B. III only   C. IV only   D. Only I, II and III   E. Only I, II and V.

Problem 20. Let y be a positive integer greater than 1, and let $0 < x < 1$. Find the value of x such that $(\log_4 y)(\log_y x) = \log_x 4$.

Algebra II Through Competitions   **Chapter 12 Logarithmic Functions**

SOLUTIONS

Problem 1. Solution: A.
Take the logarithm of both sides of the equation: $\log 3^{x+2} = \log 2^{2x-1}$
By (3.3), $(x + 2)\log 3 = (2x - 1) \log 2$.
Solving for x we get: $x = \dfrac{\log 2 + 2\log 3}{2\log 2 - \log 3} = \dfrac{\log 2 + \log 3^2}{\log 2^2 - \log 3} = \dfrac{\log(2 \cdot 3^2)}{\log(\frac{2^2}{3})} = \dfrac{\log 18}{\log \frac{4}{3}}$.

Problem 2. Solution: B.
$2^x = 3 \Rightarrow x = \log_2 3 = -1/\log_3 2$ (Rule: $\log_a N = -\log_a \dfrac{1}{N}$) $\Rightarrow$ $1/x = -\log_3 2$.

Thus $3^{\frac{1}{x}} = 3^{-\log_3 2} = 3^{\log_3 2^{-1}} = 2^{-1} = \dfrac{1}{2}$ (Rules: $\log_a x^r = r \log_a x$ and $x = a^{\log_a x}$).

Problem 3. Solution: A.
$x\log_7 3^2 = \log_7 7 \cdot 4 \quad \Rightarrow \quad x = \dfrac{\log_7 7 + \log_7 4}{2\log_7 3} \quad \Rightarrow \quad x = \dfrac{1 + b}{2a}$.

Problem 4. Solution: D.
$e^{2x-5} = 71 \quad \Rightarrow \quad \ln e^{2x-5} = \ln 71 \quad \Rightarrow \quad 2x - 5 = \ln 71 \Rightarrow \quad x = \dfrac{\ln 71 + 5}{2}$.

Problem 5. Solution: C.
$\log_7(a \cdot b) + \log_7(b \cdot c) + \log_7(a \cdot c) = 10 \quad \Rightarrow \quad \log_7(abc)^2 = 10 \quad \Rightarrow$
$2\log_7(abc)^2 = 10 \quad \Rightarrow \quad \log_7(abc) = 5 \quad \Rightarrow \quad abc = 7^5. = 16807$.

Problem 6. Solution: A
$\log_b z = \dfrac{1}{3}\log_b x + \log_b y \Rightarrow \log_b(z/x^{\frac{1}{3}}y) = 0 \Rightarrow z/x^{\frac{1}{3}}y) = 1 \Rightarrow z = x^{\frac{1}{3}}y = y\sqrt[3]{x}$.

Problem 7. Solution: D.
$3\log_5 25 + \log_{25} 625^{10} = 3\log_5 5^2 + 10\log_{25} 625 = 6 + 10 \times \log_{25} 25^2 = 6 + 20 = 26$.

Problem 8. Solution: C.
$\log_3 x + \log_3(x^3) = 8 \quad \Rightarrow \quad \log_3 x \cdot x^3 = 8 \quad \Rightarrow \quad \log_3 x^4 = 8 \quad \Rightarrow \quad 4\log_3 x = 8$
$\Rightarrow \quad \log_3 x = 2 \quad \Rightarrow \quad x = 3^2 = 9$.

Problem 9. Solution: B.
$\log(x^2-1) - \log(x-1) = 2 \Rightarrow \log(x^2-1)/(x-1) = 2 \Rightarrow \log(x+1) = 2$
$\Rightarrow (x+1) = 10^2 \Rightarrow x = 99$.

Problem 10. Solution: D.
$2\log_{10} x = \log_{10}(3x-20) + 1 \Rightarrow \log_{10} x^2 = \log_{10}(3x-20) + \log_{10} 10$
$\Rightarrow \log_{10} x^2 = \log_{10}(3x-20) \times 10 \Rightarrow x^2 = (3x-20) \times 10$
$\Rightarrow x^2 = 30x - 200 \Rightarrow x^2 - 30x + 200 = 0$
$\Rightarrow (x-10)(x-20) = 0$. So $x = 10$ or $x = 20$.
The largest solution is 20.

Problem 11. Solution: D.
Let $\log_b x = e$. Since b is a constant (not a variable), then in terms of x, the function returns y such that $b^y = x$. To find the inverse, switch x and y and solve for y. So
$$y = f^{-1}(x) = b^x$$
Making sense of this, $f(f^{-1}(x)) = \log_b b^x = x$

Problem 12. Solution: B.
$\log_b \sqrt{x} = \sqrt{\log_b x} \Rightarrow \frac{1}{2}\log_b x = \sqrt{\log_b x}$ \hfill (1)

Let $\sqrt{\log_b x} = y$. (1) becomes $\frac{1}{2}y^2 = y \Rightarrow \frac{1}{2}y^2 - y = 0 \Rightarrow y(\frac{1}{2}y - 1) = 0$.

We get $y = 0$. So $\sqrt{\log_b x} = 0 \Rightarrow \log_b x = 0 \Rightarrow x = 1$.

$\frac{1}{2}y - 1 = 0 \Rightarrow y = 2$. So $\sqrt{\log_b x} = 2 \Rightarrow \log_b x = 4 \Rightarrow x = b^4$.

We know that $1 + b^4 = 10$. So $b^4 = 9 \Rightarrow b^2 = 3$.
So $b = \sqrt{3}$ ($b = -\sqrt{3}$ is ignored).

Problem 13. Solution: A.
$\frac{e^x - e^{-x}}{2} = 2 \Leftrightarrow e^x - e^{-x} = 4$. Now let $w = e^x$ and rewrite the expression as $w + w^{-1} = 4 \Leftrightarrow$
$w^2 - 4w - 1 = 0$, and solve for w. $w = \frac{4 \pm \sqrt{20}}{2} = 2 \pm \sqrt{5}$. Since $w = e^x$, it must be positive, so $e^x = 2 + \sqrt{5}$ making $x = \ln(2 + \sqrt{5})$.

Problem 14. Solution: A.
Note that $C = A\log_{200} 5 + B\log_{200} 2 = \log_{200} 5^A + \log_{200} 2^B = \log_{200}(5^A \cdot 2^B)$, so

$200^C = 5^A \cdot 2^B$. Therefore, $5^A \cdot 2^B = 200^C = (5^2 \cdot 2^3)^C = 5^{2C} \cdot 2^{3C}$.
By uniqueness of prime factorization, $A = 2C$ and $B = 3C$. Letting $C = 1$, we get $A = 2$, $B = 3$, and $A + B + C = 6$.
The triplet $(A, B, C) = (2, 3, 1)$ is the only solution with no common factor greater than 1.

Problem 15. Solution : A.

Since $3\sqrt{3\sqrt{3\sqrt{3\sqrt{3}}}} = 3 \cdot 3^{\frac{1}{2}} \cdot 3^{\frac{1}{4}} \cdot 3^{\frac{1}{8}} \cdot 3^{\frac{1}{16}} = 3^{1+\frac{1}{2}+\frac{1}{4}+\frac{1}{8}+\frac{1}{16}} = 3^{\frac{31}{16}}$, we have $\log 3^{\frac{31}{16}} = \frac{31}{16}\log 3$ and $A = \frac{31}{16}$.

Problem 16. Solution: 6.

$$\prod_{x=5}^{15624} \log_x(x+1) = \prod_{x=5}^{15624} \frac{\log(x+1)}{\log x} = \frac{\log(6)}{\log 5} \cdot \frac{\log(7)}{\log 6} \cdots \frac{\log(15624+1)}{\log 15624} = \frac{\log 15625}{\log 5}$$
$= \log_5 15625 = \log_5 5^6 = 6$.

Problem 17. Solution: B.

Use the laws of logarithms to get $a^a \cdot \log a = \log \log 2 - \log \log a$, which is equivalent to $a^{(a^a)} = \log\left(\frac{\log 2}{\log a}\right)$. Since the two logs are equal, their arguments are also equal, so $a^{(a^a)} = \frac{\log 2}{\log a}$. But this implies that $\log a \cdot a^{(a^a)} = \log a^{a^{(a^a)}} = \log 2$, and from this it follows that $a^{(a^{(a^a)})} = 2$.

Problem 18. Solution: A.
Method 1:
$\log_7 9 = \frac{\log 9}{\log 7} = \frac{\log 3^2}{\log 7} = \frac{2\log 3}{\log 7} = \frac{2A}{B}$.

Method 2:
$\log 3 = A$ $\Rightarrow$ $2\log 3 = 2A$ $\Rightarrow$ $\log 3^2 = 2A$ $\Rightarrow$ $\log 9 = 2A$ (1)
$\log 7 = B$ (2)
(1) ÷ (2): $\frac{\log 9}{\log 7} = \frac{2A}{B}$ $\Rightarrow$ $\log_7 9 = \frac{2A}{B}$.

Problem 19. Solution: E.

$(\log(x))^3 = \log(x)^{16} \quad \Rightarrow \quad (\log(x))^3 = 16\log x$ (1)

Let $(\log(x)) = y$. (1) becomes: $y^3 = 16y \quad \Rightarrow \quad y(y^2 - 16) = 0 \Rightarrow y(y-4)(y+4) = 0$

The solutions are

(a) $y = 0$. $(\log(x)) = 0 \quad \Rightarrow \quad x = 1$

(b) $y - 4 = 0 \quad \Rightarrow \quad y = 4 \quad \Rightarrow \quad (\log(x)) = 4 \quad \Rightarrow \quad x = 10^4$.

(c) $y + 4 = 0 \quad \Rightarrow \quad y = -4 \Rightarrow \quad (\log(x)) = -4 \quad \Rightarrow \quad x = 10^{-4} = \dfrac{1}{10^4} = \dfrac{1}{10000}$.

Problem 20. Solution: 1/4.

By Change – of – Base Theorem, the given equation can be written as

$\dfrac{\log y}{\log 4} \times \dfrac{\log x}{\log y} = \dfrac{\log 4}{\log x} \quad \Rightarrow \quad \dfrac{\log x}{\log 4} = \dfrac{\log 4}{\log x} \quad \Rightarrow \quad (\log x)^2 = (\log 4)^2$

$\Rightarrow \quad (\log x - \log 4)(\log x + \log 4) = 0$.

$\log x - \log 4 = 0$ can be ignored since $0 < x < 1$.

Thus $\log x + \log 4 = 0 \Rightarrow \quad \log x = -\log 4 = \log 4^{-1} = \log \dfrac{1}{4}$.

The solution is $x = 1/4$.

Algebra II Through Competitions — Chapter 13 Rational Functions

1. DEFINITION OF RATIONAL FUNCTIONS

Ratios of polynomial functions are called rational functions.

If $p(x)$ and $q(x)$ are polynomials with $q(x) \neq 0$, then $f(x) = p(x)/q(x)$ defines a rational function.

The domain consists of all real numbers except those for which the denominator is q zero.

The simplest rational function with one variable denominator is then $f(x) = 1/x$.

When the the denominator and the numerator have no common factors (except 1 and -1), the rational expression is reduced to lowest terms or is simplified.

Example 1. Which of the following is equivalent, where it is defined, to $\dfrac{1 - \dfrac{1-x}{1+x}}{1 + \dfrac{x-1}{x+1}}$?

A. 1   B. $4x/(x+1)$ C. $1/x$ D. 0   E. $2x/(2x-2)$

Solution: A.
Note that $x - 1 = -(1 - x)$ so the numerator and denominator are the same. Alternatively, combine the fractions and simplify.

Example 2. The ratio of $2x + y$ to $2y + x$ is 5 to 4. What is the ratio of $x + 3y$ to $3x + y$?
A. $3:5$ B. $5:7$ C. $7:9$ D. $9:11$ E. $11:13$

Solution: B.
Massage the equation $5(2y + x) = 4(2x + y)$ to get $x = 2y$ from which it follows that $x + 3y = 5x/2$ and $3x + y = 7x/2$.

Example 3. Let x and y be positive integers satisfying $1/(x + 1) + 1/(y - 1) = 5/6$. Find $x + y$.
A. 2   B. 3   C. 4   D. 5   E. 6

Solution: D.

Since x and y are natural numbers, $y \geq 2$. Let $u = x + 1$ and $v = y - 1$, u and v are natural numbers with $u \geq 2$ and $v \geq 1$. One of $1/u$ and $1/v$ is at least half of $5/6$, so either $u \leq 12/5$

= 2.4 or $v \leq 12/5 = 2.4$. Consider the two cases: case one, $u = 2$ so $x = 1$ then $v = 3$ then $y = 4$; case two, $v = 2$ so $y = 3$ then $u = 3$ then $x = 2$. In either case, $x + y = 5$.

Example 4. Determine the range of the function $F(x) = \dfrac{x+3}{x}$.

A. $\{x \mid x \in \text{Reals}\}$ B. $\{x \mid x \in \text{Reals}, x \neq 0\}$ C. $\{x \mid x \in \text{Reals}, x \neq -1\}$
D. $\{x \mid x \in \text{Reals}, x \neq -3\}$ E. $\{x \mid x \in \text{Reals}, x \neq 1\}$

Solution: E.
This function has a vertical asymptote at $x = 0$, but of more importance to this question is the horizontal asymptote at $y = 1$, since $F(x) = (x + 3)/x = 1 + 3/x$. Since y can never equal 1, but can equal anything else, the range of this function is $\{x \; x \in \text{Reals}, x \neq 1\}$.

Example 5. If x is positive, what is the least value of $x + 9/x$?
A. 1   B. 2   C. 3   D. 4   E. 6

Solution: E.
Method 1 (official solution):
$$x + \dfrac{9}{x} = \dfrac{1}{x}(x^2 + 9) = \dfrac{1}{x}(x^2 - 6x + 9) + 6 = \dfrac{1}{x}(x-3)^2 + 6 \geq 6.$$

Method 2 (our solution):
By AM-GM, we have $x + \dfrac{9}{x} \geq 2\sqrt{x \times \dfrac{9}{x}} = 6$.

The greatest value 6 is obtained when $x = \dfrac{9}{x}$ or $x = 3$.

2. ASYMPTOTES OF RATIONAL FUNCTIONS

For a rational function defined by $y = f(x)$, and for real numbers a and b,

Vertical asymptote

If, as x gets closer to a, $|f(x)| \to +\infty$, the line $x = a$ is a vertical asymptote.

Horizontal asymptote

If, as $x \to +\infty$ or $x \to -\infty$, the values of $f(x)$ get closer to b, then the line $y = b$ is a horizontal asymptote.

Oblique asymptote

An asymptote that is neither horizontal nor vertical is called the oblique asymptote. In general, if the degree of $p(x)$ is exactly one more than the degree of $q(x)$, there will be an oblique asymptote.

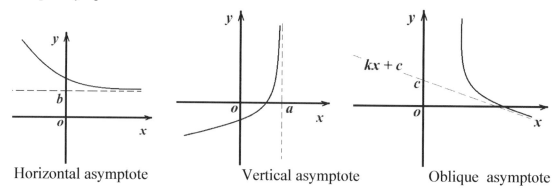

Horizontal asymptote   Vertical asymptote   Oblique asymptote

Methods To find The Asymptotes

$$f(x) = \frac{p(x)}{q(x)} = \frac{a_n x^n + a_{n-1} x^{n-1} + \ldots + a_1 x + a_0}{b_m x^m + b_{m-1} x^{m-1} + \ldots + b_1 x + b_0}, \text{ where } b_n \neq 0.$$

Vertical asymptotes

Find any vertical asymptotes by setting the denominator equal to 0 and solve for x. If a is a zero of the denominator, then the line $x = a$ is a vertical asymptote.

Other asymptotes

Case 1: If $n < m$, the line $y = 0$ (the x-axis) is a horizontal asymptote

Case 2: If $n = m$, the line $y = a_n/b_m$ is a horizontal asymptote

Case 3: If $n = m + 1$ (the degree of the numerator is one more than the degree of the denominator), the graph of $f(x)$ will have an oblique asymptote. To find it, divide the numerator by the denominator and disregard any remainder. Note that an oblique asymptote is represented by a line defined by $y = kx + c$ (note: k is not zero as that would be horizontal).

Case 4: If $n > m + 1$, the graph of $f(x)$ has no horizontal asymptote and no oblique asymptote.

Example 6. A horizontal asymptote of $f(x) = \dfrac{-3x^2}{6x^2 + 7x - 10}$ is

A. $y = \dfrac{3}{10}$ B. $x = 2$ C. $x = \dfrac{5}{6}$ D. $y = \dfrac{-1}{2}$ E. no horizontal asymptote

Solution: D.
This is the case 2 where $n = m$. The line $y = a_n/b_m = -3/6 = -1/2$ is the horizontal asymptote.

Example 7. Which of the following is not an asymptote of the function
$R(x) = \dfrac{|x|(x-2)(x+3)}{x(x+2)(x-3)}$?

A. $x = 0$ B. $x = -2$ C. $x = 3$ D. $y = 1$ E. $y = -1$

Solution: A.

The numbers $x = -2$ and $x = 3$ are zeros of the denominator but not the numerator. Also, $\lim\limits_{x \to \infty} R(x) = 1$ and $\lim\limits_{x \to -\infty} R(x) = -1$, so the function has both horizontal asymptotes. But the line $x = 0$ is not a vertical asymptote

Example 8. Which of the following are vertical asymptotes of the graph of
$y = \dfrac{x^3 - 3x^2 - 4x + 12}{x^2 - 9}$?

A. $x = -2, x = 2$ B. $x = 3$ C. $x = -3, x = 3$ D. $x = -3$ E. $x = -3, x = 3, x = -2, x = 2$

Solution: D.

$$\dfrac{x^3 - 3x^2 - 4x + 12}{x^2 - 9} = \dfrac{x^2(x-3) - 4(x-3)}{(x-3)(x+3)} = \dfrac{(x-3)(x^2-4)}{(x-3)(x+3)} = \dfrac{x^2 - 4}{x + 3}.$$

We then find the vertical asymptotes by setting the denominator equal to 0 and solve for x: $x + 3 = 0 \Rightarrow x = -3$.

3. PARTIAL FRACTION DECOMPOSITION

The partial fraction decomposition of the rational expression p/q depends on the factors of q.

Case 1: the denominator has only nonrepeated linear factors:
$q(x) = (x-a_1)(x-a_2)\cdots(x-a_n)$.

$$\frac{p(x)}{q(x)} = \frac{A_1}{x-a_1} + \frac{A_2}{x-a_2} + \cdots + \frac{A_n}{x-a_n},$$ where the numbers $A_1, A_2, \cdots, A_n$ are to be determined.

Case 2: the denominator has repeated linear factors: $q(x) = (x-a)^n$, $n \geq 2$.

$$\frac{p(x)}{q(x)} = \frac{A_1}{x-a} + \frac{A_2}{(x-a)^2} + \cdots + \frac{A_n}{(x-a)^n}.$$

Case 3: the denominator has a nonrepeated irreducible quadratic factor:
$q(x) = ax^2 + bx + c$.

$$\frac{p(x)}{q(x)} = \frac{Ax+B}{ax^2+bx+c},$$ where the numbers A and B are to be determined.

Case 4: the denominator has repeated irreducible quadratic factors: $q(x) = (ax^2+bx+c)^n$, $n \geq 2$.

$$\frac{p(x)}{q(x)} = \frac{A_1x+B_1}{ax^2+bx+c} + \frac{A_2x+B_2}{(ax^2+bx+c)^2} + \cdots + \frac{A_nx+B_n}{(ax^2+bx+c)^n},$$ where the numbers $A_1, A_2, B_1, B_2, \cdots, A_n,$ and B_n are to be determined.

Example 9. Let $\dfrac{x-3}{(x-1)(x+2)} = \dfrac{A}{x-1} + \dfrac{B}{x+2}$ be an identity in x. What is the value of $A + B$?

A. 0 B. 1 C. $\dfrac{7}{3}$ D. 2 E. None of these

Solution: B.
The partial fraction problem has unique values for A and B.
$$\frac{A}{x-1} + \frac{B}{x+2} = \frac{A(x+2)+B(x-1)}{(x-1)(x+2)} = \frac{(A+B)x+(2A-B)}{(x-1)(x+2)} = \frac{x-3}{(x-1)(x+2)},$$ so $A + B = 1$ and $2A - B = -3$. We don't need to find A and B, since the question asks for $A + B$.

Example 10. The fraction $(5x-11)/(2x^2+x-6)$ was obtained by adding the two fractions $A/(x+2)$ and $B/(2x-3)$. Find the value of $A + B$.

Algebra II Through Competitions   Chapter 13 Rational Functions

(A) −4   (B) −2   (C) 1   (D) 2   (E) 4

Solution: D.
Note that $\dfrac{5x-11}{2x^2+x-6} = \dfrac{A}{x+2} + \dfrac{B}{2x-3} = \dfrac{A(2x-3)}{(x+2)(2x-3)} + \dfrac{B(x+2)}{(x+2)(2x-3)}$. Also, note that
$5x − 11 = (2A + B)x − (3A − 2B)$, so $2A + B = 5$ and $3A − 2B = 11$.
Solving this system of equations we obtain $A = 3$ and $B = −1$, so $A + B = 2$.

Alternatively, since the degree of the numerator is less than the degree of the denominator, the value for A can be obtained by evaluating the expression $(5x−11)/(2x−3)$ at $x = −2$ (the zero of $x + 2$) and the value for B can be obtained by evaluating the expression $(5x − 11)/(x + 2)$ at $x = 3/2$ (the zero of $2x − 3$). So $A = \dfrac{-21}{-7} = 3$,
$B = \dfrac{-(7/2)}{(7/2)} = -1$, and $A + B = 2$.

Example 11. When an expression such as $\dfrac{2}{x^2-1}$ is expression as the sum $\dfrac{1}{x-1} + \dfrac{-1}{x+1}$, the fractions in the sum are called partial fractions. When $\dfrac{19x-8}{2x^2-x-21}$ is decomposed into partial fractions, what is the sum of the numerators when each fraction is reduced to lowest terms?
A. 5.   B. 7.   C. 9.   D. 12.   E. 14.

Solution: E.
$\dfrac{A}{2x-7} + \dfrac{B}{x+3} = \dfrac{A(x+3) + B(2x-7)}{(2x-7)(x+3)} = \dfrac{19x-8}{2x^2-x-21}$, so $A = 9$ and $B = 5$ and $A + B = 14$.

4. RATIONAL EQUATIONS

Example 12. Given $\dfrac{x+a}{x+b} = c$, solve for x.

A. $x = \dfrac{cb-a}{1-c}$ B. $x = c - \dfrac{a}{b}$ C. $x = \dfrac{cb+a}{c-1}$ D. $x = \dfrac{c-1}{cb-a}$ E. $x = \dfrac{cb+a}{c+1}$

Solution: A.

If $\dfrac{x+a}{x+b} = c \Rightarrow x + a = c(x+b) = cx + bc \therefore cx - x = a - bc \Rightarrow x(c-1) = a - bc$, so

$x = \dfrac{a-bc}{c-1} = \dfrac{cb-a}{1-c}$.

Example 13. A bag contains a number of marbles of which 80 are red, 24 are white, and the rest are blue. If the probability of randomly selecting a blue marble from this bag is 1 5, how many blue marbles are there in the bag?
a. 25　　　　b. 26　　　　c. 27　　　　d. 28　　　　e. 29

Solution: B.
Let x be the number of blue marbles.
$\dfrac{x}{x+80+24} = \dfrac{1}{5} \Rightarrow 5x = x + 104 \Rightarrow x = 26$.

Example 14. An airplane, flying with a tail wind, travels 1200 miles in 5 hours; the return trip, against the wind, takes 6 hours. Find the cruising speed of the plane and the speed of the wind (assume that both are constant).
A. 220 mph, 20 mph　B. 220 mph, 40 mph　C. 230 mph, 20 mph
D. 230 mph, 40 mph　E. 240 mph, 20 mph

Solution: A.
This is a $d = rt$ problem. Let c = cruising speed and w = wind speed.
1200/(c + w) = 5　　　　　　　　　　　　　　(1)
1200/(c − w) = 6　　　　　　　　　　　　　　(2)
1200/5 = c + w = 240
1200/6 = c − w = 200
From here, it's a system of equations. $c = 220$, $w = 20$.

Algebra II Through Competitions   **Chapter 13 Rational Functions**

PROBLEMS

Problem 1. Simplify $\dfrac{\dfrac{-2}{x+1}}{\dfrac{5}{x}+4}$.

A. $\dfrac{-2x}{4x^2+9x+5}$ B. $\dfrac{-2x-40}{5x+5}$ C. $\dfrac{-8x-10}{20x-40}$ D. $\dfrac{-2x}{9x+9}$ E. none of these

Problem 2. If $1-\dfrac{4}{x}+\dfrac{4}{x^2}=0$, then $\dfrac{2}{x}$ equals

A. -1 B. 1 C. 2 D. -1 or 2 E. -1 or -2

Problem 3. For all non-zero numbers x and y such that $x=\dfrac{1}{y}$, $\left(x-\dfrac{1}{x}\right)\left(y+\dfrac{1}{y}\right)$ equals

A. $2x^2$ B. $2y^2$ C. x^2+y^2 D. x^2-y^2 E. y^2-x^2

Problem 4. Mr. Goebel runs y yards in s seconds. What would his rate be, in yards per second, if he ran twice as far in 10 more seconds?

A. $\dfrac{2+y}{10+s}$ B. $\dfrac{y}{2(s+10)}$ C. $\dfrac{2y}{10s}$ D. $\dfrac{2y}{s+10}$ E. $\dfrac{2y}{s-10}$

Problem 5. Mary has d liters of punch that is d% grape juice. How many liters of grape juice must she add to make the punch $3d$% grape juice?

a. $\dfrac{d^2}{100-3d}$ b. $\dfrac{2d^2}{100-3d}$ c. $\dfrac{d}{100+3d}$ d. $\dfrac{3d^2}{100+d}$ e. none of the above

Problem 6. Find the value of $A+B$ if $\dfrac{A}{z+2}+\dfrac{B}{2z-3}=\dfrac{5z-11}{2z^2+z-6}$

A. -2 B. 0 C. 1 D. 2 E. none of these

Problem 7. Let $\dfrac{2x-11}{x^2-5x-14} = \dfrac{B}{x-7} + \dfrac{C}{x+2}$ be an identity in x. The value of $B + C$ is:
A. 4   B. –2   C. 5   D. 2   E. –4

Problem 8. Simplify the following rational function $\dfrac{x^3 - y^3}{x^4 + x^2y^2 + y^4}$.

A. $\dfrac{x-y}{x^2 + xy + y^2}$ B. $\dfrac{x-y}{x^2 - xy + y^2}$ C. $\dfrac{1}{x^2 - xy + y^2}$ D. $\dfrac{1}{x^2 + y^2}$ E. none of these

Problem 9. Compute N such that $\dfrac{N}{x-5} + \dfrac{3}{x+4} = \dfrac{10x+13}{x^2 - x - 20}$ for all $x > 2000$.
A. 4   B. 7   C. 10   D. 13   E. 16

Problem 10. The Container Company is designing an open top rectangular box, with a square base, that will hold 108 cm^3. Estimate the minimum surface area for the box.
A. 120 cm^2 B. 108 cm^2 C. 102 cm^2 D. 96 cm^2 E. 92 cm^2

Problem 11. John was contracted to work A days. For each of these A days that John actually worked, he received B dollars. For each of these A days that John did not work, he had to pay a penalty of C dollars. After the A days of contracted work was over, John received a net amount of D dollars for his work. How many of the A days of contracted work did John not work?

A. $\dfrac{AB - D}{B + C}$.   B. $\dfrac{AB + D}{B - C}$.   C. $\dfrac{AB + D}{B + C}$ D. $\dfrac{AB - D}{B - C}$.   E. None of the above.

Problem 12. How many pairs of positive integers (a, b) with $a + b \le 100$ satisfy $\dfrac{a + b^{-1}}{a^{-1} + b} = 13$?
A. 2   B. 3   C. 4   D. 5   E. 7

Problem 13. It takes Amy and Bill 15 hours to paint a house, it takes Bill and Chandra 20 hours, and it takes Chandra and Amy 30 hours. How long will it take if all three work together?

A. 9 hours and 40 minutes   B. 10 hours   C. 12 hours   D. 13 hours and 20 min
E. 14 hours.

Problem 14. A man's salary is reduced by p percent. By what percent would his salary then have to be raised to bring it back to the original amount?
A. $2p/(100-p)$ B. $(p-100)/(100-2p)$ C. $100p/(100-p)$ D. $p/(p-100)$
E. $2p/(p-100)$

Problem 15. John was contracted to work A days. For each of these A days that John actually worked, he received B dollars. For each of these A days that John didn't work, he had to pay a penalty of C dollars. After the A days of contracted work was over, John received a net amount of D dollars for his work. How many of the A days of contracted work did John not work?
A. $(AB-D)/(B+C)$ B. $(AB+D)/(B+C)$ C. $(AB-D)/(B-C)$
D. $(AB+D)/(B-C)$ E. $(AC-B)/(D-C)$

Problem 16. The non-zero real numbers a, b, c, d have the property that $\frac{ax+b}{cx+d}=1$ has no solution in x. What is the value of $\frac{a^2}{a^2+c^2}$?
A. 0   B. 1/2   C. 1   D. 2   E. an irrational number

Problem 17. Find x such that $(8+x)/(8-x) = x/(x+x)$.
A. $-8/3$ B. $-4/3$ C. $-2/3$ D. 0   E. 4.

Problem 18. (Indiana 2009 Algebra II) If $\frac{x^3+a}{x-1} = rx^2 + sx + t + \frac{u}{x-1}$, then $r+s+t+u$ is equal to
A. 1   B. 3   C. 4   D. $a+3$ E. $a+4$

Problem 19. What are the vertical asymptotes for the graph $y = (2x+14)/(x^3+7x^2-4x-28)$
A. $x=-2; x=2; x=-7$ B. $x=-2; x=2$ C. $x=-2; x=-7$
D. $x=2; x=7$ E. $x=-2; x=2; x=7$

Problem 20. What are the vertical asymptotes for the graph: $y = \dfrac{4x+20}{x^3+6x^2-x-30}$?

A. $x = 3, x = -2$ B. $x = -5, x = -3$ C. $x = -2, x = 3, x = 5$.
D. $x = -5, x = -3, x = 2$.   E. $x = -3, x = 2$.

Problem 21. The horizontal asymptote of $f(x) = \dfrac{3x^2}{x^2-1}$ is

A. $y = \dfrac{1}{3}$ B. $y = 1$ C. $x = 1$ D. $x = -1$ E. $y = 3$

Problem 22. Assuming non-zero denominators,
$\dfrac{k}{x-2} + \dfrac{w}{(x-2)^2} + \dfrac{px+f}{x^2+1} = \dfrac{3x^2+3x+7}{(x-2)^2(x^2+1)}$ for all values of x. Find the value of $k + w + p + f$.

Problem 23. The horizontal asymptote for the function $\dfrac{3x+5}{5x+10}$ is at

A. -2 B. $-5/3$ C. $1/2$ D. $3/5$

Problem 24. Consider the rational function, $f(x) = \dfrac{x-1}{x^2-2x-3}$. Find the following:
Horizontal asymptote(s); Vertical asymptote(s); x-intercept(s); and y-intercept.

SOLUTIONS

Problem 1. Solution: A.
$$\frac{\frac{-2}{x+1}}{\frac{5}{x}+4} = \frac{-2}{x+1} \cdot \frac{x}{5+4x} = \frac{-2x}{4x^2+9x+5}.$$

Problem 2. Solution: B.
$$1 - \frac{4}{x} + \frac{4}{x^2} = 1 - \frac{2}{x}(2 - \frac{2}{x}) = 0 \qquad (1)$$

Let $\frac{2}{x} = y$. (1) becomes: $1 - y(2-y) = 0 \Rightarrow y^2 - 2y + 1 = 0$

$\Rightarrow (y-1)^2 = 0 \Rightarrow y = \frac{2}{x} = 1.$

Problem 3. Solution: D.
$$x = \frac{1}{y} \Rightarrow xy = 1, \text{ so } (x-1/x)(y+1/y) = xy + x/y - y/x - 1/xy = 1 + \frac{x^2-y^2}{xy} - 1 = x^2 - y^2.$$

Problem 4. Solution: D.
Rate is distance divided by time so, Mr. Goebel's new rate is $\frac{2y}{s+10}$.

Problem 5. Solution: B.
Looking at the amount of grape juice, we get $d \cdot \frac{d}{100} + x = (d+x)\frac{3d}{100}$, where x is the amount of pure grape juice to be added. This simplifies to $d^2 + 100x = 3d^2 + 3dx$. Solving for x we get $x = \frac{2d^2}{100-3d}$.

Problem 6. Solution: D.
Multiply both sides by $(z+2)(2x-3)$. The resulting equation is now $A(2x-3) + B(z+2)$
$= 5z - 11$. When $x = \frac{3}{2}$, $B = \frac{\frac{15}{2}-11}{\frac{3}{2}+2} = -1$. When $x = -2$, $A = \frac{-10-11}{-7} = 3$. $A + B = 3 - 1$
$= 2$

Problem 7. Solution: D.
$$\frac{2x-11}{x^2-5x-14} = \frac{B}{x-7} + \frac{C}{x+2} \Rightarrow \frac{2x-11}{(x-7)(x+2)} = \frac{B(x+2)}{(x-7)(x+2)} + \frac{C(x-7)}{(x-7)(x+2)} \Rightarrow$$
$$2x - 11 = B(x+2) + C(x-7) \Rightarrow 2x - 11 = (B+C)x + (2B-7C) \Rightarrow B+C = 2.$$

Problem 8. Solution: B.
$$\frac{x^3-y^3}{x^4+x^2y^2+y^4} = \frac{(x-y)(x^2+xy+y^2)}{x^4+x^2y^2+y^4} = \frac{(x-y)(x^2+xy+y^2)}{(x^2+xy+y^2)(x^2-xy+y^2)} = \frac{(x-y)}{(x^2-xy+y^2)}$$
(Using long division one can show that $x^2 + xy + y^2$ is a factor of $x^4 + x^2y^2 + y^4$).

Problem 9. Solution: B.
Here we are decomposing a rational expression into partial fractions.
$$\frac{N}{x-5} + \frac{3}{x+4} = \frac{N(x+4)+3(x-5)}{(x-5)(x+4)} = \frac{10x+13}{x^2-x-20}$$, so for these last two fractions to be
equal, the numerators must be equal, so we have $N(x + 4) + 3(x - 5) = 10x + 13 \Leftrightarrow (N + 3)x + (4N - 15) = 10x + 13$. From this last equation we see that $N + 3 = 10$, $4N - 15 = 13$ $\Rightarrow N = 7$.

Problem 10. Solution: B.
The official solution used the calculator. We provide a solution without use of the calculator.
$$V = x^2h = 108 \Rightarrow h = \frac{108}{x^2}. \quad SA = x^2 + 4xh = x^2 + 4x\left(\frac{108}{x^2}\right) = x^2 + \frac{4 \cdot 108}{x}.$$
By AM-GM, we have $x^2 + \frac{2 \times 108}{x} + \frac{2 \times 108}{x} \geq 3\sqrt[3]{x^2 \times \frac{2 \times 108}{x} \times \frac{2 \times 108}{x}} = 3 \times 36 = 108.$

The greatest value 108 is obtained when $x^2 = \frac{2 \times 108}{x}$ or $x = 6$.

Problem 11. Solution: A.
Let x denote the number of days John did not work. Then he worked $A - x$ days and so earned $B(A - x) - Cx = D$ dollars. Solving this for x we get
$$-(B+C)x = D - AB \Rightarrow x = \frac{AB-D}{B+C}.$$

Problem 12. Solution: E.
Multiplying the given equality $a + b^{-1} = 13(b + a^{-1})$ by ab we obtain: $a(ab + 1) = 13b(ab + 1)$, or $(a - 13b)(ab + 1) = 0$. Since $ab + 1 > 0$, the given equation is equivalent to $a = 13b$. The inequality $a + b \leq 100$ means that $14b \leq 100$; therefore, the possible values of

Algebra II Through Competitions — Chapter 13 Rational Functions

the positive integer b are 1, 2, ... , 7, and there are 7 solutions: (13, 1), (26, 2), ... , (91, 7).

Problem 13. Solution: D.
If Amy, Bill and Chandra can paint a whole house at a rate of $1/a$, $1/b$, and $1/c$ of the house per hour, respectively, then $1/a + 1/b = 1/15$, $1/b + 1/c = 1/20$, and $1/c + 1/a = 1/30$. When all three work together, they can paint at a rate of $(1/a + 1/b + 1/c) = 1/2(1/15 + 1/20 + 1/30) = 3/40$. So it will take $40/3$ hours, that is, 13 hours 20 min.

Problem 14. Solution: C.
Let S denote the man's salary and let x denote the required percent. Then $S - S(p/100) + x/100 [S - S(p/100)] = S$. This is equivalent to $S(1 - p/100) + S(x/100 - xp/10000) = S$. So $1 - p/100 + x/100 - xp/10000 = 1$.
Solving for x, we get $x = 100p/(100 - p)$.

Problem 15. Solution: A.
Let x denote the number of days John did not work. Then he worked $A_i x$ days and so earned $B(A - x) - Cx = D$ dollars. Solving this for x, we get $-(B + C)x = D - AB$ and so $x = (AB - D)/(B + C)$.

Problem 16. Solution: B.
The equation is solvable if $ax + b = cx + d$ which is equivalent to $(a - c)x = d - b$. This has a solution unless $a = c$. Therefore $a = c$ and it follows that $\dfrac{a^2}{a^2 + c^2} = \dfrac{1}{2}$.

Problem 17. Solution: A.
The equation is equivalent to $2(8 + x) = 8 - x$, from which it follows that $x = -8/3$.

Problem 18. Solution: E.
When $x^3 + a$ is divided by $x - 1$, the quotient is $x^2 + x + 1$ and the remainder is $1 + a$. So $r = 1$, $s = 1$, $t = 1$, and $u = 1 + a$.
The sum of them is $1 + 1 + 1 + 1 + a = 4 + a$.

Problem 19. Solution: B.
$(2x + 14)/(x^3 + 7x^2 - 4x - 28) = 2(x + 7)/(x^2 - 4)(x + 7) = 2/(x + 2)(x - 2)$
Vertical asymptotes at $x = 2$ and $x = -2$.

Problem 20. Solution: E.

$$\frac{4x+20}{x^3+6x^2-x-30} = \frac{4(x+5)}{(x+5)(x-2)(x+3)} = \frac{4}{(x-2)(x+3)}.$$

The vertical asymptotes for the graph are $x = 2$ and $x = -3$.

Problem 21. Solution: E.
This is the case 2 where $n = m$. The line $y = a_n/b_m = 3/1 = 3$ is the horizontal asymptote.

Problem 22. Solution: 5.

$$\frac{k}{x-2} + \frac{w}{(x-2)^2} + \frac{px+f}{x^2+1} = \frac{k(x-2)(x^2+1) + w(x^2+1) + (px+f)(x-2)^2}{(x-2)^2(x^2+1)}$$

$$= \frac{k(x^3 - 2x^2 + x - 2) + wx^2 + w + (px+f)(x^2 - 4x + 4)}{(x-2)^2(x^2+1)}$$

$$= \frac{kx^3 - 2kx^2 + kx - 2k + wx^2 + w + px^3 - 4px^2 + 4px + fx^2 - 4fx + 4f}{(x-2)^2(x^2+1)}$$

$$= \frac{x^3(k+p) + x^2(-2k + w - 4p + f) + x(k + 4p) + w + 4f}{(x-2)^2(x^2+1)}.$$

Since there is no cubic term, both $k + p = 0$. So we know that $k = -p$.
So we only need to find $w + f$.
$-2k + w - 4p + f = 3$ (1)
$k + 4p = 3 \Rightarrow k + p + 3p = 3 \Rightarrow 3p = 3 \Rightarrow p = 1$ (2)
Substituting (2) into (1): $-2(-1) + w - 4 + f = 3$ $\Rightarrow$ $w + f = 5$.

Problem 23. Solution: D.
This is the case 2 where $n = m$. The line $y = a_n/b_m = 3/5$ is the horizontal asymptote.

Problem 24. Solution:
Horizontal asymptote:
This is the case 1 where $n < m$. The line $y = 0$ (the x-axis) is a horizontal asymptote
Vertical asymptotes:
We set $x^2 - 2x - 3 = 0$ $\Rightarrow$ $(x-3)(x+1) = 0$. So vertical asymptotes are $x = 3$ and $x = -1$.
x – intercept:
We set $f(x) = y = \dfrac{x-1}{x^2-2x-3} = 0 \Rightarrow x - 1 = 0$. So the x – intercept is $(1, 0)$ or $x = 1$.
y – intercept:
$f(0) = \dfrac{x-1}{x^2-2x-3} = \dfrac{0-1}{0^2 - 2\cdot 0 - 3} = \dfrac{-1}{-3} = \dfrac{1}{3}$. So the y – intercept is $(0, 1/3)$ or $y = 1/3$.

Algebra II Through Competitions **Chapter 14 Binomial Theorem**

1. THE BINOMIAL THEOREM

$$(x+y)^n = \binom{n}{0}x^n + \binom{n}{1}x^{n-1}y + \cdots + \binom{n}{n-1}xy^{n-1} + \binom{n}{n}y^n = \sum_{r=0}^{n}\binom{n}{r}x^{n-r}y^r \quad (1.1)$$

Where n is positive integer, r is nonnegative integer.

Sigma notation "$\sum$" means "the sum". For example: $a_1 + a_2 + \ldots + a_n = \sum_{k=1}^{n} a_k$.

A special case: $(1+x)^n = \binom{n}{0} + \binom{n}{1}x + \binom{n}{2}x^2 + \ldots + \binom{n}{r}x^r + \cdots + \binom{n}{n}x^n \quad (1.2)$

The binomial coefficients are the coefficients in the expansion of $(x+y)^n$.

The coefficient of the $(r+1)^{\text{th}}$ term is denoted as $C(n, r)$, C_n^r, or $\binom{n}{r}$.

Example 1. The coefficient in front of the term x^4 in $(x-1)^6$ is
A. -24 B. 1 C. 15 D. 24

Solution: C.
$(x-1)^6$
$= \binom{6}{0}x^6 + \binom{6}{1}x^5(-1) + \binom{6}{2}x^4(-1)^2 + \binom{6}{3}x^3(-1)^3 + \binom{6}{4}x^2(-1)^4 + \binom{6}{5}x(-1)^5 + \binom{6}{6}(-1)^6$
$= x^6 - 6x^5 + 15x^4 - 20x^3 + 15x^2 - 6x + 1$.
The coefficient in front of the term x^4 is 15.

2. THE GENERAL TERM OF THE BINOMIAL EXPANSION

Important note: when we say the 8th term, we mean $r + 1 = 8$ because we start to count not from 1 but from 0. So $r = 7$.

$(r+1)^{\text{th}}$ term of a binomial expansion: $T_{r+1} = \binom{n}{r}x^{n-r}y^r \quad (2.1)$

Example 2. What is the coefficient of $x^4 y^5$ in the expansion $(2x - y)^9$?
A. -16 B. 126 C. $-126(16)$ D. -126 E. 16.

Solution: C.

$n = 9$. x^4y^5 is the fifth term in the expansion of $(2x - y)^9$. The fifth term is $\binom{9}{5}(2x)^4(-y)^5 = -2^4 \times 126 \times x^4y^5$. Its coefficient is $-2^4 \times 126 = -126(16)$.

Example 3. If $(2a^3 - b^3)^{10}$ is expanded, completely simplified, and arranged in terms of descending powers of a, find the numerical coefficient of the fifth term.

Solution: 13440.

$n = 10$. The fifth term is $\binom{10}{4}(2a^3)^6(-b^3)^4 = 210 \times 2^6 \times a^{18}b^{12}$. Its coefficient is $210 \times 2^6 = 13440$.

3. FIND THE COEFFICIENT OF A TERM IN THE EXPANSION

Example 4. What is the numerical coefficient of c^2 in the expansion of $(\sqrt{c} + \sqrt{d})^{26}$?

Solution:
Method 1:
To obtain a term with a c^2, there must be 4 factors of $\sqrt{c}$ (and 22 factors of $\sqrt{d}$) in the term. With the coefficients of $\sqrt{c}$ and $\sqrt{d}$ being 1, the coefficient in the expansion then becomes $(1)^4(1)^2\left(\dfrac{26!}{4!22!}\right) = 14950$.

Method 2:
By (2.1), $(r+1)^{\text{th}}$ term of a binomial expansion:
$$T_{r+1} = \binom{n}{r}x^{n-r}y^r = \binom{26}{r}(\sqrt{c})^{26-r}(\sqrt{d})^r = \binom{26}{r}(c)^{\frac{26-r}{2}}(\sqrt{d})^r.$$

Let $(26-r)/2 = 2$, we get $r = 22$. The coefficient is $\binom{26}{22} = 14950$.

Example 5. (1963 AMC) In the expansion of $\left(a - \dfrac{1}{\sqrt{a}}\right)^7$ the coefficient of $a^{-\frac{1}{2}}$ is:

A. -7 B. 7 C. -21 D. 21 E. 35

Solution: C.
Method 1:
$$\left(a - \frac{1}{a^{1/2}}\right)^7 = a^7 - 7a^{11/2} + 21a^4 - 35a^{5/2} + 35a - 21a^{-1/2} + 7a^{-2} - a^{-7/2}.$$

Method 2:

$(r+1)$th term $= \binom{7}{r} a^{7-r}(-a^{-\frac{1}{2}})^r = (-1)^r \binom{7}{r} a^{7-r-\frac{r}{2}}$.

We need $a^{7-r-\frac{r}{2}} = a^{-\frac{1}{2}}$, $7 - \frac{3r}{2} = -\frac{1}{2}$, $r = 5$.

$\therefore$ 6th term $= \binom{7}{5} a^2 (-a)^{-5/2} = -21 a^{-1/2}$.

4. THE MULTINOMIAL COEFFICIENTS

For a multinomial $(x + y + z + ... + w)^n$, the coefficient of the term $x^a y^b z^c ... w^d$ is: $\frac{n!}{a!b!c!...d!}$, where $a + b + c + \cdots + d = n$.

The general term is T and $T = \frac{n!}{a!b!c!\cdots d!} x^a y^b z^c \cdots w^d$ (4.1)

$a, b, c, ..., d$ are nonnegative integers. $a + b + c + \cdots + d = n$.

The coefficient of the term $x^a y^b z^c ... w^d$ is: $\frac{n!}{a!b!c!...d!}$ (4.2)

The number of terms in the expansion of $(x_1 + x_2 + x_3 + ... + x_r)^n$, after the like terms combined, is $\binom{n+r-1}{n}$ or $\binom{n+r-1}{r-1}$ (4.3)

Example 6. The coefficient of x^7 in the polynomial expansion of $(1 + 2x - x^2)^4$ is
A. -8 B. 12   C. 6   D. -12 E. none of these

Solution: (A).
Method 1:

Algebra II Through Competitions   Chapter 14 Binomial Theorem

The coefficient of x^7 in $(1 + 2x - x^2)^4$ is the coefficient of the sum of four identical terms $2x(-x^2)^3$, which sum is $-8x^7$.

Method 2:
By (4.2), we have
$$T = \frac{n!}{a!b!c! \cdots d!} x^a y^b z^c \cdots w^d = \frac{4!}{a!b!c!} \cdot 1^a (2x)^b (-x^2)^c = \frac{4!}{a!b!c!} \cdot 1^a \cdot 2^b \cdot (-x)^{b+2c}.$$
We know that $a + b + c = 4$ and $b + 2c = 7$. Solving we get $c - a = 3$.
Therefore $a = 0$, $c = 3$, and $b = 1$ (note that $a = 1$, $c = 4$ will make b negative).
Therefore $T = -\frac{4!}{0!1!3!} \cdot 1^0 \cdot 2^1 x^7$ and the coefficient is $-\frac{4!}{0!1!3!} \cdot 2^1 = -8$.

5. FIND THE SUM OF COEFFICIENTS

Example 7. What is the sum of the numerical coefficients of the expression $(a + b)^8$?
A. 218   B. 128   C. 250   D. 256   E. none of the above

Solution: D.
When you expand $(a + b)^8$ you get a nine terms, each with some coefficient and a and b raised to powers from 0 to 8. Plugging 1 in for both a and b would then give you the sum of the coefficients, to go ahead and plug 1 in at the beginning, getting $(1 + 1)^8 = 2^8 = 256$.

Example 8. If $(3x - 1)^7 = a_7 x^7 + a_6 x^6 + \cdots + a_0$, then $a_7 + a_6 + \cdots + a_0$ equals
A. 0   B. 1   C. 64   D. -64 E. 128

Solution: E.
The sum of the coefficients of a polynomial $p(x)$ is equal to $p(1)$. For the given polynomial this is $(3 \cdot 1 - 1)^7 = 128$.

Example 9. What is the sum of the coefficients of the expanded form of $(2x - 3y + 3)^4$?
A. 0   B. 16   C. 81   D. 625   E. 1000

Solution: B.
Let $x = y = 1$. Then each term of the expansion is its coefficient, and the sum of these coefficients is $(2 - 3 + 3)^4 = 16$. Alternatively, expand the expression and add the coefficients.

6. FIND THE NUMBER OF TERMS IN THE EXPANSION

Example 10. The number of terms in the expansion of $[(a+3b)^2(a-3b)^2]^2$ when simplified is:
A. 4   B. 5   C. 6   (D) 7   E. 8

Solution: (B).
The binomial expansion of $(x+y)^n$ has $n+1$ terms. Thus $[(a+3b)^2(a-3b)^2]^2 = (a^2-9b^2)^4$ has $4+1=5$ terms.

Example 11. Expand $(a+b)(c+d+e)(f+g+h+i)$ and then combine the like terms, how many terms are there?

Solution:
The first part has 2 letters, second part has 3 letters, and third part has 4 letters. The total terms will be: $2 \times 3 \times 4 = 24$.

Algebra II Through Competitions **Chapter 14 Binomial Theorem**

PROBLEMS

Problem 1. What is the sum of all the coefficients of the terms of the expansion of $(3x - 4y)^9$
A. 1   B. −1   C. 144   D. −1728   E. none of these

Problem 2. The sum of the numerical coefficients in the expansion of $(x - y)^{21}$ is equal to
A. 0   B. 2   C. 221   D. 222   E. none of these

Problem 3. Determine the coefficient of x^{-3} in the expansion of $\left(\dfrac{2}{x} + \dfrac{x}{4}\right)^9$.

A. 80   B. 81   C. 82   D. 83   E. 84

Problem 4. Determine the coefficient of x^6 in the expansion of $\left(\dfrac{x^2}{4} + \dfrac{2}{x}\right)^{12}$.

A. 80   B. 81   C. 82   D. 83   E. 231/16.

Problem 5. What is the coefficient of x^6 in the polynomial $p(x) = (x + 2)^8$?
A. 48   B. 60   C. 112   D. 660   E. 1024

Problem 6. Determine the coefficient of x^7 in the expansion of $\left(\dfrac{x^2}{2} - \dfrac{2}{x}\right)^8$.

A. 12   B. − 12   C. 13   D. − 14   E. 14

Problem 7. Find the coefficient of x^3 in the expansion of $(1 + x + x^2)^{12}$.
A. 132   B. 220   C. 352   D. 552   E. 2024

Problem 8. Find the x^4 coefficient of $(x - 3)^9$.

Problem 9. If $\left(a + \dfrac{1}{a}\right)^2 = 3$, then $a^3 + \dfrac{1}{a^3}$ equals

A. 0   B. $3\sqrt{3}$ C. $\dfrac{10\sqrt{3}}{3}$ D. $6\sqrt{3}$ E. $7\sqrt{3}$

Algebra II Through Competitions **Chapter 14 Binomial Theorem**

Problem 10. Find the coefficient of the term that contains x^4 in the expansion of $(x^2 - \frac{1}{x})^5$.

A. −10 B. 10 C. −5 D. 5

Problem 11. Find the coefficients of the term that contain x^4 in the expansion of $\left(x - \frac{1}{2x}\right)^{10}$.

A. −120 B. 120 C. −15 D. 15

Problem 12. Find the coefficient of the term that contains x^3 in the expansion of $\left(x - \frac{2}{x}\right)^7$.

Problem 13. Find the coefficient of the term that contains x in the expansion of $(2x + \frac{1}{\sqrt{x}})^7$.

Problem 14. When completely simplified, one of the terms of the expansion of $\left(x + \frac{y}{2}\right)^9$ is $\frac{k}{w}x^3 y^p$ where k, w, and p are positive integers. Find the minimum value of (k + w + p).

Problem 15. What is the coefficient of x^5 in the expansion $(2x + 0.5)^{10}$?
A. 1 B. 16 C. 32 D. 252 E. 1024.

Problem 16. The coefficient in front of the term x^5 in $(x + 2)^7$ is
A. 21 B. 7 C. 42 D. 84

SOLUTIONS

Problem 1. Solution: B.
$(3 \times 1 - 4 \times 1)^9 = -1$.

Problem 2. Solution: A.
$(1 - 1)^{21} = 0$.

Problem 3. Solution: E

$n = 9$. We see that $\left(\dfrac{2}{x}\right)^6 \times \left(\dfrac{x}{4}\right)^3$ will produce the term x^{-3}.

So the term $\binom{9}{3} \times \left(\dfrac{2}{x}\right)^6 \times \left(\dfrac{x}{4}\right)^3 = 84\, x^{-3}$.

The coefficient is 84.

Problem 4. Solution: E.

$n = 9$. If $\left(\dfrac{x^2}{4}\right)^a \times \left(\dfrac{2}{x}\right)^b$ will produce the term x^6, $2a - b$ in the expression $x^{2a} \times x^{-b} = x^{2a-b}$ should be 6. Since $a + b = 12$, $a = 6$ and $b = 6$.

So the term $\binom{12}{6} \times \left(\dfrac{x^2}{4}\right)^6 \times \left(\dfrac{2}{x}\right)^6 = 924\, \dfrac{x^{12}}{4^6} \times \dfrac{2^6}{x^6} = \dfrac{231}{16} x^6$.

The coefficient is 231/16.

Problem 5. Solution: C

$n = 8$. We see that $x^6 \times 2^2$ will produce the term x^6.
So the term $\binom{8}{2} \times x^6 \times 2^2 = 28 \times 4 \times x^6$.

The coefficient is 112.

Problem 6. Solution: D.

$n = 8$. If $\left(\dfrac{x^2}{2}\right)^a \times \left(-\dfrac{2}{x}\right)^b$ will produce the term x^7, $2a - b$ in the expression $x^{2a} \times x^{-b} = x^{2a-b}$ should be 7. Since $a + b = 8$, $a = 5$ and $b = 3$.

So the term $\binom{8}{3} \times \left(\frac{x^2}{2}\right)^5 \times \left(-\frac{2}{x}\right)^3 = -56 \frac{x^{10}}{2^5} \times \frac{2^3}{x^3} = -14x^7$.

The coefficient is -14.

Problem 7. Solution: C.

By (4.2), we have $T = \frac{12!}{a!b!c!} \cdot 1^a (x)^b (x^2)^c = \frac{12!}{a!b!c!} \cdot 1^a \cdot (x)^{b+2c}$.

We know that $a + b + c = 12$ and $b + 2c = 3$. So $a - c = 9$. Since $b + 2c = 3$, $c < 2$. If $c = 1$, $a = 10$ and $b = 1$. If $c = 0$, $a = 9$, and $b = 3$. Therefore

$T = \frac{12!}{10!\,1!\,1!} + \frac{12!}{9!\,3!\,0!} = 132 + 220 = 352$.

Problem 8. Solution:

$n = 9$. We see that $x^4 \times (-3)^5$ will produce the term x^4.

So the term $\binom{9}{5} \times x^4 \times (-3)^5 = -126 \times 234 \times x^4$. The coefficient is -30618.

Problem 9. Solution: A.

$a^3 + \frac{1}{a^3} = \left(a + \frac{1}{a}\right)^3 - 3\left(a + \frac{1}{a}\right) = 3\sqrt{3} - 3\sqrt{3} = 0$.

Problem 10. Solution: B.

By (2.1), $(r+1)^{th}$ term of a binomial expansion:

$T_{r+1} = \binom{n}{r} x^{n-r} y^r = \binom{5}{r}(x^2)^{5-r}(-\frac{1}{x})^r = (-1)^r \binom{5}{r}(x^{10-3r})$.

Let $10 - 3r = 4, \therefore r = 2$.

The coefficient of the term that contains x^4 is $(-1)^2 \binom{5}{2} = 10$.

Problem 11. Solution: C

By (2.1), $(r+1)^{th}$ term of a binomial expansion:

$T_{r+1} = \binom{n}{r} x^{n-r} y^r = \binom{10}{r}(x)^{10-r}(-\frac{1}{2x})^r = (-\frac{1}{2})^r \binom{10}{r}(x^{10-2r})$.

Algebra II Through Competitions   **Chapter 14 Binomial Theorem**

Let $10 - 2r = 4$. $r = 3$. The coefficient of the term that contains x^4 is $(-\frac{1}{2})^3 \binom{10}{3} = -15$.

Problem 12. Solution:
By (2.1), $(r+1)^{th}$ term of a binomial expansion:
$$T_{r+1} = \binom{n}{r} x^{n-r} y^r = \binom{7}{r}(x)^{10-r}(-\frac{2}{x})^r = (-2)^r \binom{7}{r}(x^{7-2r}).$$

Let $7 - 2r = 3$, we get $r = 2$. Thus the coefficient is $(-2)^2 \binom{7}{2} = 84$.

Problem 13. Solution:

In the expansion of $(2x + \frac{1}{\sqrt{x}})^7$, the term with x is $\binom{7}{3}(2x)^3 \cdot (\frac{1}{\sqrt{x}})^4 = 280x$.

So the coefficient is 280.

Problem 14. Solution: 43.

By (2.1), $(r+1)^{th}$ term of a binomial expansion: $T_{r+1} = \binom{n}{r} x^{n-r} y^r = \binom{9}{r}(x)^{9-r}(\frac{y}{2})^r$.

We know that $9 - r = 3$. So $r = 6$.
$$T_{r+1} = \binom{9}{6}(x)^{9-6}(\frac{y}{2})^6 = \frac{84}{64} x^3 y^6 = \frac{21}{16} x^3 y^6.$$
$k + w + p = 21 + 16 + 6 = 43$.

Problem 15. Solution: D.
$n = 10$. We see that $(2x)^5 \times 0.5^5$ will produce the term x^5.
So the term $\binom{10}{5} \times (2x)^5 \times (0.5)^5 = 252 \, x^5$. The coefficient is 252.

Problem 16. Solution: D.
$n = 7$. We see that $x^5 \times 2^2$ will produce the term x^5.
So the term $\binom{7}{2} \times x^5 \times 2^2 = 84 \, x^5$.

The coefficient is 84.

Algebra II Through Competitions **Chapter 15 Complex Numbers**

1. DEFINITION OF COMPLEX NUMBERS
A complex number z is a number of the form:
$$z = x + yi \tag{1.1}$$
where x and y are real numbers and
$$i^2 = -1. \tag{1.2}$$

The real number $x = \text{Re}(z)$ is called the real part of the complex number z and $y = \text{Im}(z)$ is called the imaginary part of z. Complex numbers of the form iy are called purely imaginary and the complex number i is called the imaginary unit.

In the polar form: $z = r(\cos\theta + i\sin\theta)$ (1.3)
In the exponential form: $z = r\, e^{i\theta}$ (1.4)

2. PROPERTIES OF COMPLEX NUMBERS
Property 1. For two complex numbers $z_1 = x_1 + y_1 i$ and $z_2 = x_2 + y_2 i$, $z_1 = z_2$ if and only if $x_1 = x_2$ and $y_1 = y_2$.

Note that as long as the imaginary part is involved, we cannot say $z_1 < z_2$ or $z_1 > z_2$.

Property 2. $z_1 = x_1 + y_1 i$ is real if and only if $y_1 = 0$.

Property 3. $i^0 = 1$; $i^1 = i$; $i^2 = -1$; $i^3 = -i$; $i^4 = 1$.
For any positive integer n, $i^{4n} = 1$; $i^{4n+1} = i$; $i^{4n+2} = -1$; $i^{4n+3} = -i$.

Property 4. $z = \bar{z}$ if and only if z is a real number.

Property 5. $z = -\bar{z}$ if and only if z is a pure imaginary number.

$$\overline{z_1 + z_2} = \bar{z_1} + \bar{z_2} \tag{2.1}$$

$$\overline{z_1 \cdot z_2} = \bar{z_1} \cdot \bar{z_2} \tag{2.2}$$

$$\overline{\left(\frac{z_1}{z_2}\right)} = \frac{\bar{z_1}}{\bar{z_2}}, \; z_2 \neq 0 \tag{2.3}$$

Property 6. The product of two pure imaginary numbers is a real number.

Example 1. Let $i = \sqrt{-1}$ and let k represent a real number. If $k - 5i + x = -7 + 2i$ is solved for x, then $x = -10 + 7i$. Find the value of k.

207

Solution: 3.
$k - 5i + x = -7 + 2i \Rightarrow x = -7 - k + 7i$
Thus $-7 - k = -10 \Rightarrow k = 3$.

Example 2. If $a + bi = \dfrac{3-i}{1+i}$, then

A. $a = 2, b = -2$ B. $a = 3, b = 1$ C. $a = -3, b = 1$ D. $a = 1, b = -2$ E. $a = 4, b = 4$.

Solution: D.
$$\dfrac{3-i}{1+i} = \dfrac{(3-i)(1-i)}{(1+i)(1-i)} = \dfrac{3 - 3i - 1i - 1}{2} = \dfrac{2 - 4i}{2} = 1 - 2i.$$

Example 3. Solve $(a + bi)^2 = 5 + 12i$ for a and b. The number of solutions is
A. 0   B. 1   C. 2   D. 3   E. 4

Solution: C.
The beauty of complex numbers is that there will always be two solutions to any quadratic equation (if we count double solutions as two solutions), so there will be two solutions. Checking to make sure we do not have a double solution, we see that
$(a + bi)^2 = (a^2 - b^2) + 2abi = 5 + 12i$, so $a^2 - b^2 = 5$ and $ab = 6$.
We have two ways:
(1) We could sketch graphs to see that there are two distinct solutions. Note that a is plotted on the horizontal axis.

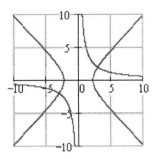

(2) We could solve this system
$a^2 - b^2 = 5 \Rightarrow a^2 = b^2 + 5$ (1)
$ab = 6 \Rightarrow a^2 b^2 = 36$ (2)
Substituting (1) into (2): $(b^2 + 5)b^2 = 36$ (3)
Let $b^2 = m$, where b is real and m is real.
(3) becomes: $(m + 5)m = 36 \Rightarrow m^2 + 5m - 36 = 0 \Rightarrow (m + 9)(m - 4) = 0$.
Solving we get $m = 4 \Rightarrow b^2 = 4 \Rightarrow b = 2$ or $b = -2$.
Thus we have two solutions: $a = 3$ and $b = 2$ or $a = -2$ and $b = -2$.

Example 4. If $i = \sqrt{-1}$, then what is the value of i^{2009}?
A. 1   B. i C. -1 D. $-i$ E. None of the above
Solution: B.

$i^{2009} = i^{2008+1} = i$.

Example 5. Let $S = i^{2n-2} + i^{2n+2}$, where n is an integer and $i^2 = -1$. The total number of possible distinct values of S is
A. 1   B. 2   C. 3   D. 4   E. 5

Solution: B.
$S = i^{2n-2} + i^{2n+2} = i^{2n-2} / i^2 + i^{2n} \times i^2 = -i^{2n} - i^{2n} = -2i^{2n} = -2(i^2)^n = -2 \times (-1)^n$.
S can be 2 (when n is odd) and -2 (when n is even).

3. CONJUGATE OF A COMPLEX NUMBER

The conjugate of $z = x + yi$ is expressed as $\bar{z} = x - yi$ \hfill (3.1)

Example 6. Given that z is a complex number and $i^2 = -1$. Solve for z given $2z + 1 = 2 - iz$
A. $0.6 - i$ B. $0.8 + 0.6i$ C. $0.8 - 0.6i$ D. $0.6 - 0.8i$ E. none of these

Solution: E.
$z(2 + i) = 1 = \dfrac{1}{2+i} \cdot \dfrac{2-i}{2-i} = \dfrac{2-i}{5} = 0.4 - 0.2i$.
None of the given choices represent z.

Example 7. If a polynomial, $F(x)$, has real coefficients with zeros at 2, $1 + i$, $3 - i$, then this polynomial must have a degree of :
A. at least 5   B. exactly 6   C. exactly 3   D. at least 6   E. none of the above

Solution: A.

If three of the zeros are 2, $1+ i$, $3- i$, and the coefficients are real, then the conjugates of the imaginary roots must also be zeros. Thus the minimum degree of the polynomial will be five.

Example 8. Let $i^2 = -1$. If z represents a complex number, then $\bar{z}$ represents its complex conjugate. If $z = 5 - 4i$, then $4\bar{z} = k + wi$ where k and w represent real numbers. Find the value of $(k + w)$.

Solution: 36.

If $z = 5 - 4i$ $\Rightarrow$ $\bar{z} = 5 + 4i$ $\Rightarrow$ $4\bar{z} = 4 \cdot 5 + 4 \cdot 4i = 20 + 16i$
$k + w = 20 + 16 = 36$.

Example 9. The polynomial equation $x^3 + kx^2 + wx + p = 0$ with k, w, and p representing integers has 5 and $3 + i$ as two of its roots for x. If $i^2 = -1$, find the value of w.

Solution: 40.
Since $3 + i$ is a root, $3 - i$ is the root as well.
By Vieta's Theorem,

$x_1 + x_2 + x_3 = -\dfrac{a_1}{a_0}$ $\Rightarrow$ $5 + (3 + i) + (3 - i) = -k$ $\Rightarrow$ $k = -11$.

$x_1 x_2 x_3 = -\dfrac{a_3}{a_0}$ $\Rightarrow$ $5 \times (3 + i) \times (3 - i) = -p$ $\Rightarrow$ $p = -50$.

Since 5 is a root, we have $5^3 - 11 \times 5^2 + 5w - 50 = 0$ $\Rightarrow$ $w = 40$.

4. MODULUS OF A COMPLEX NUMBER

The modulus or the absolute value of the complex number $z = x + yi$ is expressed as

$|z| = \sqrt{x^2 + y^2}$ (4.1)

$|z| = |-z| = |\bar{z}|$.(a nonnegative real number). (4.2)

$z \cdot \bar{z} = |z|^2$ (The product of z and its conjugate is a nonnegative real number). (4.3)

$|z_1 \cdot z_2| = |z_1| \cdot |z_2|$ (4.4)

$\left|\dfrac{z_1}{z_2}\right| = \dfrac{|z_1|}{|z_2|}, z_2 \neq 0$ (4.5)

$z + \bar{z} = (x + yi) + (x - yi) = 2x$ (a nonnegative real number). (4.6)

$z - \bar{z} = (x + yi) - (x - yi) = 2yi$ (a real number if $y = 0$). (4.7)

Example 10. Let $i = \sqrt{-1}$, $\left|\dfrac{-2 + 3i}{1 - i}\right| = \dfrac{\sqrt{k}}{w}$, where k and w represent positive integers. Find the smallest possible value of $(k + w)$.

Solution: 28.

$$\left|\frac{-2+3i}{1-i}\right| = \frac{|-2+3i|}{|1-i|} = \frac{\sqrt{(-2)^2+3^2}}{\sqrt{1^2+(-1)^2}} = \frac{\sqrt{13}}{\sqrt{2}} = \frac{\sqrt{13}}{\sqrt{2}} \cdot \frac{\sqrt{2}}{\sqrt{2}} = \frac{\sqrt{26}}{2}.$$

$k + w = 26 + 2 = 28$.

Example 11. Let $i = \sqrt{-1}$. If $|21 + ki| = 75$ and $k < 0$, find the value of k.

Solution: -72.

$|21 + ki| = \sqrt{21^2 + k^2} = 75 \quad \Rightarrow \quad 21^2 + k^2 = 75^2 \Rightarrow$
$k^2 = 75^2 - 21^2 = (75-21)(75+21) = 54 \times 96 = 72^2 = (-72)^2$.
So $k = -72$.

5. RULES OF OPERATIONS:

$z_1 + z_2 = (x_1 + y_1 i) + (x_2 + y_2 i) = (x_1 + x_2) + (y_1 + y_2)i$ \hfill (5.1)

$z_1 - z_2 = (x_1 + y_1 i) - (x_2 + y_2 i) = (x_1 - x_2) + (y_1 - y_2)i$ \hfill (5.2)

$z_1 \cdot z_2 = (x_1 + y_1 i)(x_2 + y_2 i) = (x_1 x_2 - y_1 y_2) + (x_1 y_2 + x_2 y_1)i$ \hfill (5.3)

$\dfrac{z_1}{z_2} = \dfrac{x_1 x_2 + y_1 y_2}{x_2^2 + y_2^2} + \dfrac{-x_1 y_2 + x_2 y_1}{x_2^2 + y_2^2} i.$ \hfill (5.4)

$\lambda(z_1 + z_2) = \lambda z_1 + \lambda z_2;$ \hfill (5.5)

$\lambda_1(\lambda_2 z) = (\lambda_1 \lambda_2)z;$ \hfill (5.6)

$(\lambda_1 + \lambda_2)z = \lambda_1 z + \lambda_2 z$ \hfill (5.7)

$z^m \cdot z^n = z^{m+n}$ \hfill (5.8)

$\dfrac{z^m}{z^n} = z^{m-n}$ \hfill (5.8)

$(z^m)^n = z^{mn}$ \hfill (5.10)

$(z_1 \cdot z_2)^n = z_1^n \cdot z_2^n$ \hfill (5.11)

$\left(\dfrac{z_1}{z_2}\right)^n = \dfrac{z_1^n}{z_2^n}.$ \hfill (5.12)

$z^n = r^n[\cos(n\theta) + i\sin(n\theta)]$ **(DeMoivre's Theorem)** \hfill (5.13)

Example 12. If $f(x) = \dfrac{4x^3 + 3x^2 + 2x + 1}{2x - 2}$ and $i = \sqrt{-1}$ then the value of $f(i)$ is:

A. $i - 1$ \quad B. $1 + i$ \quad C. $-i$ \quad D. i \quad E. none of these

Solution: D.

$$f(i)=\frac{4i^3+3i^2+2i+1}{2i-2}=\frac{-4i-3+2i+1}{2i-2}=\frac{-2i-2}{-2+2i}=\frac{-i-1}{-1+i}=\frac{-(1+i)}{-(1-i)}=\frac{1+i}{1-i}=\frac{(1+i)^2}{(1-i)(1+i)}$$

$$=\frac{1+2i+i^2}{1^2+1^2}=\frac{1+2i-1}{2}=\frac{2i}{2}=i.$$

Note: the official answer key is C.

Example 13. Let $z = 6 + 7i$, where $i^2 = -1$. If $1/z = a + bi$, what is the value of a?

A. 1/6 B. –1/7 C. 7/13 D. –7/85 E. 6/85

Solution: E.

$z = 6 + 7i \quad \Rightarrow \quad \frac{1}{z}=\frac{1}{6+7i}=\frac{6-7i}{(6+7i)6-7i)}=\frac{6-7i}{6^2-(7i)^2}=\frac{6+7i}{6^2+7^2}=\frac{6}{85}+\frac{7}{85}i.$

$a = 6/85$.

Example 14. If $x+y=i$ and $x\times y=i$, where $i=\sqrt{-1}$, determine the value of x^3+y^3.

A. $3 + i$ B. $i + 1$ C. 6 D. 2 E. $3 - i$

Solution: E.

$x^3+y^3=(x+y)^3-3xy(x+y)=i^3-3i\times i=-i+3.$

Example 15. Let $i=\sqrt{-1}$. Which of the following is equivalent to $\frac{3-2i}{4+5i}$?

A. $\frac{3}{4}+\frac{2}{5}i$ B. $\frac{5-4i}{2+3i}$ C. $\frac{2}{41}-\frac{23}{41}i$ D. $\frac{7}{81}+\frac{21}{81}i$ E. $\frac{21}{80}-\frac{71}{80}i$

Solution: C.

$$\frac{3-2i}{4+5i}=\frac{(3-2i)(4-5i)}{(4+5i)(4-5i)}=\frac{12-23i-10}{4^2+5^2}=\frac{2}{41}-\frac{23}{41}i.$$

Example 16. If $i=\sqrt{-1}$, what is the value of $(1-i)^{20}$?

A. 1024 B. –1024 C. 1024i D. –1024i E. None of these

Solution: B.

By the DeMoivre's Theorem, $z^n=r^n[\cos(n\theta)+i\sin(n\theta)]$.

$(1-i)^{20}=\left[-\sqrt{2}(\cos 135^\circ+i\sin 135^\circ)\right]^{20}=(-\sqrt{2})^{20}[\cos(135^\circ\times 20)+i\sin(135^\circ\times 20)$

$$= 2^{10}[\cos(7\times 360° + 180) + i\sin(7\times 360° + 180) = 2^{10}(\cos 180° + i\sin 180°)$$
$$= 2^{10}(-1+0) = -1024$$

6. QUADRATIC EQUATIONS

The two solutions of the quadratic equation $ax^2 + bx + c = 0$, $a \neq 0$.

$$x = \frac{-b \pm i\sqrt{-(b^2 - 4ac)}}{2a} \qquad (6.1)$$

Example 17. Find the roots of $y^{-2} - 2y^{-1} + 2 = 0$

A. $\dfrac{1 \pm i}{2}$ B. $1 \pm i$ C. $-1 \pm i$ D. $\dfrac{-1 \pm i}{2}$ E. none of these

Solution: A.
$y^{-2} - 2y^{-1} + 2 = 0 \Rightarrow 1 - 2y + 2y^2 = 0$.

$$y = \frac{-(-2) \pm \sqrt{(-2)^2 - 4(2)(1)}}{2 \times 2} = \frac{2 \pm \sqrt{-4}}{4} = \frac{2 \pm 2i}{4} = \frac{1 \pm i}{2}.$$

Example 18. Find all real and imaginary solutions of the equation $3ix^2 + 7x - 4i = 0$.

A. $x = i, x = \dfrac{-4i}{3}$ B. $x = -i, x = \dfrac{-4i}{3}$ C. $x = -i, x = \dfrac{4i}{3}$

D. $x = i, x = \dfrac{4i}{3}$ E. $x = -6i, x = \dfrac{8i}{3}$

Solution: D.
$$x = \frac{-7 \pm \sqrt{7^2 - 4(3i)(-4i)}}{6i} = \frac{-7 \pm \sqrt{49 - 48}}{6i} = \frac{-7 \pm 1}{6i}$$

$$x_1 = \frac{-7+1}{6i} = \frac{-6i}{6i^2} = i \text{ and } x_2 = \frac{-7-1}{6i} = \frac{-8i}{6i^2} = \frac{4i}{3}.$$

Example 19. If $x^2 - (1-2i)x = (\dfrac{1}{2} + i)$, find the complete solution.

A. $\left\{\dfrac{1\pm 2i}{2}\right\}$ B. $\left\{\dfrac{1\pm 3i}{3}\right\}$ C. $\left\{\dfrac{1+i}{i},\dfrac{1+3i}{2}\right\}$ D. $\left\{\dfrac{1-2i}{2},\dfrac{1+3i}{2}\right\}$ E. none of these

Solution: E.
$$x = \dfrac{-(-2+4i)\pm\sqrt{(-2+4i)^2 - 4(2)(-1-2i)}}{4} = \dfrac{2-4i\pm\sqrt{-4}}{4}$$
$$x = \dfrac{2-4i\pm 2i}{4},\quad x = \dfrac{1-i}{2} \text{ or } x = \dfrac{1-3i}{2}.$$

Example 20. If the complex number $2i + 1$ is a solution to the quadratic equation $x^2 + bx + c = 0$, then what is $b + c$?
A. 2   B. 3   C. 4   D. 5   E. 6

Solution: B.
Since $2i + 1$ is a solution, $-2i + 1$ is also a solution.
By Vieta's Theorem,

$x_1 + x_2 = -\dfrac{b}{a}$ $\Rightarrow$ $(2i + 1) + (-2i + 1) = -b$ $\Rightarrow$ $b = -2$.

$x_1 \cdot x_2 = \dfrac{c}{a}$ $\Rightarrow$ $(2i + 1) \times (-2i + 1) = c$ $\Rightarrow$ $c = 5$.

$b + c = -2 + 5 = 3$.

Note: the conjugate of $2i + 1$ is $-2i + 1$, not $2i - 1$.

Algebra II Through Competitions   **Chapter 15 Complex Numbers**

PROBLEMS

Problem 1. The standard form of $i^6 - i^{10} - i^{15}$ is:
A. i^6 B. i C. $-i$ D. $-1 + i$ E. $1 - i$

Problem 2. If $f(x) = \left(\dfrac{x^4 - x^3 + x^2 - x + 1}{x}\right)^3$ and $i = \sqrt{-1}$, then $f(i)$ equals

A. i B. -1 C. $-i$ D. 1 E. $2i$

Problem 3. If $f(x) = \dfrac{x^2 - x + 1}{x + 1}$ and $i = \sqrt{-1}$, then what does $f(1 - i)$ equal?

A. $-i$ B. $\dfrac{1 - 2i}{5}$ C. $\dfrac{1 - 3i}{5}$ D. $\dfrac{3 - 2i}{5}$ E. $1 + i$.

Problem 4. Which of the following expressions, if any, is different from the other 3? ($i = \sqrt{-1}$, $c \in$ Reals).

A. $\dfrac{1}{2} - \dfrac{i + c}{i - 1}$ B. $\dfrac{c}{2} + \dfrac{i + ci}{2}$ C. $\dfrac{i}{2} + \dfrac{c}{i + 1}$ D. $\dfrac{-1 + (2c + 1) \cdot i}{2(1 + i)}$ E. All are the same.

Problem 5. Evaluate $\left(\dfrac{i^{123}}{i^6}\right)^{-5}$ given that $i = \sqrt{-1}$.

A. $-i$ B. i C. -1 D. 1 E. 0

Problem 6. Find $i + 2i^2 + 3i^3 + 4i^4 + \ldots + 2006i^{2006}$ where $i = \sqrt{-1}$.
A. $-1004 - 1003i$ B. $-1004 + 1003i$ C. 0 D. $1004 - 1003i$ E. $1004 + 1003i$

Problem 7. If $i^2 = -1$, then the sum $i^0 + i^1 + i^2 + i^3 + \ldots + i^{2009} + i^{20110}$, is
A. 0 B. 1 C. -1 D. i E. $-i$

Problem 8. Which of the following expressions, if any, is different from the other 3? $i = \sqrt{-1}$.

A. $i - \dfrac{4}{1 + 5i}$ B. $\dfrac{1 + 9i}{5 - i}$ C. $\dfrac{23i - 2}{13}$ D. $\dfrac{i - 9}{1 + 5i}$ E. All are the same.

Algebra II Through Competitions **Chapter 15 Complex Numbers**

Problem 9. $f(x) = \dfrac{x^{17} + x^9}{x^5 - 1}$ and $i = \sqrt{-1}$. The value of $f(i)$ is:
A) $i - 1$ B) $1 + i$ C) $1 - i$ D) $2i$ E) none of these

Problem 10. If $f(x) = \dfrac{x^5 + x^2}{x^3 + 1}$, then $f(i)$, where $i = \sqrt{-1}$, is equal to:
A. -2 B. -1 C. 0 D. 1 E. None of these

Problem 11. If $i^2 = -1$, then what is the value of $\left(\dfrac{1+i}{\sqrt{2}}\right)^{2011}$?

A. 1 B. $\dfrac{1+i}{\sqrt{2}}$ C. $\dfrac{1-i}{\sqrt{2}}$ D. i E. $-i$.

Problem 12. If $i = \sqrt{-1}$, what is the value of $\left(\dfrac{1-i}{1+i}\right)^{2012}$?

A. 1 B. -1 C. i D. $-i$ E. $\dfrac{1}{1+i}$

Problem 13. If z is a complex number satisfying $z^2 - z + 1 = 0$, then z^9 is equal to
A. 1 B. -1 C. i D. $-i$ E. None of these

Problem 14. If z is a complex number satisfying $z^2 - z + 1 = 0$, then z^{24} is equal to
A. 1 B. -1 C. i D. $-i$ E. None of these

Problem 15. If $i^2 = -1$, then what is the value of i^{2010}?
A. 1 B. i C. -1 D. $-i$ E. None of these

Problem 16. If $S = i^n + i^{-n}$ (where $i^2 = -1$) and n is an integer, then the total number of possible distinct value for S?
A. 1 B. 2 C. 3 D. 4 E. ∞

Problem 17. For what value of b is $\dfrac{2+i}{3-bi}$ a real number?

A. –3/2   B. –1/2   C. 1/2   D. 3   E. There is no such value

Problem 18. How many real solutions does the equation $x^9 - x = 0$ have?
A. 1   B. 2   C. 3   D. 4   E. 9

Problem 19. Let $i = \sqrt{-1}$. Then $i^{13} - i^7 = ki$ where k is a real number. Find the value of k.

Problem 20. Let $i = \sqrt{-1}$. In an arithmetic sequence of complex numbers, the first term is $3 + 7i$, and the second term is $4 + 5i$. If the ninth term is written in the form $x + yi$ where x and y are real numbers, find the value of $(x - y)$.

Problem 21. If $i = \sqrt{-1}$, one of the roots of $x^3 + (-7+i)x^2 - (54+22i)x + 120i = 72$ is $2i$. Find the real root.

Problem 22. Let $i = \sqrt{-1}$. Let $a \pm bi$ and k be the roots for x of the cubic equation $x^3 + wx^2 + 85x - 156 = 0$. Find the value of w.

Problem 23. Let $i = \sqrt{-1}$, let $\bar{z}$ be the complex conjugate of z, and let x and y be real numbers. If $z = 2 + 3i$, find the value of $x + y$ when $z + 3\bar{z}$ is written in $x + yi$ form.

Problem 24. Which of the following polynomial equations has roots 1, 3 + 2i, and 3 – 2i?
A. $x^3 - 7x^2 + 19x - 13 = 0$ B. $x^3 - 8x^2 + 21x - 14 = 0$ C. $x^3 + 5x^2 - 14x + 8 = 0$
D. $x^3 + 4x^2 - 15x + 10 = 0$ E. $x^3 + 6x^2 - 12x + 5 = 0$

Problem 25. Among the zeros of a polynomial with rational coefficients are –1, 3, 1 + $\sqrt{5}$, and 3 – 2i. The least degree of the polynomial can have is
A. 2   B. 4   C. 6   D. 8   E. 10

Problem 26. Find a function with integer coefficients that has the zeros: 2 and 1 – 3i.
A. $y = x^2 - 6x - 16$ B. $y = x^2 - x + 2 + 3xi - 6i$
C. $y = x^3 - 4x^2 + 4x + 32$ D. $y = x^3 - 4x^2 + 14x - 20$ E. None of these

Problem 27. Let $i = \sqrt{-1}$. Find the value of $|7 + 4i|$.

Problem 28. The polynomial equation $x^3 + kx^2 + wx + p = 0$ with k, w, and p representing integers has 5 and $1+ i$ as two of its roots for x. If $i = \sqrt{-1}$, find the value of p.

SOLUTIONS:

Problem 1. Solution: B.
$i^6 - i^{10} - i^{15} = i^2 - i^2 - i^3 = i.$

Problem 2. Solution: A.
$$f(i) = \left(\frac{i^4 - i^3 + i^2 - i + 1}{i}\right)^3 = \left(\frac{1 + i - 1 - i + 1}{i}\right)^3 = \left(\frac{1}{i}\right)^3 = \frac{1}{i^3} \cdot \frac{i}{i} = \frac{i}{1} = i.$$

Problem 3. Solution: B.
$$f(1-i) = \frac{(1-i)^2 - (1-i) + 1}{(1-i) + 1} = \frac{(1 - 2i - 1) - 1 + i + 1}{2 - i} = \frac{-i}{2 - i} = \frac{-i(2+i)}{(2-i)(2+i)} = \frac{1 - 2i}{5}.$$

Problem 4. Solution: C.
Simplifying each by rationalizing all denominators we get:

A. $\dfrac{i + c(1+i)}{2}$ B. $\dfrac{i + c(1+i)}{2}$ C. $\dfrac{i + c(1-i)}{2}$ D. $\dfrac{i + c(1+i)}{2}$ C. is different.

Problem 5. Solution: A.
$$\left(\frac{i^{123}}{i^6}\right)^{-5} = \left(\frac{i^{(120+3)(5)}}{i^{(4+2)(5)}}\right)^{-1} = \left(\frac{i^{(120+3)(4+1)}}{i^{(4+2)(4+1)}}\right)^{-1} = \left(\frac{i^3}{i^2}\right)^{-1} = i^{-1} = \frac{1}{i} = \frac{i}{i^2} = -i.$$

Problem 6. Solution: B.
By the Property 3, $i^1 = i$; $i^2 = -1$; $i^3 = -i$; $i^4 = 1$; $i^{4n} = 1$; $i^{4n+1} = i$; $i^{4n+2} = -1$; $i^{4n+3} = -i$.
$i + 2i^2 + 3i^3 + 4i^4 + \ldots + 2006i^{2006} = [i + 2i^2 + 3i^3 + 4i^4 + \ldots + 2004i^{2004}] + 2005i^{2005} + 2006i^{2006} = (i + 2i^2 + 3i^3 + 4i^4) \times \dfrac{2004}{4} + 2005i^{2005} + 2006i^{2006} = (i - 2 - 3i + 4) \times 501 + 2005i^{2004+1} + 2006i^{2004+2} = (2 - 2i) \times 501 + 2005i + 2006i^2 = 1002 - 1002i + 2005i - 2006 = -1004 + 1003i.$

Problem 7. Solution: D.
$i^0 + i^1 + i^2 + i^3 + \ldots + i^{2009} + i^{2010} = 1 + [i^1 + i^2 + i^3 + \ldots + i^{2008}] + i^{2009} + i^{2010}$
$= 1 + [i^1 + i^2 + i^3 + i^4] \times \dfrac{2008}{4} + i^{2009} + i^{2010}$
$= 1 + (i - 1 - i + 1) \times 502 + i^{2008+1} + i^{2008+2} = 1 + 0 \times 502 + i + i^2 = 1 + i - 1 = i.$

Algebra II Through Competitions   **Chapter 15 Complex Numbers**

Problem 8. Solution: E.

A. $i - \dfrac{4}{1+5i} = i - \dfrac{4(1-5i)}{(1+5i)(1-5i)} = i - \dfrac{2-10i}{13} = \dfrac{13i - 2 + 10i}{13} = \dfrac{-2 + 23i}{13}$

B. $\dfrac{1+9i}{5-i} = \dfrac{(1+9i)(5+i)}{(5-i)(5+i)} = \dfrac{5 + 46i - 9}{26} = \dfrac{-4 + 46i}{26} = \dfrac{-2 + 23i}{13}$

C. $\dfrac{23i - 2}{13} = \dfrac{-2 + 23i}{13}$

D. $\dfrac{i-9}{1+5i} = \dfrac{(i-9)(1-5i)}{(1+5i)(1-5i)} = \dfrac{i - 9 - 5i^2 + 45i}{26} = \dfrac{-9 + 5 + 46i}{26} = \dfrac{-2 + 23i}{13}$.

So the answer is E. All are the same.

Problem 9. Solution: C.

$f(i) = \dfrac{i^{17} + i^9}{i^5 - 1} = \dfrac{i^{4\times 4 + 1} + i^{4\times 2 + 1}}{i^{4+1} - 1} = \dfrac{i + i}{i - 1} = \dfrac{2i(-1-i)}{(-1+i)(-1-i)} = \dfrac{2i(-1-i)}{2} = -i - i^2 = 1 - i$.

Problem 10. Solution: B.

$f(i) = \dfrac{i^5 + i^2}{i^3 + 1} = \dfrac{i - 1}{-i + 1} = \dfrac{i - 1}{-(i - 1)} = -1$.

Problem 11. Solution: C.

$\left(\dfrac{1+i}{\sqrt{2}}\right)^{2011} = \left(\dfrac{\sqrt{2}}{2} + \dfrac{\sqrt{2}}{2}i\right)^{2011} = (\cos 45° + i\sin 45°)^{2011}$.

By the DeMoivre's Theorem, $z^n = r^n[\cos(n\theta) + i\sin(n\theta)]$.

$(\cos 45° + i\sin 45°)^{2011} = \cos(45° \times 2011) + i\sin(45° \times 2011)$
$= \cos(251 \times 360° + 135°) + i\sin(251 \times 360° + 135°) = \cos 135° + i\sin 135° =$
$-\dfrac{1}{\sqrt{2}} + \dfrac{1}{\sqrt{2}}i = \dfrac{-1+i}{\sqrt{2}}$.

Problem 12. Solution: A.

$\left(\dfrac{1-i}{1+i}\right) = \dfrac{(1-i)^2}{(1+i)(1-i)} = \dfrac{1 + i^2 - 2i}{2} = -i = (\cos 270° + i\sin 270°)$.

By the DeMoivre's Theorem, $z^n = r^n[\cos(n\theta) + i\sin(n\theta)]$.

$(\cos 270° + i\sin 270°)^{2012} = \cos(270° \times 2012) + i\sin(270° \times 2012)$
$= \cos(1509 \times 360°) + i\sin(1508 \times 360°) = \cos 360° + i\sin 360° = 1 + 0 = 1$.

Problem 13. Solution: B.
Since $z^2 - z + 1 = 0$, so $z^2 + 1 = z$, and $z^2 = z - 1$.
$z^9 = z(z^2)^4 = z(z-1)^4 = z[(z-1)^2]^2 = z(z^2+1-2z)^2 = z(z-2z)^2 = z(-z)^2$
$= z(z^2) = z(z-1) = z^2 - z = z - 1 - z = -1$.

Problem 14. Solution: A.
Method 1:
Since $z^2 - z + 1 = 0$, so $z^2 + 1 = z$, and $z^2 = z - 1$.
$z^{24} = (z^2)^{12} = (z-1)^{12} = [(z-1)^2]^6 = (z^2+1-2z)^6 = (z-2z)^6 = (-z)^6 = z^6$
$= (z^2)^3 = (z-1)^3 = (z-1)^2(z-1) = -z(z-1) = -z^2 + z = -(z-1) + z = 1$.

Method 2:
Solving $z^2 - z + 1 = 0$, $z = \dfrac{1 \pm \sqrt{3}}{2}$. We use $z = \dfrac{1+\sqrt{3}}{2}$ in our calculation.

$\dfrac{1+\sqrt{3}}{2} = \dfrac{1}{2} + i\dfrac{\sqrt{3}}{2} = (\cos 60° + \sin 60°)$.

By the DeMoivre's Theorem, $z^n = r^n[\cos(n\theta) + i\sin(n\theta)]$.

$z^{24} = (\dfrac{1+\sqrt{3}}{2})^{24} = (\cos 60° + i\sin 60°)^{24} = \cos(60° \times 24) + i\sin(60° \times 24)$
$= \cos(4 \times 360°) + i\sin(4 \times 360°) = \cos 360° + i\sin 360° = 1 + 0 = 1$.

Problem 15. Solution: C.
Method 1:
$i^{2010} = i^{2008+2} = i^2 = -1$.

Method 2:
$i = 0 + i = \cos(90°) + i\sin(90°)$.

By the DeMoivre's Theorem, $z^n = r^n[\cos(n\theta) + i\sin(n\theta)]$.
$i^{2010} = = (\cos 90° + i\sin 90°)^{2010} = \cos(90° \times 2010) + i\sin(90° \times 2010)$
$= \cos(502 \times 360° + 180) + i\sin(502 \times 360° + 180) = \cos 180° + i\sin 180° = -1 + 0 = -1$.

Problem 16. Solution: C.

$S = i^n + i^{-n} = i^n + \dfrac{1}{i^n} = \dfrac{i^{2n}+1}{i^n} = \dfrac{(i^{2n}+1)i^{3n}}{i^n \cdot i^{3n}} = i^{5n} + i^{3n} = i^n + i^{3n} = i^n(1+i^{2n})$
$= i^n[1+(i^2)^n] = i^n[1+(-1)^n]$.

When n is odd, $S = 0$.

When n is even, $S = i^n[1+(-1)^n] = 2i^n$. $S = 2$ when n is a multiple of 4 and $S = -2$ otherwise. So the total number of values for S is 3.

Problem 17. Solution: A.

$\dfrac{2+i}{3-bi} = \dfrac{(2+i)(3+bi)}{(3-bi)(3+bi)} = \dfrac{6-b+i(2b+3)}{9+b^2}$.

The imaginary part will disappear when $2b+3 = 0$ or $b = -\dfrac{3}{2}$.

Problem 18. Solution: B

$x^9 - x = 0 \qquad x(x^8 - 1) = 0$.

So $x = 0$ or $x^8 = 1$.

For $x^8 = 1$, we have $(x^4)^2 - 1 = 0 \quad \Rightarrow \quad (x^4 - 1)(x^4 + 1) = 0$.

Only $(x^4 - 1) = 0$ has real solutions.

Factoring we get $(x^2 - 1)(x^2 + 1) = 0$.

Similarly we know that real solutions are from $(x^2 - 1) = 0$, which are 1 and -1.

So there are three real solutions.

Problem 19. Solution: 2.

$i^{13} - i^7 = ki \quad \Rightarrow \quad i^{4\times 3+1} - i^{4+3} = ki \quad \Rightarrow \quad i - i^3 = ki \quad \Rightarrow i + i = ki$

$\Rightarrow \quad 2i = ki \quad \Rightarrow \quad k = 2$.

Problem 20. Solution: 20.

The common difference is $(4+5i) - (3+7i) = 1 - 2i$.

$a_9 = 3 + 7i + (9-1)(1-2i) = 3 + 7i + 8 - 16i = 11 - 9i$.

$x - y = 11 + 9 = 20$.

Problem 21. Solution: 12.

Let a be the real root. $a^3 + (-7+i)a^2 - (54+22i)a + 120i = 72$, or

$a^3 - 7a^2 + ia^2 - 54a - 22ia + 120i - 72 = 0$.

Algebra II Through Competitions   Chapter 15 Complex Numbers

Since a is the real root, we must have $ia^2 - 22ia + 120i = 0$ or $a^2 - 22a + 120 = 0$

Solving for a: $a = 12$ or $a = 10$.

When $a = 10$, $a^3 - 7a^2 - 54a - 72 = 10^3 - 7 \cdot 10^2 - 54 \times 10 - 72 \neq 0$.

When $a = 12$, $a^3 - 7a^2 - 54a - 72 = 12^3 - 7 \cdot 12^2 - 54 \times 12 - 72 = 0$.

Thus the real solution is 12.

Problem 22. Solution: -18.

By Vieta's Theorem, for $a_0 x^3 + a_1 x^2 + a_2 x^1 + a_3 = 0$,

$$x_1 + x_2 + x_3 = -\frac{a_1}{a_0} \quad \Rightarrow \quad k + (a+bi) + (a-bi) = -w \quad \Rightarrow \quad k + w + 2a = 0 \quad (1)$$

$$x_1 x_2 + x_2 x_3 + x_3 x_1 = \frac{a_2}{a_0} \quad \Rightarrow \quad k \times (a+bi) + k \times (a-bi) + (a+bi)(a-bi) = 85 \quad \Rightarrow$$

$$2ak + a^2 + b^2 = 85 \quad (2)$$

$$x_1 x_2 x_3 = -\frac{a_3}{a_0} \quad \Rightarrow \quad k \times (a+bi) \times (a-bi) = -(-156) \quad \Rightarrow \quad k(a^2 + b^2) = 156 \quad (3)$$

Combining (2) and (3), we have $k(85 - 2ak) = 156$.

We also know that k is a positive integer from (3).

Thus $k(85 - 2ak) = 156 = 1 \times 156 = 2 \times 78 = 3 \times 52 = 4 \times 39 = 6 \times 26 = 12 \times 13$.

Only when $k = 12$, a is an integer and $a = 3$. At this value $12 + w + 2 \times 3 = 0 \Rightarrow w = -18$.

Problem 23. Solution: 2.

Since $z = 2 + 3i$, $\bar{z} = 2 - 3i$.

$z + 3\bar{z} = 2 + 3i + 3(2 - 3i) = 2 + 3i + 6 - 9i = 8 - 6i$. Thus $x + y = 2$.

Problem 24.

Solution: A.

By Vieta's Theorem, for $a_0 x^3 + a_1 x^2 + a_2 x^1 + a_3 = 0$,

$$x_1 + x_2 + x_3 = -\frac{a_1}{a_0} \quad \Rightarrow \quad 1 + (3 + 2i) + (3 - 2i) = 7.$$

Only A is correct.

Problem 25. Solution:

The polynomial has the following roots: $1 - \sqrt{5}$, and $3 + 2i$. So its least degree is 6.

Problem 26. Solution: D.

The third root is $1 + 3i$.

The function is $(x-2)[x-(1-3i)][(x-(1+3i)] = (x-2)(x-1+3i)(x-1-3i)$
$= (x-2)[(x-1)^2 - 9i^2] = (x-2)(x^2 - 2x + 10) = x^3 - 2x^2 - 2x^2 + 4x + 10x - 20$
$= x^3 - 4x^2 + 14x - 20$.

Problem 27. Solution: $\sqrt{65}$.

$|7 + 4i| = \sqrt{7^2 + 4^2} = \sqrt{65}$.

Problem 28. Solution: -10.

By Vieta's Theorem, for $a_0 x^3 + a_1 x^2 + a_2 x^1 + a_3 = 0$,

$$x_1 x_2 x_3 = -\frac{a_3}{a_0} \quad \Rightarrow \quad 5 \times (1+i) \times (1-i) = -p \quad \Rightarrow \quad p = -5(1+1) = -10.$$

Algebra II Through Competitions Chapter 16 Arithmetic and Geometric Sequences

1. DEFINITIONS:

Arithmetic sequence
If any two consecutive terms in a sequence $a_1, a_2, a_3, ..., a_n, ...$, have the same difference, the sequence is called an arithmetic sequence (or arithmetic progression).

Geometric sequence
If any two consecutive terms in a sequence $a_1, a_2, a_3, ..., a_n, ...$, have the same ratio, the sequence is called a geometric sequence (or geometric progression).

Terms
An element of an arithmetic or geometric sequence is called a term of the sequence, written as $a_1, a_2, a_3, ...,$.
a_1 is called the first term.
a_n is called the general term or n^{th} term.
The same difference is called the common difference (d).
The same ratio is called the **common ratio** (q or r).
The **sum of the first n terms** is expressed as S_n. For example, S_{12} means the sum of the first twelve terms.

2. PROPERTIES OF ARITHMETIC SEQUENCE:
(1). The common difference (d)

$$d = a_{n+1} - a_n \tag{2.1}$$

$$d = \frac{a_m - a_n}{m - n} \tag{2.2}$$

$$\frac{d}{2} = \frac{\frac{S_m}{m} - \frac{S_n}{n}}{m - n} \tag{2.3}$$

Example 1. If the first term of an arithmetic sequence is 14 and if the twelfth term of this arithmetic sequence is 23, find the common difference of the arithmetic sequence. Express your answer as a common fraction reduced to lowest terms.

Solution: 9/11.

Algebra II Through Competitions Chapter 16 Arithmetic and Geometric Sequences

By the formula $d = \dfrac{a_m - a_n}{m - n}$ $\Rightarrow$ $d = \dfrac{a_{12} - a_1}{12 - 1} = \dfrac{23 - 14}{11} = \dfrac{9}{11}$.

Example 2. In an arithmetic sequence a_n, the sum of first m terms is 30, the sum of first $2m$ terms is 100. Find S_{3m}, the sum of first $3m$ terms.

(A)   130   (B) 170   (C) 210   (D) 260

Solution: (C).
Using formula (2.3):

$\dfrac{S_{3m}}{3m} - \dfrac{S_{2m}}{2m} = \dfrac{S_{2m}}{2m} - \dfrac{S_m}{m}$ $\Rightarrow$ $\dfrac{2S_{3m} - 3S_{2m}}{6m} = \dfrac{S_{2m} - 2S_m}{2m}$ $\Rightarrow$ $2S_{3m} - 3S_{2m} = 3(S_{2m} - 2S_m)$

$\Rightarrow$ $S_{3m} = 3S_{2m} - 3S_m$.

Substituting $S_m = 30$, $S_{2m} = 100$ into the equation above yields
$S_{3m} = 3S_{2m} - 3S_m = 3 \times 100 - 3 \times 30 = 300 - 90 = 210$.

(2). The nth term

The nth term is expressed as $a_n = a_1 + (n-1)d$ (2.4)

Other forms:
$a_n = a_m + (n-m)d$ (2.5)
$a_n = (a_2 - a_1)n + (2a_1 - a_2)$ (2.6)
$a_n = S_n - S_{n-1}$ (n is positive integer and $n > 1$)   (2.7)

If $m + n = p + q$ where $m, n, p,$ and q are positive integers, then

$$a_m + a_n = a_p + a_q \qquad (2.8)$$

If $m + n = 2q$, then $a_m + a_n = 2a_q$ (2.9)

Algebra II Through Competitions Chapter 16 Arithmetic and Geometric Sequences

If $a_m = p$, $a_n = q$, then $a_{m+n} = \dfrac{mp - nq}{m - n}$ (2.10)

For three consecutive terms a, b, and c: $b = \dfrac{a+c}{2}$ (2.11)

$a_1 = S_1$ (2.12)

Example 3. In an arithmetic sequence a_n, if $a_1 + a_9 = 10$, find a_5.

(A) 5   (B) 6   (C) 8   (D) 10

Solution: (A)
By formula (2.9): $a_1 + a_9 = 2a_5$. Therefore $a_5 = 5$.

Example 4. The sixteenth term of an arithmetic progression is 40. The fifty-fifth term of this arithmetic progression is 157. Find the eighty-first term of this arithmetic progression.

Solution: 235.

By the formula $d = \dfrac{a_m - a_n}{m - n}$ $\Rightarrow$ $d = \dfrac{a_{55} - a_{16}}{55 - 16} = \dfrac{157 - 40}{39} = 3$

By the formula $a_n = a_m + (n - m)d$ $\Rightarrow$ $a_{81} = a_{55} + (81 - 55)3 = 235$.

Example 5. In an arithmetic sequence a_n, the sum of first n terms is S_n. If $a_2 = 3$, $a_6 = 11$, find S_7.

A. 13   B. 35   C. 49   D. 63

Solution: C.
Method 1:
$S_7 = \dfrac{7(a_1 + a_7)}{2} = \dfrac{7(a_2 + a_6)}{2} = \dfrac{7(3 + 11)}{2} = 49$.

Note that by (2.8) we have $a_1 + a_7 = a_2 + a_6$.

Method 2:

Algebra II Through Competitions Chapter 16 Arithmetic and Geometric Sequences

$$\begin{cases} a_2 = a_1 + d = 3 \\ a_6 = a_1 + 5d = 11 \end{cases} \Rightarrow \begin{cases} a_1 = 1 \\ d = 2 \end{cases}, \quad a_7 = 1 + 6 \times 2 = 13.$$

Therefore $S_7 = \dfrac{7(a_1 + a_7)}{2} = \dfrac{7(1+13)}{2} = 49$.

(3). The sum of n terms in the sequence:

$$S = \frac{(a_1 + a_n)n}{2} \tag{2.13}$$

$$S = na_1 + \frac{(n-1)d}{2} n \tag{2.14}$$

Other forms:

$$S_{m+n} = (m+n)\frac{S_m - S_n}{m-n} \tag{2.15}$$

In an arithmetic sequence a_n, the sum of first n terms is S_n.

If the number of terms is $2n - 1$, then $S_{odd} - S_{even} = a_n$ (2.16)

If the number of terms is $2n$, then $S_{even} - S_{odd} = nd$ (2.17)

Example 6. Find the value of $a_2 + a_4 + a_6 + \cdots + a_{98}$ if $a_1, a_2, a_3, \ldots$ is an arithmetic progression with common difference 1, and $a_1 + a_2 + a_3 + \cdots + a_{98} = 137$.

Solution:
Method 1:
Adding 49 to both sides of the equation: $a_1 + a_2 + a_3 + \cdots + a_{98} = 137$, we get
$(a_1 + 1) + a_2 + (a_3 + 1) + a_4 + \cdots + (a_{97} + 1) + a_{98} = 137 + 49 = 186$.
Notice that $a_2 = a_1 + 1$, $a_4 = a_3 + 1$, $\cdots$, $a_{98} = a_{97} + 1$
So $(a_1 + 1) + a_2 + (a_3 + 1) + a_4 + \cdots + (a_{97} + 1) + a_{98} = 137 + 49 = 186$ can be re-written as
$2(a_2 + a_4 + a_6 + \cdots + a_{98}) = 186$
Therefore $a_2 + a_4 + a_6 + \cdots + a_{98} = 93$.

Method 2:
$a_1 + a_2 + \cdots + a_{98} = \dfrac{98}{2}(a_1 + a_{98}) = 49(a_1 + a_{98}) = 137$.

Algebra II Through Competitions Chapter 16 Arithmetic and Geometric Sequences

$$\therefore a_2 + a_4 + \cdots + a_{98} = \frac{49}{2}(a_2 + a_{98}) = \frac{49}{2}(a_1 + 1 + a_{98}) = \frac{49}{2}(a_1 + a_{98}) + \frac{49}{2}$$

$$= \frac{137}{2} + \frac{49}{2} = 93$$

Method 3:
Let $S_o = a_1 + a_3 + a_5 + \cdots + a_{97}$
$S_e = a_2 + a_4 + a_6 + \cdots + a_{98}$
By formula (2.17), we have $S_e - S_o = nd$. We know that the common difference is 1 and the number of terms is $2n$, so $n = 49$.

Therefore $\begin{cases} S_e - S_o = 49 \\ S_o + S_e = 137 \end{cases}$

Solving by adding the two equations together, we get $S_e = 93$.

(4). Important relationship between a_n and S_n

If a_n is an arithmetic sequence, then $S_{2n-1} = (2n-1)a_n$ \hfill (2.18)

or $a_n = \dfrac{S_{2n-1}}{2n-1}$ \hfill (2.19)

Example 7. If a_n is an arithmetic sequence and $a_9 = 23$, $a_{18} = 54$, find S_{18}, the sum of first 18 terms.

Solution:
By the formula (2.7): $S_{18} = S_{17} + a_{18}$.
By the formula (2.18): $S_{17} = 17 \times 23$.
$S_{18} = S_{17} + a_{18} = 17 \times 23 + 54 = 445$.

Example 8. (ARML) In a certain arithmetic progression, the ratio of the sum of the first r terms to the sum of the first s terms is equal to the ratio of r^2 to s^2. Compute the ratio of the 8th term to the 23rd term.

Solution:
Method 1 (Official Solution):

Algebra II Through Competitions Chapter 16 Arithmetic and Geometric Sequences

Applying the formula for the sum of terms of an arithmetic progression,

$$\frac{\frac{r}{2}[2a+(r-1)d]}{\frac{s}{2}[2a+(s-1)d]} = \frac{r^2}{s^2} \Rightarrow \frac{2a+(r-1)d}{2a+(s-1)d} = \frac{r}{s}.$$

Cross multiplying and simplifying leads to $2a = d$. Now the ratio of

$$\frac{8th\text{ term}}{23rd\text{ term}} = \frac{a+7d}{a+22d} = \frac{15a}{45a} = \frac{1}{3}.$$

Method 2 (our solution):

By the formula $a_n = \dfrac{S_{2n-1}}{2n-1}$, we have

$$a_8 = \frac{S_{2\times 8-1}}{2\times 8-1} = \frac{S_{15}}{15} \tag{1}$$

$$a_{23} = \frac{S_{2\times 23-1}}{2\times 23-1} = \frac{S_{45}}{45} \tag{2}$$

We also have $\dfrac{S_{15}}{S_{45}} = (\dfrac{15}{45})^2 = \dfrac{1}{9}$ \hfill (3)

(1) ÷ (2): $\dfrac{a_8}{a_{15}} = 3\times(\dfrac{S_{15}}{S_{45}}) = 3\times\dfrac{1}{9} = \dfrac{1}{3}.$

Method 3 (our solution):

Since $\dfrac{S_r}{S_s} = \dfrac{r^2}{s^2}$, $\dfrac{S_{15}}{S_{45}} = \dfrac{15^2}{45^2} = \dfrac{1}{9}.$

$$\frac{a_8}{a_{23}} = \frac{2a_8}{2a_{23}} = \frac{a_1+a_{15}}{a_1+a_{45}} = \frac{\frac{15}{2}(a_1+a_{15})}{\frac{45}{2}(a_1+a_{45})} \cdot \frac{45}{15} = 3\cdot\frac{S_{15}}{T_{15}} = 3\cdot\frac{1}{9} = \frac{1}{3}.$$

3. PROPERTIES OF GEOMETRIC SEQUENCE :

(1). The common ratio (q or r).

$$\frac{a_n}{a_{n-1}} = q \qquad (q\neq 0,\ n\geq 2) \tag{3.1}$$

Algebra II Through Competitions Chapter 16 Arithmetic and Geometric Sequences

(2). The nth term

$$a_n = a_1 \cdot q^{n-1} \qquad (a_1, q \neq 0) \qquad (3.2)$$

$$a_n = a_m \cdot q^{n-m} \quad (a_1, q \neq 0) \qquad (3.3)$$

$$a_n^2 = a_{n-k} \cdot a_{n+k} \qquad (n \geq k) \qquad (3.4)$$

If $m + n = p + q$, $a_m \cdot a_n = a_p \cdot a_q$ \qquad (3.5)

If $m + n = 2t$, $a_m \cdot a_n = a_t^2$ \qquad (3.6)

(3). The sum of the first n terms

$$S_n = \frac{a_1(1-q^n)}{1-q} \qquad q \neq 1 \qquad (3.7)$$

$$S_n = \frac{a_1 - a_n q}{1 - q} \qquad (3.8)$$

$$S_n = na_1 \qquad q = 1 \qquad (3.9)$$

Example 9. By adding the same constant to each of 20, 50, 100 a geometric progression results. The common ratio is

A. $\frac{5}{3}$ B. $\frac{4}{3}$ C. $\frac{3}{2}$ D. $\frac{1}{2}$ E. $\frac{1}{3}$

Solution: A.
Let a denote the constant that is added to 20, 50, and 100.
$$\frac{20+a}{50+a} = \frac{50+a}{100+a}$$
$\therefore a = 25$.
The common ratio is $r = \frac{5}{3}$.

Example 10. The second and fourth terms of a geometric sequence are 2 and 6. Which of the following is a possible first term?

(A) $-\sqrt{3}$ (B) $-\frac{2\sqrt{3}}{3}$ (C) $-\frac{\sqrt{3}}{3}$ (D) $\sqrt{3}$ (E) 3

Algebra II Through Competitions Chapter 16 Arithmetic and Geometric Sequences

Solution: (B).
Let the sequence be denoted as $a, ar, ar^2, ar^3, \ldots$, with $ar = 2$ and $ar^3 = 6$. Solving for r, we get $r^2 = 3$ and $r = \sqrt{3}$ or $r = -\sqrt{3}$. Therefore $a = \dfrac{2\sqrt{3}}{3}$ or $a = -\dfrac{2\sqrt{3}}{3}$.

(4). The sum of n terms in the infinite geometric sequence:

In the equation $S_n = \dfrac{a_1(1 - q^n)}{1 - q}$, if n increases while $-1 < q < 1$, that is $|q| < 1$, q^n decreases in size and becomes zero, and the sum S_n approaches the following specific value:

$$S = \dfrac{a_1}{1-q} \qquad (|q| < 1) \qquad (3.10)$$

If $q > 1$ then as n increases, q^n increases in size without bound.
If $q < -1$ then as n increases q^n oscillates between positive and negative values but again increases in size without bound.
In either situation, as n increases, S_n does not approach a specific value, so we can say that the sum of the infinite geometric series does not exist.

Example 11. Find the sum of the first ten terms of the following geometric series
$$\dfrac{1}{2} + \dfrac{1}{6} + \dfrac{1}{18} + \dfrac{1}{54} + \dfrac{1}{162} + \cdots$$

(A) $\dfrac{1}{2}(1 + \dfrac{1}{3^{10}})$ B) $\dfrac{1}{2}(1 - \dfrac{1}{3^{10}})$ C) $\dfrac{3}{4}(1 + \dfrac{1}{3^{10}})$ D) $\dfrac{3}{4}(1 - \dfrac{1}{3^{10}})$ E) none of these

Solution: D.

$a_1 = \dfrac{1}{2}$, $q = \dfrac{\frac{1}{6}}{\frac{1}{2}} = \dfrac{1}{3}$, and $n = 10$.

By the formula, $S_n = \dfrac{a_1(1-q^n)}{1-q}$, $S_{10} = \dfrac{\frac{1}{2}[1 - (\frac{1}{3})^{10}]}{1 - \frac{1}{3}} = \dfrac{\frac{1}{2}(1 - \frac{1}{3^{10}})}{\frac{2}{3}} = \dfrac{3(1 - \frac{1}{3^{10}})}{4}$.

Algebra II Through Competitions Chapter 16 Arithmetic and Geometric Sequences

Example 12. Find the sum of the following infinite series:
$$\frac{2}{7}-\frac{3}{7^2}+\frac{2}{7^3}-\frac{3}{7^4}+\frac{2}{7^5}-\frac{3}{7^6}+\cdots$$
A. $\frac{5}{6}$ B. $\frac{7}{24}$ C. $\frac{11}{48}$ D. $\frac{13}{49}$ E. $\frac{3}{8}$

Solution: C.
$$\frac{2}{7}-\frac{3}{7^2}+\frac{2}{7^3}-\frac{3}{7^4}+\frac{2}{7^5}-\frac{3}{7^6}+\cdots = \frac{2}{7}+\frac{2}{7^3}+\frac{2}{7^5}+\cdots-\frac{3}{7^2}-\frac{3}{7^4}-\frac{3}{7^6}+\cdots$$
$$=\frac{\frac{2}{7}}{1-\frac{1}{7^2}}-\frac{\frac{3}{7^2}}{1-\frac{1}{7^2}}=\frac{14}{48}-\frac{3}{48}=\frac{11}{48}.$$

Example 13. Simplify: $x^{\frac{1}{2}}\cdot x^{-\frac{1}{4}}\cdot x^{\frac{1}{8}}\cdot x^{-\frac{1}{16}}\cdots$
A. x B. $\sqrt{x}$ C. $\sqrt[3]{x}$ D. x^2 E. $\sqrt[5]{x}$

Solution: C.
$x^{\frac{1}{2}}\cdot x^{-\frac{1}{4}}\cdot x^{\frac{1}{8}}\cdot x^{-\frac{1}{16}}\cdots = x^{\frac{1}{2}-\frac{1}{4}+\frac{1}{8}-\frac{1}{16}+\cdots}$ The exponent in this last expression is an infinite geometric series with sum $\dfrac{\frac{1}{2}}{1-(-\frac{1}{2})}=\dfrac{\frac{1}{2}}{\frac{3}{2}}=\dfrac{1}{3}$, so $x^{\frac{1}{2}}\cdot x^{-\frac{1}{4}}\cdot x^{\frac{1}{8}}\cdot x^{-\frac{1}{16}}\cdots = x^{\frac{1}{3}}=\sqrt[3]{x}$.

Algebra II Through Competitions Chapter 16 Arithmetic and Geometric Sequences

PROBLEMS

Problem 1.   Find the sum of the distinct terms of an arithmetic sequence whose first term is 3, whose fifth term is 19, and whose last term is 103.

Problem 2.   An arithmetic sequence of 15 terms has a sum of 3060. The common difference and each member of the arithmetic sequence are all positive even integers. If a is the first term of any such arithmetic sequence, find the sum of all distinct a in which at least one of the first four terms of the arithmetic sequence is the square of an integer.

Problem 3.   The fifth term of an arithmetic sequence is -18, and the sum of the first thirty-two terms is 1448. Find the ninth term.

Problem 4.   Given $x + x^2 + x^3 + x^4 + \ldots = 1.5$, find x.
A. 2/3   B. 3/5   C. 3/4   D. 5/6   E. none of these.

Problem 5.   A hot-air balloon rises 80 feet in the first minute of flight. If in each succeeding minute the balloon rises only 90% as far as in the previous minute, what will be its maximum altitude if it is allowed to rise without limit?
A. 88.8 feet   B. 800 feet   C. 888.8 feet   D. 900 feet   E. 905 feet.

Problem 6.   Find the sum of all proper fractions whose denominators are less than or equal to 100. (Include unreduced fractions in the sum.)
A. 2075   B. 1050   C. 1175   D. 1275   E. 2475

Problem 7.   The natural numbers are grouped as indicated: {1}, {2,3}, {4, 5, 6}, {7, 8, 9, 10}, $\cdots$ with m numbers in the mth group. Find the sum of the numbers in the 110th group.
A. 665,555   B. 55,000   C. 55,555   D. 450,000   E. 700,565

Problem 8.   The first four terms in an arithmetic sequence are $x + y$, $x - y$, xy, x/y in that order. What is the fifth term?
A. $-15/8$ B. $-6/5$ C. 0   D. 27/20   E. 123/40

Algebra II Through Competitions Chapter 16 Arithmetic and Geometric Sequences

Problem 9. Let $x > 0$. The terms of the sequence $\sqrt[3]{x}$, $\sqrt{x}$, $\sqrt[3]{x^2}$, ... are in geometric progression. What will be the fourth term?
A. $\sqrt[4]{x^3}$ B. $\sqrt[5]{x^4}$ C. $\sqrt[6]{x^5}$ D. x E. $\sqrt{x^3}$

Problem 10. If $f(1) = 2$ and $f(n+1) = (f(n))^2$, what is the value of $f(4)$?
A. 4.   B. 16.   C. 64.   D. 256.   E. 65,536.

Problem 11. What is the sum of all the odd numbers from 1 to 2008?
A. 1,008,016   B. 1,010,025   C. 1,008,518   D. 2,015,028   E. 2,017,036

Problem 12. Rubber ball is dropped from a height of 4 meters onto a concrete floor. It bounces to 75% of its original height and drops back down, bouncing many times in ever smaller bounces. If it did this forever how much total distance did the ball cover?
A. 10m   B. 24m   C. 28m   D. 42m   E. more than 42m

Problem 13. Define $s = \sum_{d=0}^{\infty} 32(3/4)^d$. Which of the following is true of s?
A. $0 \le s \le 32$ B. $32 < s \le 64$ C. $64 < s \le 128$ D. $128 < s \le 256$ E. $256 < s < 1$

Problem 14. Determine $i + i^2 + i^3 + \cdots + i^{50}$, where $i^2 = -1$.
A. 1   B. $i - 1$ C. -1 D. $-i$ E. cannot be determined.

Problem 15. The sum of 67 consecutive integers is 2010. What is the least of these numbers?
A. -3 B. -1 C. 1   D. 3   E. 4.

Problem 16. Find the sum $12 + 17 + 22 + 27 + 32 + \ldots + 97$ of all the two-digit numbers whose units digit is either 2 or 7.
A. 972   B. 981   C. 990   D. 999   E. 1008.

Problem 17. Find the largest possible integer n such that $1 + 2 + 3 + \cdots n \le 200$.
A. 14   B. 17   C. 19   D. 21   E. 23.

Problem 18. In a 10-team baseball league, each team plays each of the other teams 18 times. No game ends in a tie, and, at the end of the season, each team is the same positive

Algebra II Through Competitions Chapter 16 Arithmetic and Geometric Sequences

number of games ahead of the next best team. What is the greatest number of games that the last place team could have won?

A. 27　　　　B. 36　　　　C. 54　　　　D. 72　　　　E. 90

Problem 19. (Indiana 2010 Mathematics Contest Algebra II)

$$1 - \frac{1}{2} + \frac{1}{4} - \frac{1}{8} + \frac{1}{16} - \frac{1}{32} + \cdots + \frac{1}{2^{1,000}} =$$

A) $2 - \frac{1}{2^{1,000}}$　　B) $2 + \frac{1}{2^{1,000}}$　　C) $\frac{1}{3}(2 - \frac{1}{2^{1,000}})$　　D) $\frac{1}{3}(2 + \frac{1}{2^{1,000}})$　　E) none of these

Problem 20. (Indiana 2011 Mathematics Contest Algebra II) What is the sum of the following infinite geometric series? $2 + \left(-\frac{1}{2}\right) + \left(\frac{1}{8}\right) + \left(-\frac{1}{32}\right) \cdots$

A) $1\frac{3}{8}$　　B) $1\frac{2}{5}$　　C) $1\frac{1}{2}$　　D) $1\frac{3}{5}$　　E) $1\frac{5}{8}$

Problem 21.　The sequence a_n satisfies $a_n = a_{n-1} + d$, where d is a constant. If the sum of the first 100 terms is 50, and the sum of the first 200 terms is 90, then the sum of the first 300 terms is ?

A. 50　　　　B. 90　　　　C. 100　　　　D. 120　　　　E. 140

Problem 22.　Find the sum of the following infinite series:

$$\frac{1}{7} - \frac{5}{7^2} + \frac{1}{7^3} - \frac{5}{7^4} + \frac{1}{7^5} - \frac{5}{7^6} + \cdots$$

A. $\frac{1}{7}$　　B. $\frac{1}{24}$　　C. $\frac{2}{49}$　　D. 0　　E. $\frac{1}{5}$

Algebra II Through Competitions Chapter 16 Arithmetic and Geometric Sequences

SOLUTIONS

Problem 1. Solution: 1378.

By the formula $d = \dfrac{a_m - a_n}{m - n} \Rightarrow d = \dfrac{a_5 - a_1}{5 - 1} = \dfrac{19 - 3}{4} = 4$.

Use the formula again: $d = \dfrac{a_m - a_n}{m - n} \Rightarrow 4 = \dfrac{a_n - a_1}{n - 1} \Rightarrow 4 = \dfrac{103 - 3}{n - 1} \Rightarrow n = 26$.

$S = na_1 + \dfrac{(n-1)d}{2}n \Rightarrow S_{26} = 26 \times 3 + \dfrac{(26-1) \times 4}{2} \times 26 = 78 + 1300 = 1378$.

Problem 2. Solution: 574.

$S = na_1 + \dfrac{(n-1)d}{2}n \Rightarrow 3060 = 15a + \dfrac{(15-1)d}{2} \times 15 \Rightarrow a + 7d = 204$.

Solving this equation we get:
$a = 8, 22, 36, 50, 64, 78, 92, 106, 120, 134, 148, 162, 176$, and 190.
All distinct a in which at least one of the first four terms of the arithmetic sequence is the square of an integer: $a = 8, 22, 36, 64, 120, 134$, and 190. The sum is 574.

Problem 3. Solution: 4.

By the formula $S = na_1 + \dfrac{(n-1)d}{2}n \Rightarrow 1448 = 32a_1 + \dfrac{(32-1)d}{2} \times 32$

$\Rightarrow 4a_1 + 62d = 181$ (1)

By the formula $a_n = a_m + (n - m)d \Rightarrow a_5 = a_1 + (5-1)d \Rightarrow a_1 + 4d = -18$ (2)

Solving the system of equations (1) and (2): $d = \dfrac{11}{2}$.

$a_9 = a_5 + (9-5) \times \dfrac{11}{2} = -18 + 22 = 4$.

Problem 4. Solution: B.
Assuming $|x| < 1$, $x + x^2 + x^3 + x^4 + \ldots = x/(1 - x) = 3/2$.
$x = 3/5$, which agrees with our assumption.

Problem 5. Solution: B.
The total height in feet that the balloon rises is $80 + 80(0.9) + 80(0.9)^2 + \ldots$
This is an infinite geometric series whose sum is $80/(1 - 0.9) = = 800$.

Algebra II Through Competitions Chapter 16 Arithmetic and Geometric Sequences

Problem 6. Solution: E.
If we write the fraction in order as $(1/2) + (1/3 + 2/3) + (1/4 + 2/4 + 3/4) + \ldots + (1/100 + 2/100 + \ldots 99/100)$, we see that this sum is $(1/2) + (3/3) + (6/4) + (10/5) + (15/6) + \ldots + \left(\frac{10099}{2}/100\right)$, using the formula for the sum of n consecutive integers $1 + 2 + 3 + \ldots + n = \frac{n(n+1)}{2}$. If you now factor one-half from each term you get $\frac{1}{2}(1 + 2 + 3 + 4 + \ldots + 99) = \frac{1}{2}\left(\frac{99 \cdot 100}{2}\right) = 2475$.

Problem 7. Solution: A.
The 110th group has 110 numbers beginning with 5996. So the sum S is
$$S = \frac{110}{2}(5996 + 6105) = 665{,}555.$$

Problem 8. Solution: E.
$$x - y - (x + y) = xy - (x - y) \quad \Rightarrow \quad xy - x + 3y = 0 \quad \Rightarrow \quad y = \frac{x}{x+3}.$$

$$x + \frac{x}{x+3},\ x - \frac{x}{x+3},\ \frac{x^2}{x+3},\ x+3.$$

$$x - \frac{x}{x+3} - x + \frac{x}{x+3} = x + 3 - \frac{x^2}{x+3} \quad \Rightarrow \quad x = -\frac{9}{8}.$$

Let the fifth term be z.
$$2(x+3) = \frac{x^2}{x+3} + z \quad \Rightarrow \quad z = 2(x+3) - \frac{x^2}{x+3} = 2(x+3) - \frac{x^2 - 3^2 + 3^2}{x+3}$$
$$= 2(x+3) - (x-3) - \frac{9}{x+3} = 2(-\frac{9}{8} + 3) - (-\frac{9}{8} - 3) - \frac{9}{-\frac{9}{8}+3} = \frac{123}{40}.$$

Problem 9. Solution: C.
We see that $\sqrt{x}/\sqrt[3]{x} = x^{\frac{1}{2} - \frac{1}{3}} = x^{\frac{1}{6}}$ and $\sqrt[3]{x^2}/\sqrt{x} = x^{\frac{2}{3} - \frac{1}{2}} = x^{\frac{1}{6}}$. So the common ratio is $x^{\frac{1}{6}}$.

Let the fourth term be z.

$$z/\sqrt[3]{x^2} = x^{\frac{1}{6}} \quad \Rightarrow \quad z = \sqrt[3]{x^2} \times x^{\frac{1}{6}} = x^{\frac{2}{3}} \times x^{\frac{1}{6}} = x^{\frac{2}{3} + \frac{1}{6}} = x^{\frac{5}{6}} = \sqrt[6]{x^5}.$$

Problem 10. Solution: D.
$f(2) = f(1+1) = f(1)^2 = 4$ and $f(3) = f(2+1) = f(2)^2 = 16$, and

$f(4) = f(3+1) = (f(3))^2 = 16^2 = 256$.

Problem 11. Solution: A.
There are 1004 odd numbers with average value of 1004. Answer = $1004^2 = 1,008,016$

Problem 12. Solution: C.
$$4 + 8 \times 0.75 + 8 \times 0.75^2 + 8 \times 0.75^3 + \cdots = 4 + 8\sum_{k=1}^{\infty} 0.75^k = 4 + 8 \times \frac{0.75}{0.25} = 28.$$

Problem 13. Solution: C.
$$s = \sum_{d=0}^{\infty} 32(3/4)^d = 32\sum_{d=0}^{\infty} (3/4)^d = 32[1 + \frac{3}{4} + (\frac{3}{4})^2 + (\frac{3}{4})^3 + \cdots] = 32 \times \frac{1}{1-\frac{3}{4}} = 128.$$

Problem 14. Solution: B.
The common ratio is $i^2/i = i$.
By the formula, $S_n = \frac{a_1(1-q^n)}{1-q}$,
$$S_{50} = \frac{i(1-i^{50})}{1-i} = \frac{i(1-i^{4\times12+2})}{1-i} = \frac{i(1-i^2)}{1-i} = \frac{2i}{1-i} = \frac{2i(1+i)}{(1-i)(1+i)} = \frac{2i-2}{2} = i-1.$$

Problem 15. Solution: A.
We have $x + (x + 1) + (x + 2) + \ldots + (x + 66) = 67x + (1 + 2 + 3 + \ldots + 66) = 67x + 66(67)/2 = 2010$. We find that $x = -3$.

Problem 16. Solution: B.
Subtract 7 from each entry, then divide the resulting terms by 5 to get 1, 2, 3, . . . 18, so there are 18 numbers in the sum. The average value is $(12 + 97) \div 2$, so the total sum is $18 \cdot (12 + 97) \div 2 = 9(109) = 981$.

Problem 17. Solution: C.
The sum $1 + 2 + 3 + \cdots + n$ can be found easily to be $n(n + 1)/2$. Thus $n(n + 1)/2 \leq 200$ and $n(n + 1)$ _ 400. So n is close to 20. Check $n = 20$ to see that $1 + 2 + 3 + \cdot + 20 = 10 \cdot 21 = 210$, so we try $n = 19$ and see that $1 + 2 + \cdots + 19 = 190$.

On the other hand, since $1 + 2 + \cdots + n = n(n + 1)/2 \leq 200$ it follows that $n(n + 1) = n^2 + n \leq 400$. No matter what the value of n, $n^2 < n^2 + n < (n + 1)2$. Since $400 = 20^2$, n cannot be 20, but $n + 1$ can. So the correct value is 19.

Algebra II Through Competitions Chapter 16 Arithmetic and Geometric Sequences

Problem 18. Solution: D.

The number of games played is $18(9)10/2 = 810$. If n is the number of wins of the last-place team, and d is the common difference of wins between successive teams, then $n + (n + d) + (n + 2d) + \cdots + (n + 9d) = 10n + 45d = 810$, so $2n + 9d = 162$. Now n is the maximum when d is a minimum (but not zero, because there are no ties). The smallest integral value of d for which n is integral is $d = 2$. Thus $n = 72$.

Problem 19. Solution: D.

The common ratio is $-\frac{1}{2}/1 = -\frac{1}{2}$ and the first term is 1.

By the formula, $S_n = \frac{a_1(1-q^n)}{1-q}$, $S_{1001} = \frac{1[1-(-\frac{1}{2})^{1001}]}{1-(-\frac{1}{2})} = \frac{1+(\frac{1}{2})^{1001}}{\frac{3}{2}} = \frac{1}{3}(2+\frac{1}{2^{1000}})$.

Problem 20. Solution: D.

The common ratio is $-\frac{1}{2}/2 = -\frac{1}{4}$ and the first term is 2.

By the formula, $S = \frac{a_1}{1-q}$, $S = \frac{2}{1-(-\frac{1}{4})} = \frac{2}{\frac{5}{4}} = \frac{8}{5} = 1\frac{3}{5}$.

Problem 21. Solution: D.

We know that a_n is an arithmetic sequence.

Using formula (2.3): $\frac{\frac{S_{3m}}{3m} - \frac{S_{2m}}{2m}}{3m-2m} = \frac{\frac{S_{2m}}{2m} - \frac{S_m}{m}}{2m-m}$ $\Rightarrow$ $\frac{\frac{2S_{3m}-3S_{2m}}{6m}}{m} = \frac{\frac{S_{2m}-2S_m}{2m}}{m}$ $\Rightarrow$

$2S_{3m} - 3S_{2m} = 3(S_{2m} - 2S_m)$ $\Rightarrow$ $S_{3m} = 3S_{2m} - 3S_m$.

Substituting $S_m = 50$, $S_{2m} = 90$ into the equation above yields

$S_{3m} = 3S_{2m} - 3S_m = 3 \times 90 - 3 \times 50 = 3(40) = 120$.

Problem 22. Solution: B.

$\frac{1}{7} - \frac{5}{7^2} + \frac{1}{7^3} - \frac{5}{7^4} + \frac{1}{7^5} - \frac{5}{7^6} + \cdots = \frac{1}{7} + \frac{1}{7^3} + \frac{1}{7^5} \cdots - \frac{5}{7^2} - \frac{5}{7^4} - \frac{5}{7^6} + \cdots$

$= \frac{\frac{1}{7}}{1-\frac{1}{7^2}} - \frac{\frac{5}{7^2}}{1-\frac{1}{7^2}} = \frac{7}{48} - \frac{5}{48} = \frac{2}{48} = \frac{1}{24}$.

Algebra II Through Competitions   **Chapter 17 Sequences and Series**

1. TERMS

Sequences

A sequence is a function whose domain is the set of natural numbers.

The function: $f(n) = 2n - 1 \Rightarrow$ The sequence: $1, 3, 5, \ldots, 2n - 1, \ldots$

$2n - 1$ is called the general term or the nth term.

The function $f(n) = 2n. \Rightarrow$ The sequence: $2, 4, 6, \ldots, 2n, \ldots$

$2n$ is called the general term or the nth term.

Where n is a natural number.

Series

Let $a_1, a_2, a_3, \ldots, a_n$ be the terms of a sequence. Then S_n is defined as the sum of the first n terms: $S_n = a_1 + a_2 + a_3 + \ldots + a_n$.

The sum of the first n terms is called the series.

A finite series is defined as $S_n = a_1 + a_2 + a_3 + \ldots + a_n. = \sum_{i=1}^{n} a_i$

An infinite series is defined as $S_n = a_1 + a_2 + a_3 + \ldots + a_n + \ldots = \sum_{i=1}^{\infty} a_i$

The symbol $\sum$, the Greek capital letter sigma, indicates a sum.

Properties of Sequences

1. $\sum_{i=1}^{\infty} c = c + c + c + \cdots + c = nc$

2. $\sum_{i=1}^{\infty} ca_i = c \sum_{i=1}^{\infty} a_i$

3. $\sum_{i=1}^{\infty} (a_i \pm b_i) = \sum_{i=1}^{\infty} a_i \pm \sum_{i=1}^{\infty} b_i$

Example 1. Find the value of: $\sum_{k=2}^{k=6} (3k + 1)$.

Solution: 65.

$$\sum_{k=2}^{k=6}(3k+1) = 3\cdot 2+1+3\cdot 3+1+3\cdot 4+1+3\cdot 5+1+3\cdot 6+1 = 3(2+3+4+5+6)+5 = 65.$$

Example 2. $\sum_{m=1}^{3}\left(\sum_{n=1}^{3}n\right) =$

A. 6   B. 12   C. 18   D. 54   E. NOTA

Solution: C.
1 + 2 + 3 = 6 and the outer summation gives 6 + 6 + 6 = 18.

Example 3. If the sum of the first n terms of a sequence is $n^2 + 5n$, what is the 2010th term of the sequence?

A. 4042   B. 4024   C. 2042   D. 4240   E. 2240

Solution: B.
$a_{2010} = S_{2010} - S_{2009} = (2010^2 + 5\times 2010) - (2009^2 + 5\times 2009) = 2010^2 - 2009^2 + 5\times 2010 - 5\times 2009 = (2010-2009)(2010+2009) + 5\times (2010-2009) = 4019 + 5 = 4024.$

Example 4. A function f from the integers to the integers is defined as follows:
$$f(n) = \begin{cases} n+3 & \text{if } n \text{ is odd} \\ n/2 & \text{if } n \text{ is even} \end{cases}.$$
Suppose k is odd and $f(f(f(k))) = 27$. What is the sum of the digits of k?

A. 3.   B. 6.   C. 9.   D. 12.   E. 15.

Solution: B.
Since k is odd, $f(k) = k + 3$. So $k + 3$ is even, $f(k+3) = f(f(k)) = \frac{k+3}{2}$. If $\frac{k+3}{2}$ is odd, then $27 = f(f(f(k))) = f\left(\frac{k+3}{2}\right) = \frac{k+3}{2} + 3$, which implies that $k = 45$. This is not possible because $f((f(45))) = f(f(48)) = f(24) = 12$. Hence $\frac{k+3}{2}$ must be even, and $27 = f(f(f(k))) = f\left(\frac{k+3}{2}\right) = \frac{k+3}{4}$, which implies that $k = 105$. Checking, we find that $f((f(105))) = f(f(108)) = f(54) = 27$. Hence the sum of the digits of k is 1 + 0 + 5 = 6.

Algebra II Through Competitions   **Chapter 17 Sequences and Series**

2. COMMONLY USED METHODS TO SOLVE SERIES PROBLEMS

(1). The Formula Method

This is the most basic and important method used to find the sum of the first n terms in a sequence.

(1). Sum of terms in an arithmetic sequence: $S_n = \dfrac{n(a_1 + a_n)}{2} = na_1 + \dfrac{n(n-1)}{2}d$

(2). Sum of terms in a geometric sequence: $S_n = \begin{cases} na_1 & (q=1) \\ \dfrac{a_1(1-q^n)}{1-q} = \dfrac{a_1 - a_n q}{1-q} & (q \neq 1) \end{cases}$

(3). $S_n = \sum\limits_{k=1}^{n} k = 1 + 2 + \ldots + n = \dfrac{1}{2}n(n+1)$

(4). $S_n = \sum\limits_{k=1}^{n} k^2 = 1^2 + 2^2 + \ldots + n^2 = \dfrac{1}{6}n(n+1)(2n+1)$

(5). $S_n = \sum\limits_{k=1}^{n} k^3 = 1^3 + 2^3 + \ldots + n^3 = (1 + 2 + \ldots + n)^2 = [\dfrac{1}{2}n(n+1)]^2$

(6). $S_n = 1 + 3 + 5 + \ldots + (2n-1) = n^2$

(7). $S_n = 2 + 4 + 6 + \ldots + 2n = n(n+1)$.

Example 5. Find $\sum\limits_{n=1}^{\infty} \dfrac{4^n - 8}{5^n}$.

A. -5 B. 2   C. 6   D. 15   E. NOTA

Solution: B.

$\sum\limits_{n=1}^{\infty} \dfrac{4^n - 8}{5^n} = \sum\limits_{n=1}^{\infty} (\dfrac{4}{5})^n - 8\sum\limits_{n=1}^{\infty} (\dfrac{1}{5})^n = \dfrac{\frac{4}{5}}{1 - \frac{4}{5}} - \dfrac{8 \times \frac{1}{5}}{1 - \frac{1}{5}} = 2$.

Example 6. Find the sum of $1 \cdot 2 + \cdot 3 \cdot 4 + \cdots + (2n-1)(2n)$.

Solution:

Algebra II Through Competitions   **Chapter 17 Sequences and Series**

$$1 \cdot 2 + 3 \cdot 4 + \cdots + (2n-1)(2n) = \sum_{m=1}^{n}(2m-1)(2m) = \sum_{m=1}^{n}(4m^2 - 2m) = 4\sum_{m=1}^{n}m^2 - 2\sum_{m=1}^{n}m$$

$$= \frac{4n(n+1)(2n+1)}{6} - \frac{2n(n+1)}{2} = \frac{2n(n+1)(2n+1)}{3} - \frac{3n(n+1)}{3}$$

$$= \frac{n(n+1)(4n+2-3)}{3} = \frac{n(n+1)(4n-1)}{3}.$$

Example 7. Find the sum of $1 \cdot 2 \cdot 3 + 2 \cdot 3 \cdot 4 + \cdots + n(n+1)(n+2)$.

Solution:
$$1 \cdot 2 \cdot 3 + 2 \cdot 3 \cdot 4 + \cdots + n(n+1)(n+2)$$

$$= \sum_{m=1}^{n} m(m+1)(m+2) = \sum_{m=1}^{n}(m^3 + 3m^2 + 2m) = \sum_{m=1}^{n}m^3 + 3\sum_{m=1}^{n}m^2 + 2\sum_{m=1}^{n}m$$

$$= \left[\frac{n(n+1)}{2}\right]^2 + \frac{3n(n+1)(2n+1)}{6} + \frac{2n(n+1)}{2} = \frac{n(n+1)}{4}[n(n+1) + 2(2n+1) + 4]$$

$$= \frac{n(n+1)}{4}(n^2 + 5n + 6) = \frac{1}{4}n(n+1)(n+2)(n+3).$$

Example 8. When the sum of the first k terms of $1^2 + 2^2 + 3^2 + \ldots + n^2 + \ldots$ is subtracted from the sum of the first k terms of $1(2) + 3(4) + 5(6) + \ldots + (2n-1)(2n) + \ldots$, the result is 111720. Find the value of k.

Solution: 48.

$$\sum_{n=1}^{k}[2n(2n-1) - n^2] = \sum_{n=1}^{k}[3n^2 - 2n] = \frac{3}{6}n(n+1)(2n+1) - \frac{2}{2}n(n+1) = n(n+1)(n - 1/2).$$

It follows that $k(k+1)(2k-1) = 2 \times 111720$. Thus $k = 48$.

Example 9. If n is a positive integer, find the value of n such that
$$\frac{1(2) + 3(4) + 5(6) + \cdots + (2n-1)(2n)}{1(2)(3) + 2(3)(4) + 3(4)(5) + \cdots + n(n+1)(n+2)} = \frac{29}{150}.$$

Solution: 22.

$$\frac{1(2)+3(4)+5(6)+\cdots+(2n-1)(2n)}{1(2)(3)+2(3)(4)+3(4)(5)+\cdots+n(n+1)(n+2)} = \frac{\sum_{n=1}^{\infty}(2n-1)(2n)}{\sum_{n=1}^{\infty}n(n+1)(n+2)} = \frac{29}{150}.$$

Using the results of obtained from examples 6 and 7:

$$\frac{\frac{1}{3}n(n+1)(4n-1)}{\frac{1}{4}n(n+1)(n+2)(n+3)} = \frac{29}{150} \quad \Rightarrow \quad \frac{\frac{1}{3}(4n-1)}{\frac{1}{4}(n+2)(n+3)} = \frac{29}{150} \quad \Rightarrow$$

$$29n^2 - 655n + 374 = 0 \quad \Rightarrow \quad (29n-17)(n-22) = 0.$$

Since n is a positive integer, $n = 22$.

(2). The Method of Subtraction with Order Shifting

This method is very effective when we want to find the sum of the first several terms of the sequence $\{a_n \cdot b_n\}$, where $\{a_n\}$ is an arithmetic sequence and $\{b_n\}$ is a geometric sequence.

Example 10. If $S = \frac{3}{2} + \frac{5}{4} + \frac{7}{8} + \cdots + \frac{2n+1}{2^n} + \cdots$, where $n = 1, 2, 3, \ldots$, then find the value of S.

A. 5   B. 6   C. 8   D. 16   E. NOTA

Solution: A.

We have $S = \frac{3}{2} + \frac{5}{4} + \frac{7}{8} + \cdots + \frac{2n+1}{2^n} + \cdots$ (1)

Multiply (1) by 2: $2S = 3 + \frac{5}{2} + \frac{7}{4} + \cdots$ (2)

(2) − (1): $S = 3 + 1 + \frac{1}{2} + \frac{1}{4} + \cdots = 3 + S_1$, where $S_1 = 1 + \frac{1}{2} + \frac{1}{4} + \cdots = \frac{1}{1-\frac{1}{2}} = 2$.

Thus $S = 3 + S_1 = 3 + 2 = 5$.

Example 11. Find the value of $S_n = \frac{1}{2} + \frac{3}{2^2} + \frac{5}{2^3} + \cdots + \frac{2n-3}{2^{n-1}} + \frac{2n-1}{2^n}$.

Solution:
We know that $S_n = \frac{1}{2} + \frac{3}{2^2} + \frac{5}{2^3} + \cdots + \frac{2n-3}{2^{n-1}} + \frac{2n-1}{2^n}$ (1)

Multiplying both sides of (1) by $\frac{1}{2}$:

$\frac{1}{2}S_n = \frac{1}{2^2} + \frac{3}{2^3} + \cdots + \frac{2n-3}{2^n} + \frac{2n-1}{2^{n+1}}$ (2)

(1) – (2): $S_n - \frac{1}{2}S_n = \frac{1}{2} + \frac{1}{2} + \frac{1}{2^2} + \cdots + \frac{1}{2^{n-1}} - \frac{2n-1}{2^{n+1}}$ $\Rightarrow$ $S_n = 3 - \frac{2n+3}{2^{n+2}}$.

Example 12. Find the sum of the first n terms of the sequence $\{n\}$.

Solution:
Let $S_n = 1 + 2 + \ldots + (n-1) + n$ (1)
We rewrite (1) in the reversing order:
$S_n = n + (n-1) + \ldots + 2 + 1$ (2)
(1) + (2): $2S_n = (1+n) + (2+n-1) + n \ldots + [(n-1) + 2] + (n+1)$
Or $2S_n = n(n+1)$ $\Rightarrow$ $S_n = \frac{1}{2}n(n+1)$.

(3). The Method of Regroup

Example 13. Consider the sequence 1, –2, 3, –4, 5, –6,…, $n - (-1)^{n-1}$. What is the average of the first 300 terms of the sequence?
A. –1   B. 0.5   C. 0   D. –0.5   E. 1

Solution: D.
$1 - 2 + 3 - 4 + 5 - 6 + \ldots 299 - 300 = (1-2) + (3-4) + (5-6) + \ldots + (299-300) = -1 \times 300/2 = -150$.
The average is $-150/300 = -0.5$.

Example 14. Give the base ten, common fraction representation for $0.\overline{123}_{four}$.

Solution:

$0.\overline{123}_{four} = \frac{1}{4} + \frac{2}{4^2} + \frac{3}{4^3} + \frac{1}{4^4} + \frac{2}{4^5} + \frac{3}{4^6} + \cdots$, which equals

$$\frac{\frac{1}{4}}{1-\frac{1}{64}} + \frac{\frac{2}{16}}{1-\frac{1}{64}} + \frac{\frac{3}{64}}{1-\frac{1}{64}} = \frac{16}{63} + \frac{8}{63} + \frac{3}{63} = \frac{27}{63} = \frac{3}{7}.$$

Example 15. Find $\sum_{n=2}^{20} \frac{1}{n^2-n}$.

Solution: C.

$$\sum_{n=2}^{20} \frac{1}{n^2-n} = \sum_{n=2}^{20} [\frac{1}{n-1} - \frac{1}{n}] = \left(\frac{1}{1} - \frac{1}{2}\right) + \left(\frac{1}{2} - \frac{1}{3}\right) + \cdots \left(\frac{1}{19} - \frac{1}{20}\right) = 1 - \frac{1}{20} = \frac{19}{20}$$

(4). Finding the Pattern

Example 16. Given $a_3 = 5$, $a_5 = 8$ and $a_n + a_{n+1} + a_{n+2} = 7$ for all positive integers, n, compute a_{2001}.
A. 5   B. 6   C. 8   D. −6   E. −8

Solution: A.
We know that $a_n + a_{n+1} + a_{n+2} = 7$, so $a_{n+2} = 7 - (a_n + a_{n+1})$. Using this relation, we can generate the next several terms. They are:
$$a_6 = 7 - (a_4 + a_5) = 7 - (-6 + 8) = 5$$
$$a_7 = 7 - (a_5 + a_6) = 7 - (8 + 5) = -6$$
$$a_8 = 7 - (a_6 + a_7) = 7 - (5 + -6) = 8$$
As you can see, the value will continue to cycle in groups of three. Since 2001 is divisible by 3, $a_{2001} = a_3 = a_6 = 5$.

Example 17. Let $f(x) = \frac{1}{1-x}$. Define $f_1(x) = f(x)$ and $f_{n+1}(x) = f(f_n(x))$ for $n \geq 1$. What is $f_{2011}(x)$?

A. x B. $\frac{1}{1-x}$ C. $\frac{x-1}{x}$ D. $\frac{x}{x-1}$ E. $\frac{x}{1-x}$.

Solution: B.

$$f_{n+1}(x) = f(f_n(x)) \Rightarrow f_2(x) = f_{1+1}(x) = f(f_1(x)) = f(f(x)) = f(\frac{1}{1-x}) = \frac{1}{1-\frac{1}{1-x}} = \frac{x-1}{x}.$$

$$f_3 = f(f_2(x)) = f(\frac{x-1}{x}) = \frac{1}{1-\frac{x-1}{x}} = x. \quad f_4 = f(f_3(x)) = f(x) = \frac{1}{1-x}.$$

So we see that pattern that every three terms are repeated.

$$f_{2011} = f_{2010+1} = f_1(x) = f(x) = \frac{1}{1-x}.$$

Example 18. Define a sequence by $b_1 = 2$ and $b_{n+1} = \frac{1+b_n}{1-b_n}$, for $n > 0$. What is the value of b_{2008}?

A. 2008   B. 2   C. –3   D. $-\frac{1}{2}$ E. $\frac{1}{3}$.

Solution: E.

$$b_{n+1} = \frac{1+b_n}{1-b_n} \quad \Rightarrow \quad b_2 = \frac{1+b_1}{1-b_1} = \frac{1+2}{1-2} = -3; \quad b_3 = \frac{1+b_2}{1-b_2} = \frac{1+(-3)}{1-(-3)} = -\frac{2}{4} = -\frac{1}{2}.$$

$$b_4 = \frac{1-b_3}{1+b_3} = \frac{1+(-\frac{1}{2})}{1-(-\frac{1}{2})} = \frac{\frac{1}{2}}{\frac{3}{2}} = \frac{1}{3}; \quad b_5 = \frac{1+b_4}{1-b_4} = \frac{1+\frac{1}{3}}{1-\frac{1}{3}} = 2.$$

We see that the pattern is that every four terms repeated. $b_{2008} = b_4 = \frac{1}{3}$.

Algebra II Through Competitions   **Chapter 17 Sequences and Series**

PROBLEMS

Problem 1. Find the value of $\sum_{n=1}^{4}\left(n\left(\sum_{k=1}^{3}(k+1)^2\right)\right)$.

Problem 2. Evaluate: $\sum_{n=7}^{17}(7n+17)$.

A) 264   B) 999   C) 1010   D) 1111   E) NOTA

Problem 3. Find the value of $\sum_{x=3}^{4}(3x+4^x)$.

Problem 4. Suppose $f(0) = 3$ and $f(n) = f(n-1) + 2$. Let $T = f(f(f(f(5))))$. What is the sum of the digits of T?

A. 6   B. 7   C. 8   D. 9   E. 10.

Problem 5. Suppose that for any integer n,
$$f(n) = \begin{cases} n-1, & \text{if } n \text{ is even;} \\ 2n, & \text{if } n \text{ is odd.} \end{cases}$$
If $k \in N$, and $f(f(f(k))) = 21$, find the sum of the digits in k.

A. 3   B. 4   C. 5   D. 7   E. 11

Problem 6. Let the function $A(x, y)$ be defined by the following rule:
  $A(0, n) = n + 1$,
  $A(m, 0) = A(m-1, 1)$,
  $A(m, n) = A(m-1, A(m, n-1))$ where m and n are positive integers.
Find the numerical value of $A(2, 3)$.

A. 5   B. 6   C. 7   D. 8   E. 9

Problem 7. When the sum of the first k terms of the series $1^2 + 2^2 + 3^3 + \ldots + n^2 + \ldots$ is subtracted from the sum of the first k terms of the series $1(2) + 2(3) + 3(4) + \ldots + n(n+1) + \ldots$, the result is 210. Find the value of k.

Problem 8. Find the sum of $2^2 + 5^2 + 8^2 + \cdots + (3n-1)^2$.

Problem 9. Multiplying two different numbers from the following n numbers 1, 2, 3,..., n. What is S, the sum of all products?

Problem 10. The sum of the squares of the first N natural number is given by the formula $S_N = \dfrac{N(N+1)(2N+1)}{6}$. Find the sum of the first 50 terms of series
$(1)(2) + (2)(3) + (3)(4) + \ldots + (N)(N+1)$.

Problem 11. Find the sum of the first n terms of $\dfrac{2}{2}, \dfrac{4}{2^2}, \dfrac{6}{2^3}, \ldots, \dfrac{2n}{2^n}, \ldots$.

Problem 12. The decimal form of $S = \dfrac{1}{10} + \dfrac{2}{100} + \dfrac{3}{1000} + \dfrac{4}{10000} + \cdots$ will never contain which of the following digits?
A. 9 B. 6 C. 8 D. 7 E. all digits are possible

Problem 13. Which of the following recursive equations generates the sequence: {1, 2, 5, 8, 2, 5, 8, 2,...}?
A. $x_{t+1} = 15 - x_t - x_{t-1}$ B. $x_{t+1} = 3x_t - 1$ C. $x_{t+1} = 3x_{t-1} + 2$
D. $x_{t+1} = \dfrac{9}{2}x_t - \dfrac{1}{2}x_t^2 - 2$ E. $x_{t+1} = \dfrac{1}{2}(9 + 5x_t - 9x_{t-1})$

Problem 14. For any nonzero a and b, define a sequence as follows: $x_1 = a$, $x_2 = b$, $x_{n+2} = \dfrac{x_{n+1} + 1}{x_n}$ for $n = 1, 2, 3 \ldots$. Find x_6.

A. a B. b C. $\dfrac{a+1}{b}$ D. $\dfrac{a+b+1}{ab}$ E. $\dfrac{b+1}{a}$

Problem 15. A positive integer is said to be a palindromic number if it is equal to itself when its digits are reversed. Let $P_2, P_3, P_4,\ldots$ denote the number of palindromic numbers with two digits, three digits, four digits, and so on. What is true regarding the following sequence of ratios?
$\dfrac{P_3}{P_2}, \dfrac{P_4}{P_3}, \dfrac{P_5}{P_4}, \dfrac{P_6}{P_5} \ldots$
A. The sequence always increases. B. The sequence always decreases.
C. The sequence is constant. D. The sequence is composed of a finite number of values that neither increase nor always decrease.

E. The sequence is composed of an infinite number of values that neither increase nor always decrease.

Problem 16. Suppose that $a_1, a_2, \ldots a_5$ are positive, single-digit integers and that $\sum_{n=1}^{5} a_n 10^n = 624380$. Find $\sum_{n=1}^{5}(-1)^n a_{6-n}$.

A. 13 B. −16 C. 7 D. −13 E. 16

Problem 17. The Lucas numbers are defined by $L_0 = 2$, $L_1 = 1$ and $L_n = L_{n-1} + L_{n-2}$ when $n > 1$. Find L_{10}.

A. 55 B. 113 C. 66 D. 123 E. 76

Problem 18. For integers n and m, let gcd(n, m) denote the greatest common divisor of n and m. Compute $\sum_{n=1}^{30} \gcd(n,30)$.

A. 232 B. 245 C. 465 D. 66 E. 135

Problem 19. In the Fibonacci sequence, where $a_1 = 1$, $a_2 = 1$ and $a_n = a_{n-1} + a_{n-2}$ for all $n \geq 3$ which are true?

I. a_{2009} is even. II. a_{2009} is odd. III. a_{2010} is even. IV. a_{2010} is odd.

A. I & III B. I & IV C. II & III D. II & IV E. cannot be determined.

Problem 20. If the sum of the first $n + 1$ terms of a sequence is $(n + 9)(n + 10)/2$, what is the 2012th term of the sequence?

A. 2018 B. 2019 C. 2020 D. 2011 E. 2012.

Problem 21. Let a_n be the nth term of a sequence, and define $S_n = \sum_{k=1}^{n} a_k$. If $S_n = 4n^2 + 2n + 8$, find the value of a_{428}.

A. 428 B. 855 C. 1710 D. 3422 E. NOTA

Problem 22. Define a sequence by $b_1 = 2$ and $b_{n+1} = \dfrac{1-b_n}{1+b_n}$, for $n \geq 1$, what is the value of b_{2010}?

A. −2010 B. 2010 C. 2000 D. −1 E. $-\dfrac{1}{3}$

Problem 23. Simplify the product $\left(1+\dfrac{1}{\alpha}\right)\left(1+\dfrac{1}{\alpha^2}\right)\left(1+\dfrac{1}{\alpha^4}\right)\ldots(1+\dfrac{1}{\alpha^{2^{2012}}})$.

A. $\dfrac{1-\dfrac{1}{\alpha^{2^{2012}}}}{1-\dfrac{1}{\alpha}}$
B. $\dfrac{1-\dfrac{1}{\alpha^{2^{2013}}}}{1-\dfrac{1}{\alpha}}$
C. $\dfrac{1-\dfrac{1}{\alpha^{2^{2014}}}}{1-\dfrac{1}{\alpha}}$
D. $\dfrac{1-\dfrac{1}{\alpha^{2^{2015}}}}{1-\dfrac{1}{\alpha}}$
E. NOTA.

Problem 25. Let $P_n = 1+2+\cdots+n$. Evaluate $\dfrac{\sum_{n=1}^{2012} P_n}{(2012)(2013)}$.

A) $\dfrac{1007}{3}$
B) $\dfrac{1009}{3}$
C) $\dfrac{1012}{3}$
D) $\dfrac{1015}{3}$
E) NOTA

Problem 26. Calculate $S = 1 + \dfrac{1}{1+2} + \dfrac{1}{1+2+3} + \cdots + \dfrac{1}{1+2+3+\cdots 100}$.

SOLUTIONS:

Problem 1. Solution: 290

$$\sum_{n=1}^{4}\left(n\left(\sum_{k=1}^{3}(k+1)^2\right)\right) = \sum_{n=1}^{4}\left(n\left((1+1)^2+(2+1)^2+(3+1)^2\right)\right) = \sum_{n=1}^{4}(n(4+9+16)) = 29\sum_{n=1}^{4}(n)$$
$$= 29(1+2+3+4) = 29\times 10 = 290.$$

Problem 2. Solution: 1111.

$$\sum_{n=7}^{17}(7n+17) = \frac{17-7+1}{2}(66+136) = \frac{11}{2}\times 202 = 11\times 101 = 1111.$$

Problem 3. Solution: 341.

$$\sum_{x=3}^{4}(3x+4^x) = 3\cdot 3 + 4^3 + 3\cdot 4 + 4^4 = 21 + 4^3(1+4) = 341$$

Problem 4. Solution: C.
In fact, $f(n) = 2n + 3$, and $T = f(f(f(f(5)))) = f(f(f(13))) = f(f(29)) = f(61) = 125$.
The sum of the digits is $1 + 2 + 5 = 8$.

Problem 5. Solution: A.
If k is odd, $f(k) = 2k$. So $2k$ is even, and $f(2k) = f(f(k)) = 2k - 1$. Since $2k - 1$ is odd, then $21 = f(f(f(k))) = f(2k-1) = 2(2k-1)$, which implies that k is not integer. So k is even, $f(k) = k - 1$. Since $k - 1$ is odd. $f(k-1) = f(f(k)) = 2(k-1)$. So $2(k-1)$ is even, then $21 = f(f(f(k))) = f(2(k-1)) = 2(k-1) - 1$. Thus $k = 12$. The sum of the digits is $1 + 2 = 3$.

Problem 6. Solution: E.
$A(0, n) = n + 1 \quad \Rightarrow \quad A(0, 1) = 1 + 1 = 2,$
$\qquad\qquad\qquad\qquad\qquad A(0, 2) = 2 + 1 = 3,$
$\qquad\qquad\qquad\qquad\qquad A(0, 3) = 4.$

$A(m, 0) = A(m - 1, 1) \Rightarrow \quad A(1, 0) = A(0, 1) = 2,$
$\qquad\qquad\qquad\qquad\qquad A(2, 0) = A(1, 1) = 3.$

$A(m, n) = A(m - 1, A(m, n - 1)) \quad \Rightarrow \quad A(1, 1) = A(0, A(1, 0)) = A(0, 2) = 3,$
$\qquad\qquad\qquad\qquad\qquad\qquad\qquad\qquad A(1, 2) = A(0, A(1, 1)) = A(0, 3) = 4,$
$\qquad\qquad\qquad\qquad\qquad\qquad\qquad\qquad A(1, 3) = 5.$
$\qquad\qquad\qquad\qquad\qquad\qquad\qquad\qquad A(1, 4) = 6.$

$$A(1, 5) = 7.$$
$$A(1, 6) = 8.$$
$$A(1, 7) = 9.$$
$A(2, 1) = A(2 - 1, A(2, 1 - 1)) = A(1, A(2, 0)) = A(1, 3) = 5.$
$A(2, 2) = A(1, A(2, 1)) = A(1, 5) = A(0, A(1, 5 - 1)) = A(0, A(1, 4)) = A(0, 6) = 7.$
$A(2, 3) = A(1, A(2, 2)) = A(1, 7) = 9.$

Problem 7. Solution:
$$\sum_{n=1}^{k} n(n+1) - n^2 = \sum_{n=1}^{k} n = \frac{k(k+1)}{2} = 210.$$
It follows that $k^2 + k - 420 = 0$. Thus $(k - 20)(k + 21) = 0$, so $k = 20$.

Problem 8. Solution:
$$2^2 + 5^2 + 8^2 + \cdots + (3n - 1)^2 = \sum_{m=1}^{n}(3m - 1)^2 = \sum_{m=1}^{n}(9m^2 - 6m + 1)$$
$$= 9\sum_{m=1}^{n} m^2 - 6\sum_{m=1}^{n} m + \sum_{m=1}^{n} 1 = 9 \times \frac{1}{6} n(n+1)(2n+1) - 6 \times \frac{1}{2} n(n+1) + n = \frac{n}{2}(6n^2 + 3n - 1)$$

Problem 9. Solution:
From $(a + b + c + \cdots)^2 = (a^2 + b^2 + c^2 + \cdots) + 2(ab + ac + \cdots + bc + \cdots)$, we have
$(1 + 2 + 3 + \cdots + n)^2 = (1^2 + 2^2 + 3^2 + \cdots + n^2) + 2S$.

Thus $S = \frac{1}{2}\{(1 + 2 + 3 + \cdots + n)^2 - (1^2 + 2^2 + 3^2 + \cdots + n^2)\}$

$= \frac{1}{2}\{\frac{1}{4} n^2(n+1)^2 - \frac{1}{6} n(n+1)(2n+1)\} = \frac{1}{24}(n-1)n(n+1)(3n+2)$.

Problem 10. Solution:
Each term of the given series is in the form $N(N + 1)$, which may be rewriten as $N^2 + N$.
We are given that $1^2 + 2^2 + 3^2 + \cdots + N^2 = \frac{N(N+1)(2N+1)}{6}$. We also can show that

$1 + 2 + 3 + \cdots + N = \frac{N(N+1)}{2}$. Now we have

$(1 + 1^2) + (2 + 2^2) + (3 + 3^2) + \cdots + (N + N^2) = \frac{N(N+1)(2N+1)}{6} + \frac{N(N+1)}{2}$. Thus the given series has a sum of $\frac{N(N+1)(2N+4)}{6}$ or $\frac{N(N+1)(N+2)}{3}$. Thus if $N = 50$, the sum is $\frac{(50)(51)(52)}{3}$ or 44,200.

Problem 11. Solution:

We see that the general term of the given sequence $\{\frac{2n}{2^n}\}$ is the the product of the general terms of the arithmetic sequence $\{2n\}$ and the geometric sequence $\{\frac{1}{2^n}\}$.

Let $S_n = \frac{2}{2} + \frac{4}{2^2} + \frac{6}{2^3} + \cdots + \frac{2n}{2^n}$ \hfill (1)

Both sides of (1) multiplying by $\frac{1}{2}$:

$\frac{1}{2}S_n = \frac{2}{2^2} + \frac{4}{2^3} + \frac{6}{2^4} + \cdots + \frac{2n}{2^{n+1}}$ \hfill (2)

(1) − (2): $(1-\frac{1}{2})S_n = \frac{2}{2} + \frac{2}{2^2} + \frac{2}{2^3} + \frac{2}{2^4} + \cdots + \frac{2}{2^n} - \frac{2n}{2^{n+1}} = 2 - \frac{1}{2^{n-1}} - \frac{2n}{2^{n+1}}$

Therefore $S_n = 4 - \frac{n+2}{2^{n-1}}$.

Problem 12. Solution: C

$S = \frac{1}{10} + \frac{2}{100} + \frac{3}{1000} + \frac{4}{10000} + \cdots$ \hfill (1)

$10 \times (1)$: $10S = 1 + \frac{2}{10} + \frac{3}{100} + \frac{4}{1000} + \frac{5}{10000} + \cdots$ \hfill (2)

(2) − (1): $9S = 1 + \frac{1}{10} + \frac{1}{100} + \frac{1}{1000} + \frac{1}{10000} + \cdots = \frac{1}{1-\frac{1}{10}} = \frac{10}{9}$.

$S = \frac{10}{81} = 0.\overline{123456790}$. So S never contains the digit 8.

Problem 13. Solution: D.
Trying to go directly from the sequence to the formula is difficult, but since there are 5 choices, we can try each one. Clearly (A) fails as the third term is 12. (B) starts fine but fails on the 4th term, which is 14. (C) grows without limits, so it fails, and (D) works.

Problem 14. Solution: A.

$$x_3 = \frac{x_2+1}{x_1} = \frac{b+1}{a}, \quad x_4 = \frac{x_3+1}{x_2} = \frac{\frac{x_2+1}{x_1}+1}{x_2} = \frac{\frac{x_2+1+x_1}{x_1}}{x_2} = \frac{x_2+1+x_1}{x_1 x_2},$$

$$x_5 = \frac{x_4+1}{x_3} = \frac{\frac{x_2+1+x_1}{x_1 x_2}+1}{\frac{x_2+1}{x_1}} = \frac{\frac{x_2+1+x_1+x_1 x_2}{x_1 x_2}}{\frac{x_2+1}{x_1}} = \frac{x_2+1+x_1+x_1 x_2}{x_2(x_2+1)} = \frac{(x_2+1)(x_1+1)}{x_2(x_2+1)} = \frac{x_1+1}{x_2}$$

$$x_6 = \frac{x_5+1}{x_4} = \frac{\frac{x_1+1}{x_2}+1}{\frac{x_2+1+x_1}{x_1 x_2}} = \frac{\frac{x_2+1+x_1}{x_2}}{\frac{x_2+1+x_1}{x_1 x_2}} = x_1 = a.$$

Problem 15. Solution: D
The number of palindromes with n digits can be found by
$9 \times 10^{\frac{n-1}{2}}$, if n is odd and $9 \times 10^{\frac{n-2}{2}}$, if n is even.

$\frac{P_3}{P_2} = \frac{9 \times 10^{\frac{3-1}{2}}}{9 \times 10^{\frac{2-2}{2}}} = \frac{90}{9} = 10$; $\frac{P_4}{P_3} = \frac{9 \times 10^{\frac{4-2}{2}}}{90} = \frac{90}{90} = 1$; $\frac{P_5}{P_4} = \frac{9 \times 10^{\frac{5-1}{2}}}{90} = \frac{900}{90} = 10$; $\frac{P_6}{P_5} = \frac{9 \times 10^{\frac{6-2}{2}}}{900} = \frac{900}{900} = 1$. So D is the answer.

Problem 16. Solution: D

$\sum_{n=1}^{5} a_n 10^n = 10 a_1 + 100 a_2 + 1000 a_3 + 10000 a_4 + 100000 a_5 = 624380$
$= 10 \times 8 + 100 \times 3 + 1000 \times 4 + 10000 \times 2 + 100000 \times 6$

$\sum_{n=1}^{5} (-1)^n a_{6-n} = -a_5 + a_4 + -a_3 + a_2 - a_1 = -6 + 2 - 4 + 3 - 8 = -13.$

Problem 17. Solution: D.
The terms are 2, 1, 3, 4, 7, 11, 18, 29, 47, 76, 123 (L_{10}).

Problem 18. Solution: E.

There are 7 numbers relatively prime to 30 (7, 11, 13, 17, 19, 23, and 29). The gcd is 1 and the sum is 7.

$$\sum_{n=1}^{30} \gcd(n,30) = 1+2+3+2+5+6+2+3+10+6+2+15+2+6+10$$
$$+3+2+6+5+2+3+2+30 = 128$$

The answer is $128 + 7 = 135$.

Problem 19. Solution: B.
$a_1 = 1$, $a_2 = 1$, $a_3 = a_1 + a_2 = 1 + 1 = 2$.
This is the Fibonacci sequence:
$(1, 1, 2), (3, 5, 8), (13, 21, 34), \ldots$
The pattern is that there is one even number every three numbers.
$a_{2009} = a_{2007+2}$.
So a_{2009} and a_2 have the same parity, which is odd, and a_{2010} and a_3 have the same parity, which is even.

Problem 20. Solution: C.
$a_{2012} = S_{2011} - S_{2010} = (2011 + 9)(2011 + 10)/2 - (2010 + 9)(2010 + 10)/2$
$= (2021 - 2019)(2020)/2 = 2020$.

Problem 21. Solution:
$a_{428} = S_{428} - S_{427} = 4(428^2 - 427^2) + 2(428 - 427) + 8 - 8 = 3422$.

Problem 22. Solution: E.

$$b_{n+1} = \frac{1-b_n}{1+b_n} \quad \Rightarrow \quad b_2 = \frac{1-b_1}{1+b_1} = \frac{1-2}{1+2} = -\frac{1}{3}; \quad b_3 = \frac{1-b_2}{1+b_2} = \frac{1-(-\frac{1}{3})}{1+(-\frac{1}{3})} = \frac{\frac{4}{3}}{\frac{2}{3}} = 2.$$

$$b_4 = \frac{1-b_3}{1+b_3} = \frac{1-2}{1+2} = -\frac{1}{3}; \quad b_5 = \frac{1-b_4}{1+b_4} = \frac{1-(-\frac{1}{3})}{1+(-\frac{1}{3})} = 2.$$

We see that the pattern is that every two terms repeated. $b_{2010} = b_2 = -\frac{1}{3}$.

Problem 23. Solution: B.

Call the product P and consider what happens when we multiply by $1-\dfrac{1}{\alpha}$. We have

$$\left(1-\frac{1}{\alpha}\right)P = \left(1-\frac{1}{\alpha}\right)\left(1+\frac{1}{\alpha}\right)\cdots = \left(1-\frac{1}{\alpha^2}\right)\left(1+\frac{1}{\alpha^2}\right)\cdots = 1-\frac{1}{\alpha^{2^{2013}}}.$$

Thus we have $P = \dfrac{1-\dfrac{1}{\alpha^{2^{2013}}}}{1-\dfrac{1}{\alpha}}$.

Problem 25. Solution: A.

Since $P_n = 1+2+\ldots+n = \dfrac{n(n+1)}{2}$ we have that

$$\sum_{n=1}^{2012} P_n = \sum_{n=1}^{2012} \frac{n(n+1)}{2} = \sum_{n=1}^{2012} n^2 + \frac{1}{2}\sum_{n=1}^{2012} n = \frac{1}{2}\frac{n(n+1)(2n+1)}{6} + \frac{1}{2}\frac{n(n+1)}{2}\bigg|_{n=2012}$$

$$= \frac{(2012)(2013)(4025)}{12} + \frac{(2012)(2013)}{4}.$$

Thus $\dfrac{\sum_{n=1}^{2012} P_n}{(2012)(2013)} = \dfrac{4025}{12} + \dfrac{1}{4} = \dfrac{1007}{3}$.

Problem 26.

$$\frac{1}{1+2+3+\cdots n} = \frac{1}{\dfrac{(1+n)n}{2}} = \frac{2}{n(n+1)} = 2\left(\frac{1}{n}-\frac{1}{n+1}\right).$$

$$1+\frac{1}{1+2}+\frac{1}{1+2+3}+\cdots+\frac{1}{1+2+3+\cdots 100} = 1+2(\frac{1}{2}-\frac{1}{3})+2(\frac{1}{3}-\frac{1}{4})+\cdots+2(\frac{1}{100}-\frac{1}{101})$$
$$=$$
$$1+2\cdot\frac{1}{2}-2\cdot\frac{1}{3}+2\cdot\frac{1}{3}-2\cdot\frac{1}{4}+\cdots+2\cdot\frac{1}{100}-2\cdot\frac{1}{101} = 1+2\cdot\frac{1}{2}-2\cdot\frac{1}{101} = \frac{3}{2}-\frac{1}{101} = \frac{303-2}{202} = \frac{301}{202}$$

.

Algebra II Through Competitions   Chapter 18 Combinations and Permutations

1. TWO IMPORTANT TERMS

(1.1). Permutations

A permutation is an arrangement or a listing of objects in which the order is important.

(1). We have n different elements, and we would like to arrange r of these elements with no repetition, where $1 \leq r \leq n$.

The number of such permutations is $\quad P(n,r) = \dfrac{n!}{(n-r)!}$ \hfill (1.1)

(2). We have n different elements, and we would like to arrange all n of these elements with no repetition.

We let $r = n$ in (1) to get $P(n, n) = n!$ \hfill (1.2)

These n distinct objects can be permutated in $n!$ permutations.

The symbol ! (factorial) is defined as follows: $0! = 1$ \hfill (1.3)
and for integers $n \geq 1$, $n! = n \cdot (n-1) \cdots 1$ \hfill (1.4)

Example 1. How many three-digit numbers are there using the digits 1, 5, 9? No digit can be used twice in any such three-digit number.

Solution:
We have three digits and we want form three digits numbers, so we have $3! = 3 \cdot 2 \cdot 1 = 6$ numbers $\{1,5,9\}, \{1,9,5\}, \{5,1,9\}, \{5,9,1\}, \{9,1,5\}, \{9,5,1\}$.

Example 2. How many three-digit numbers are there using the digits 1, 5, 7, 9? No digit can be used twice in any such three-digit number.

Solution:
We have four digits and we want form three digits numbers, so we have
$P(4,3) = 4 \cdot 3 \cdot 2 = 24$ numbers.

(1.2). Combinations

Definition:
A combination is an arrangement or a listing of things in which order is not important.

Let n, r be non-negative integers such that $0 \leq r \leq n$. The symbol $\binom{n}{r}$ (read "n choose m") is defined and denoted by
$$\binom{n}{r} = \frac{P(n,r)}{P(r,r)} = \frac{n!}{r!(n-r)!} \qquad (1.5)$$

Remember: $\binom{n}{0} = 1$, $\binom{n}{1} = n$, and $\binom{n}{n} = 1$

Since $n - (n - r) = r$, we have $\binom{n}{r} = \binom{n}{n-r}$ \qquad (1.6)

Unlike permutations, combinations are used when the order of the terms does not matter. If we have n different elements, and it doesn't matter which order we arrange the elements, the number of combinations to arrange m elements where $1 \leq m \leq n$, is $\binom{n}{m}$

Example 3. At one of Governor Pat McCrory's parties, each man shook hands with everyone except his spouse, and no handshakes took place between women. If 13 married couples attended, how many handshakes were there among these 26 people?
A. 78 B. 185 C. 234 D. 312 E. 325

Solution: C.
Each man shakes hands with all women but his wife, which gives 13×12 handshakes. There are also $\binom{13}{2} = 78$ handshakes between men; totally, there are 78 × 3 = 234 handshakes.

2. TWO IMPORTANT RULES

(2.1). The product rule (Fundamental Counting Principle) (step work)

When a task consists of k separate steps, if the first step can be done in n_1 ways, the second step can be done in n_2 ways, and so on through the kth step, which can be done in n_k ways, then the total number of possible results for completing the task is given by the product:
$$N = n_1 \times n_2 \times n_3 \times \ldots \times n_k \qquad (2.1)$$

Example 4. There are 8 girls and 6 boys in the Math Club at Central High School. The Club needs to form a delegation to send to a conference, and the delegation must contain exactly two girls and two boys. The number of possible delegations that can be formed from the membership of the Club is
A. 48   B. 420   C. 576   D. 1680   E. 2304

Solution: B.
In order to form the delegation, we need to go through the following two steps:
Step 1: Selecting two girls: The number of combinations to select 2 girls out of 8 is $\binom{8}{2}$.

Step 2: Selecting two boys: The number of combinations to select 2 boys out of 6 is $\binom{6}{2}$.

By the product rule, $\binom{8}{2} \times \binom{6}{2} = 28 \times 15 = 420$.

Example 5. A city council is composed of 6 men and 5 women. Four members are to be chosen as delegates. In how many ways can exactly 2 men and 2 women be chosen?
A. 45   B. 6   C. 115   D. 90   E. 150

Solution: E.
We select 2 men from 6 men and 2 women from 5 women: $\binom{6}{2} \times \binom{5}{2} = 15 \times 10 = 150$.

Example 6. How many integers from 101 to 999 contain the digit 8?
A. 270   B. 252   C. 243   D. 219   E. 180.

Solution: B.
Consider all three-digit numbers and count the numbers without 8. We have 900 three-digit numbers, and $8 \times 9 \times 9 = 648$ numbers do not contain 8. Thus, $900 - 648 = 252$ numbers have the digit 8.

(2.2). The sum rule (case work)
If an event E_1 can happen in n_1 ways, event E_2 can happen in n_2 ways, event E_k can happen in n_k ways, and if any event $E_1, E_2,...$ or E_k happens, the job is done, then the total ways to do the job is
$$N = n_1 + n_2 + \cdots + n_k \qquad (2.2)$$

Example 7. Mr. and Mrs. Zesa want to name baby Zesa so that its monogram (first, middle, and last initials) will be in alphabetical order with no letters repeated. How many such monograms are possible?
A. 276   B. 300   C. 552   D. 600   E. 15600

Solution: B.
The last initial is fixed at Z. If the first initial is *A*, the second initial must be one of *B, C, D, . . ., Y*, so there are 24 choices for the second. If the first initial is *B*, there are 23 choices for the second initial: of *C, D, E, . . ., Y*. Continuing in this way we see that the number of monograms is $24 + 23 + 22 + \cdots + 1$.

Using the formula $1 + 2 + \cdots + n = \dfrac{n(n+1)}{2}$, we get the answer $\dfrac{24 \cdot 25}{2} = 300$.

3. THREE IMPORTANT THEOREMS

THEOREM 1: (Grouping)

(a). Let the number of different objects be n. Divide n into r groups $A_1, A_2, ..., A_r$ such that there are n_1 objects in group A_1, n_2 objects in group A_2, ..., n_r objects in the group A_r, where $n_1 + n_2 + \cdots + n_r = n$. The number of ways to do so is

$$N = \dfrac{n!}{n_1! n_2! \cdots n_r!} \tag{3.1}$$

Example 8. In how many different ways can all the letters in INDIANA be arranged in a line? Assume that duplicate letters are indistinguishable.
A. 5040   B. 2520   C. 1260   D. 630   E. none of these

Solution: D.
$N = \dfrac{7!}{2!2!2!} = 630$.

THEOREM 2: (Combinations with Repetitions)

(a). n identical balls are put into r labeled boxes and the number of balls in each box is not limited. The number of ways is

$$\binom{n+r-1}{n} \text{ or } \binom{n+r-1}{r-1} \tag{3.3}$$

(b) The number of terms in the expansion of $(x_1 + x_2 + x_3 + ... + x_r)^n$, after the like terms combined, is

$$\binom{n+r-1}{n} \text{ or } \binom{n+r-1}{r-1} \qquad (3.4)$$

(c). Let n be a positive integer. The number of positive integer solutions to $x_1 + x_2 + \cdots + x_r = n$ is

$$\binom{n-1}{r-1}. \qquad (3.5)$$

(d). Let n be a positive integer. The number of non-negative integer solutions to $y_1 + y_2 + \cdots + y_r = n$ is

$$\binom{n+r-1}{n} \text{ or } \binom{n+r-1}{r-1} \qquad (3.6)$$

Example 9: How many distinct ordered triples of the form (x, y, z) exist if x, y, and z are non-negative and if $x + y + z = 5$?

Solution:

By (3.6), we have $\binom{n+r-1}{r-1} = \binom{5+3-1}{3-1} = \binom{7}{2} = 21$.

Example 10: A baking company produces four different cookies: Chocolate Chip Cookies, Peanut Butter Cookies, Oatmeal Cookies, and Blueberry Cookies. (a) If a package contains 8 cookies, how many different packages are possible? (b) If a package contains 8 cookies with at least one cookie of each kind, how many different packages are possible?

Solution:
(a). From (3.6), we write $y_1 + y_2 + \cdots + y_r = n \Rightarrow y_1 + y_2 + y_3 + y_4 = 8$

There are $N = \binom{n+r-1}{r-1} = \binom{8+4-1}{3} = \binom{11}{3} = 165$ different packages possible.

(b). By (3.5), we write $y_1 + y_2 + \cdots + y_r = n \Rightarrow y_1 + y_2 + y_3 + y_4 = 8$

There are $N = \binom{n-1}{r-1} = \binom{8-1}{4-1} = \binom{7}{3} = 35$ different packages possible.

THEOREM 3: (Circular Permutations)

The number of circular permutations (arrangements in a circle) of n distinct objects is
$$N = (n-1)! \tag{3.7}$$

We can think of this as n people being seated at a round table. Since a rotation of the table does not change an arrangement, we can put person A in one fixed place and then consider the number of ways to seat all the others. Person B can be treated as the first person to seat and M the last person to seat. The number of ways to arrange persons A to M is the same as the number of ways to arrange persons B to M in a row. So the number of ways to seat n people around a round table, or arranging n distinct objects around a circle, is $N = (n-1)!$.

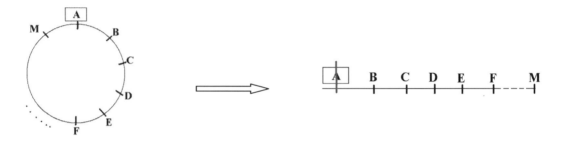

Example 11: In how many ways can four married couples be seated at a round table if no two men, as well as no husband and wife are to be in adjacent seats?

Solution: 12.

There are $(4-1)! = 3!$ to seat four women. After the ladies are seated, person M_4 (whose wife is not shown in the figure below) has two ways to sit. After he is seated in any one of the two possible seats, the other men have only one way to sit in the remaining seats. The solution is $3! \times 2 = 12$.

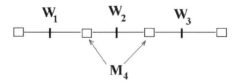

Algebra II Through Competitions   Chapter 18 Combinations and Permutations

PROBLEMS

Problem 1. How many integers between 199 and 301 are divisible by 4 or 10?
A. 26   B. 31   C. 35   D. 37   E. 39

Problem 2. Josh and nine of his friends volunteered to help clean Mr. Camp's vacant lot. Mr. Camp needed 2 mowers, 5 twig collectors and 3 to rake. In how many ways can these jobs be assigned to Josh and his friends?
A. 5040   B. 50,400   C. 15210   D. 25,200   E. 2520

Problem 3. Mary typed a six-digit number, but the two 1's she typed did not show. What appeared instead was 2002. How many different six-digit numbers could she have typed?
A. 5   B. 8   C. 10   D. 15   E. 18

Problem 4. If S and T are sets such that S has 2 more elements than T and 96 more subsets than T, how many elements are in S?
A. 3   B. 5   C. 7   D. 9   E. 11

Problem 5. For how many three-element sets of positive integers $\{a,b,c\}$ is it true that $a \times b \times c = 2310$?
A. 32   B. 36   C. 40   D. 43   E. 45

Problem 6. A standard digital clock displays hours with numbers ranging from one to twelve as opposed to the European digital clock that displays hours with numbers ranging from one to 24. The difference between the least and greatest numbers that are squares of integers that can be displayed on a standard digital clock is
A. 1035.   B. 1056.   C. 1104.   D. 1125.   E. 1196.

Problem 7. Suppose that you are asked to change a twenty dollar bill and that you have 20 one dollar bills, 4 five dollar bills and 2 ten dollar bills at your disposal. If x, y, and z denote, respectively, the number of one dollar, five dollar, and ten dollar bills exchanged for the twenty dollar bill, how many different ordered triples (x, y, z) are possible?
A. 6.   B. 7.   C. 8.   D. 9.   E. 10.

Problem 8. Two integers are said to be partners if both are divisible by the same set of prime numbers. The number of positive integers greater than 1 and less than 25 that have no partners in this set of integers is
A. 9.   B. 10.   C. 11.   D. 12.   E. 13.

Problem 9. An urn contains marbles of four colors, red, yellow, blue and green. All but 25 are red, all but 25 are yellow, and all but 25 are blue. All but 36 are green. How many of the marbles are green?
A. 1   B. 2   C. 3   D. 4   E. 5.

Problem 10. How many two-element subsets $\{a, b\}$ of $\{1, 2, 3, \ldots, 16\}$ satisfy ab is a perfect square?
A. 4   B. 5   C. 6   D. 7   E. 8.

Problem 11. Two points A and B are 4 units apart are given in the plane. How many lines in the plane containing A and B are 2 units from A and 3 units from B?
A. 0   B. 1   C. 2   D. 3   E. 4

Problem 12. How many four-digit numbers between 6000 and 7000 are there for which the thousands digits equal the sum of the other three digits?
A. 20   B. 22   C. 24   D. 26   E. 28.

Problem 13. How many 5-digit numbers can be built using the digits 1, 2, and 3 if each digit must be used at least once?
A. 60   B. 90   C. 120   D. 150   E. 243

Problem 14. The diagram below represents the only possible paths for trips between cities A, B, C, and D. For instance, there are only five trip paths from city B to city C. How many different round-trip paths are there between A and D such that each round-trip passes **ONLY ONCE** through the location represented by the asterisk (that is, a round trip goes from A to B, to C, to D, to C, to B, to A)?

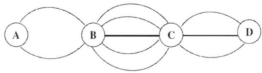

A) 144   B) 324   C) 576   D) 864   E) none of these

Problem 15. With four cents, two nickels and one dime, how many different amounts of money can be made using one or more of these coins?
(A) 24   (B) 20   (C) 6   (D) 19   (E) 22

Problem 16. If you toss a fair coin four times, in how many ways can you obtain at least one head?
A. 5   B. 4   C. 9   D. 6   E. 15.

Problem 17. A city council is composed of 6 men and 3 women. Four members are to be chosen as delegates. In how many ways can exactly 2 men and 2 women be chosen?
A. 35   B. 40   C. 45   D. 60   E. 90.

Problem 18. How many different pairs (m, n) can be formed using numbers from the list of integers $\{1, 2, 3, \cdots, 20\}$ such that $m < n$ and $m + n$ is even ?
A. 55   B. 90   C. 100   D. 50   E. 45.

Problem 19. How many 5 digit numbers that are divisible by 3 can be formed using 0, 1, 2, 3, 4, 5 if no repetitions allowed?
A. 216   B. 120   C. 240   D. 126   E. 96.

Problem 20. A set of signal flags consists of one red flag, one blue flag, one yellow flag, and one green flag. Signals are sent by fying one or more flags on a vertical pole in a specified order. How many signals of this type are possible?
A. 64   B. 36   C. 48   D. 72   E. 52.

Problem 21. A swim team has eight members, only two of which are boys. The coach wants to take a delegation from the team to a special swimming camp. If the delegation must have either five or six members, and must include at least one boy, how many ways are there to select the delegation?
A. 55   B. 66   C. 77   D. 88   E. 99

Problem 22. 16 college students are going to the beach in 4 identical vans. Each van can hold exactly four students. How many ways can we distribute the students in the vans? (Two distributions are different if there are two students who ride together in one distribution but not in the other. If the same groups of students are together, it does not matter which van they ride in).

(A) $16!$ (B) $(4!)^3$ (C) $\binom{16}{4}^3$ (D) $\binom{16}{4}\binom{12}{4}\binom{8}{4}$ (E) None of the above

Problem 23. A pizza restaurant offers 8 different toppings and 4 different cheeses. A deluxe pizza contains your choice of 5 different toppings and your choice of 2 different cheeses. How many different deluxe pizzas are possible?
A. 10   B. 48   C. 336   D. 80640   E. $8^5 \cdot 4^2$.

Problem 24. A baseball coach is determined the batting order for the team. The team has 9 members, but the coach does not want pitcher to be one of the first four to bat. How many batting orders are possible?
A. 362,880   B. 40,320   C. 50,400   D. 320,000   E. 201,600

Algebra II Through Competitions   Chapter 18 Combinations and Permutations

SOLUTIONS

Problem 1. Solution: B.
From 200 to 300, there are 26 multiples of 4 and 11 multiples of 10. However, this overcounts 6 numbers that are multiples of both 4 and 10. Thus, there are 26 + 11 − 6 = 31 numbers from 200 to 300 that are divisible by 4 or 10.

Problem 2. Solution: E.
$\binom{10}{2}\binom{8}{5}\binom{3}{3} = 45 \cdot 56 \cdot 1 = 2520$.

Problem 3. Solution:
Case 1: two 1's are separated:

There are 5 positions and we have two 1's, so we have $\binom{5}{2} = 10$ such 6-digit numbers.

Case 2: two 1's are next to each other:

There are 5 positions so we have $\binom{5}{1} = 5$ such 6-digit numbers. 10 + 5 = 15.

Problem 4. Solution: C.
Let the number of elements in S and T be m and n respectively. We know that the number of subsets of a set with m elements is 2^m, so we have $m - n = 2$ and $2^m - 2^n = 96$.
Let $m = n + 2$ so that $2^m - 2^n = 2^{n+2} - 2^n = 2^n(2^2 - 1) = 96 \Rightarrow 2^n \cdot 3 = 96$
$\Rightarrow 2^n = 32 \Rightarrow n = 5 \Rightarrow m = 7$

Problem 5. Solution: C.
Since $2310 = 2 \cdot 3 \cdot 5 \cdot 7 \cdot 11$, there are five factors to choose from. To get only three numbers whose product is 2310, we would need combinations for either 3-1-1 of these factors or 2-2-1. $\binom{5}{3}\binom{1}{1}\binom{1}{1} = 10 \cdot 1 \cdot 1 = 10$ and $\binom{5}{3}\binom{3}{2}\binom{1}{1} = 10 \cdot 3 \cdot 1 = 30$. The answer: 40.

Problem 6. Solution: D
The least integer is 100 and the greatest is 1225 and their difference is 1125.

Problem 7. Solution: D.
There are 9 ordered triples, (0, 0, 2), (0, 2, 1), (5, 1, 1), (10, 0, 1), (0, 4, 0), (5, 3, 0), (10, 2, 0), (15, 1, 0) and (20, 0, 0).

Problem 8. Solution: C.
The 11 such integers with no partners in the set are 5, 7, 11, 13, 14, 15, 17, 19, 21, 22, 23.

Problem 9. Solution: A.
Let r, g, y, b denote the number of red, green, yellow and blue marbles respectively. Then the equations translate into $g + b + y = 25$, $r + g + y = 25$, $r + b + g = 25$ and $r + b + y = 36$. Adding all four equations together yields $3(r + g + y + b) = 111$, which means that $r + g + y + b = 37$. Subtract the fourth equation from this one to get $g = 1$.

Problem 10. Solution: E.
Any two-element subset of {1, 4, 9, 16} satisfies the condition. There are just two other sets, {2, 8} and {3, 12}. A four element set has $\binom{4}{2} = 6$ two-element subsets, so there are $6 + 2 = 8$ (unordered) pairs whose product is a perfect square.

Problem 11. Solution: C.
In order that a line be 2 units from A, it must be tangent to the circle of radius 2 about the point A and similarly for B. The circles of radius 2 about A and radius 3 about B intersect, so there are just two lines tangent to both circles.

Problem 12. Solution: E.
The question is equivalent to 'how many three digit numbers abc (0 allowed as the hundreds digit) satisfy $a + b + c = 6$. By (3.6), we have $\binom{6+3-1}{3-1} = \binom{8}{2} = 28$.

Problem 13. Solution: D.
There are two patterns 1 - 1 - 3, where we use one digit three times, and 2 - 2 - 1, where we use two digits twice. There are $\binom{3}{1}\binom{5}{2} \cdot 2 = 60$ in the first group, and $\binom{3}{1} \cdot 5 \cdot \binom{4}{2} = 90$ in the second group for a total of $60 + 90 = 150$.

Problem 14. Solution: E.
By the Fundamental Counting Principle, we get $2 \times 5 \times 3 \times 3 \times 5 \times 2 = 900$ paths.

Problem 15. Solution: A.
We can get the amount from 1¢ to $1 + 1 + 1 + 1 + 5 + 5 + 10 = 24$¢.

Problem 16. Solution: E
We have 4 cases: one head, HTTT: $\frac{4!}{3!1!} = 4$; two heads, HHTT: $\frac{4!}{2!2!} = 6$; three heads, HTTT: $\frac{4!}{3!1!} = 4$; and 4 heads, HHHH: one way. Total: $4 + 6 + 4 + 1 = 15$ ways.
Method 2: The total number of ways is 2^4. There is one case that has no head. So $2^4 - 1 = 15$.

Problem 17. Solution: C.
We select 2 men from 6 men and 2 women from 3 women: $\binom{6}{2} \times \binom{3}{2} = 15 \times 3 = 45$.

Problem 18. Solution: B.
We divide the numbers to two groups: {1, 3, 5, 7, 9, 11, 13, 15, 17, 19}, and {2, 4, 6, 8, 10, 12, 14, 16, 18, 20}. If we select two numbers from any of the group, we will get an even sum. So we have $\binom{10}{2} + \binom{10}{2} = 90$ such pairs.

Problem 19. Solution: A.
If the sum of the digits of a number is divisible by 3, the number is divisible by 3. We have two cases: 54321 and 54210. There are $5! = 120$ permutations for 54321 and $120 - 4! = 120 - 24 = 96$ permutations for 54210. The answer is $120 + 96 = 216$.

Problem 28. Solution: A.
We can put one flag, two flags, three flags, and 4 flags as a set. Considering orders (red-blue and blue-red are different), the number of signals is
$\binom{4}{1} + \binom{4}{2} \times 2! + \binom{4}{3} \times 3! + \binom{4}{4} \times 4! = 4 + 12 + 24 + 24 = 64$.

Problem 21. Solution: C.

If the delegation must have five members with one boy or two boys:
$$\binom{2}{1}\binom{6}{4}+\binom{2}{2}\binom{6}{3}=30+20=50.$$
If the delegation must have six members with one boy or two boys:
$$\binom{2}{1}\binom{6}{5}+\binom{2}{2}\binom{6}{4}=12+15=27.$$ The answer is 50 + 27 = 77.

Problem 22. Solution: D.

We first put 4 students out of 16 to one van and we have $\binom{16}{4}$ ways to do so. We then put 4 students out of 16 - 4 = 12 to one van and we have $\binom{12}{4}$ ways to do so. Third, we put 4 students out of 8 to one van and we have $\binom{8}{4}$ ways to do so. Last we put 4 students left to one van left and we have $\binom{4}{4}$ ways to do so.

Total we have $\binom{16}{4}\binom{12}{4}\binom{8}{4}\binom{4}{4}=\binom{16}{4}\binom{12}{4}\binom{8}{4}\times 1=\binom{16}{4}\binom{12}{4}\binom{8}{4}$ ways.

Problem 23. Solution: C.

You can choose 5 different toppings out of 8 and 2 different cheeses out of 4. The number of different deluxe pizzas is $\binom{8}{5}\binom{4}{2}=56\times 6=336$.

Problem 24. Solution: E.
We select four members from 8 members (excluding the pitcher) to the first four to bat. Note that these four members can change the order. So we have $\binom{8}{4}4!=1680$ ways to do so. We then arrange the rest of members. We have 5! ways. The answer is then $1680\times 5!=1680\times 120=201,600$.

Algebra II Through Competitions — Chapter 19 Probability

1. PROPERTIES OF PROBABILITY:

(1). The probability of an event A is between 0 and 1.
(2). The probability of an impossible event is 0.
(3). The probability of a certain event is 1.
(4). The probability that an event A will occur $P(A)$ is equal to one minus the probability that it will not occur $P(\overline{A})$.

2. BASIC PROBABILITY FORMULA

$$\text{Probability} = \frac{\text{number of ways that a certain outcome can occur}}{\text{total number of possible outcomes}}$$

Example 1: Suppose you rolled two standard dice and the total number of pips on the top faces was 6. What is the probability that one of the dice had 2 pips on its top face?
A) 1/6   B) 1/36   C) 1/5   D) 2/5   E) 1/3.

Solution: D.
If we look at all of the possible ways to get a 6, we have (1,5), (2,4), (3,3), (4,2), and (5,1). Two of these five have a 2. Thus the probability is $P = 2/5$.

Example 2: A fair die is rolled three times. The probability that you get a larger number each time is:
A. 17/216   B. 9/216   C. 7/17   D. 9/26   E. none of these

Solution: E.
When a die is rolled 3 times, there are a total of $6 \times 6 \times 6 = 216$ possible outcomes. If a 1 is rolled first, then there are 10 possible outcomes for the next two digits such that the third digit is greater than the second, which is greater than the first. If a 2 is rolled first, then there are 6 possible outcomes for the next two digits where the digits increase. If a 3 is rolled first, there are 3 ways for the digits to increase, and if a 4 is rolled first, there is 1 way for digits to increase. So there are a total of $10 + 6 + 3 + 1 = 20$ favorable outcomes. Thus the probability is $P = 20/216 = 5/54$. The answer is E.

Example 3: A class with 7 women and 5 men has 4 students chosen at random. If all have an equal chance of being picked, what is the chance that the group will have 2 men and 2 women?
A. 1/3   B. 14/33   C. 7/33   D. 7/99   E. 1225/3456

Solution: B.
Method 1:
Total number of possible outcomes (select 4 students from 7 + 5 = 12 students) is $\binom{12}{4} = 495$.

The number of favorable ways (select 2 women from 7 women and select 2 men from 5 men): $\binom{7}{2} \times \binom{5}{2} = 210$.

Probability = $\dfrac{\text{number of ways that a certain outcome can occur}}{\text{total number of possible outcomes}}$ = 210/495 = 14/33.

Method 2:
We select one man first and our chance is $\dfrac{5}{12}$. We select second man and our chance is $\dfrac{4}{11}$. Then we select one woman and our chance is $\dfrac{7}{10}$. Last we select second woman and our chance is $\dfrac{6}{9}$. Then we have the probability: $\dfrac{5}{12} \times \dfrac{4}{11} \times \dfrac{7}{10} \times \dfrac{6}{9}$.

We note that the order that the two men and two women are selected can be changed. There are $\dfrac{4!}{2! \times 2!} = 6$ ways of ordering them. So the answer is $6 \times \dfrac{5}{12} \times \dfrac{4}{11} \times \dfrac{7}{10} \times \dfrac{6}{9} = \dfrac{14}{33}$.

3. BASIC GEOMETRIC PROBABILITY FORMULA

"Geometric probability" is exactly the same as basic probability, except that we are dealing with the geometric figures instead of the "numbers".

The basic probability formula becomes:

$$P = \dfrac{\text{measure of geometric figure representing desired outcomes in the event}}{\text{measure of geometric figure representing all outcomes in the same space}}$$

Some examples of **geometric measures** are lengths, areas, angle measures, and volumes. For example, a probability determined by comparing the area of a given section to that of a total available region, is:

$$P = \frac{\text{measure of area of favorable region}}{\text{measure of area of total region}}$$

Example 4: A point E is chosen at random from within square $ABCD$. Express as a decimal to the nearest hundredth the probability that $\triangle ABE$ is obtuse.

Solution: 0.39.
If E is inside the semicircle that has AB as its diameter, then $\triangle ABE$ will be obtuse.

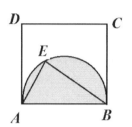

$$P = \frac{\text{Area of the semicircle}}{\text{Area of square}} = \frac{\frac{1}{2}\pi r^2}{(2r)^2} = \frac{\pi}{8} = 0.39$$

Example 5: (2006 NC Math Contest) Five concentric circles with radii 1, 2, 3, 4 and 5 are drawn on a flat surface. The circles with radii 1 through 4 divide the circle of radius 5 into five regions: a circle of radius 1 and 4 annuli (rings). The first annulus has inner radius 1 and outer radius 2, the second has inner radius 2 and outer radius 3, and so on. A point is chosen at random in the circle of radius 5. What is the probability that the point lies in the annulus whose inner radius is 3 and whose outer radius is 4?
A. 3/25   B. 1/5   C. 7/25   D. 9/25   E. None of A through D is correct.

Solution: C.
The total area is $\pi(5)^2$. The area of the favorable region is $\pi(4^2 - 3^2)$
The probability is $\frac{\pi(4^2 - 3^2)}{\pi 5^2} = \frac{7}{25}$.

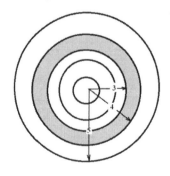

4. GENERAL ADDITION RULE OF PROBABILITY:

(1). If A and B are any two events: $P(A \text{ or } B) = P(A \cup B) = P(A) + P(B) - P(A \cap B)$

(2). If A and B are mutually exclusive ($P(A\cap B)=0$):

$$P(A \text{ or } B) = P(A\cup B) = P(A) + P(B)$$

Note: Any two events that cannot both occur at the same time are called mutually exclusive.

(3). General addition rule of probability (three events):

$$P(A\cup B\cup C) = P(A) + P(B) + P(C) - P(A\cap B) - P(B\cap C) - P(C\cap A) + P(A\cap B\cap C)$$

Example 6: A box contains 5 red marbles and 7 blue marbles. If you draw two marbles at random from the box, what is the probability that you have drawn two marbles of the same color?

A. $\dfrac{31}{66}$ B. $\dfrac{5}{33}$ C. $\dfrac{7}{22}$ D. $\dfrac{5}{11}$ E. none of these

Solution: A.

The probability to select two red marbles is $P(A) = \dfrac{\binom{5}{2}}{\binom{12}{2}} = \dfrac{10}{66}$

The probability to select two blue marbles is $P(B) = \dfrac{\binom{7}{2}}{\binom{12}{2}} = \dfrac{21}{66}$.

The probability that you have drawn two marbles of the same color is

$$P(A \text{ or } B) = P(A\cup B) = P(A) + P(B) = \dfrac{10}{66} + \dfrac{21}{66} = \dfrac{31}{66}.$$

Example 7: Alex and Bob take a two-round math test. There are 10 problems written on 10 cards, one problem per card. Alex can answer correctly 8 of the problems and Bob can answer correctly 6 of them. Each person needs to answer correctly at least two of the three problems randomly selected from these 10 problems in order to advance to the next round. What is the probability that at least one of them will advance to the next round?

Solution:

Method 1:

Let the event that Alex will advance to the next round be *A*. Let the event that Bob will advance to the next round be *B*.

$$P(A) = \frac{\binom{8}{2} \times \binom{2}{1} + \binom{8}{3}}{\binom{10}{3}} = \frac{14}{15}$$

$$P(B) = \frac{\binom{6}{2} \times \binom{4}{1} + \binom{6}{3}}{\binom{10}{3}} = \frac{2}{3}.$$

The probability that at least one of them will advance to the next round

$$P(A \cup B) = P(A) + P(B) - P(A \cap B) = \frac{2}{3} + \frac{14}{15} - \frac{2}{3} \times \frac{14}{15} = \frac{44}{45}.$$

Method 2:

Let the event that Alex will not advance to the next round be *C*. Let the event that Bob will not advance to the next round be *D*.

$$P(C) = \frac{\binom{2}{2}\binom{8}{1}}{\binom{10}{3}} = \frac{1}{15} \text{ and } P(D) = \frac{\binom{4}{2}\binom{6}{1} + \binom{4}{3}}{\binom{10}{3}} = \frac{1}{3}.$$

We know that event *A* and even *B* are independent.

The probability that no one will advance to the next round is

$$P(C \cap D) = P(C) \times P(D) = \frac{1}{15} \times \frac{1}{3}$$

The probability that at least one of them will advance to the next round $1 - P(\overline{A} \cdot \overline{B}) = 1 -$
$$1 - P(C \cap D) = P(C) \times P(D) = 1 - \frac{1}{45} = \frac{44}{45}.$$

5. GENERAL MULTIPLICATION RULE OF PROBABILITY:

(1). If *A* and *B* are any two events:

$P(A \text{ and } B) = P(A \cap B) = P(A) \times P(B|A)$, where $P(B|A)$ is the probability of B happening under the condition of event A.

(2). If A and B are two independent events:
If the outcome of event A does not affect the outcome of event B, A and B are called independent events. $P(B|A) = P(B)$
$P(A \text{ and } B) = P(A \cap B) = P(A) \times P(B)$

Example 8: A coin is biased so that the probability of obtaining a head is 0.25. Another coin is biased where the probability of obtaining a head is 0.6. If both coins are tossed, find the probability of obtaining at least one head.
A. 11/20   B. 7/10   C. 3/10   D. 9/10   E. none of the above

Solution: B.
The only case not to be considered is if both are tails, and this occurs with probability of
$P(T_1 \text{ and } T_2) = P(T_1) \times P(T_2) = (1 - 0.25) \times (1 - 0.6) = 0.75 \times 0.4 = 0.3$.
Thus the probability of obtaining at least one head is $1 - 0.3 = 0.7 = 7/10$.

Example 9: Five men and five women go into a movie theater and occupy a row of ten seats. If they choose their seats randomly, the probability that all of the men are sitting together and all of the women are sitting together is:
A) 1/126 B) 2/191 C) 1/191 D) 5/126 E) 1/2

Solution: A.
Method 1: Let the first person sit down. The 4 remaining people of the same gender must be next to choose their seats to ensure that they sit together. The second person has 4 seats to choose from out of 9 remaining seats, while the third person has 3 seats to choose from 8 remaining seats, etc. When the first 5 people are seated, the remaining 5 people, who are of the opposite gender, are forced to sit together. So the probability that the first 5 people sit together is $1 \cdot \frac{4}{9} \cdot \frac{3}{8} \cdot \frac{2}{7} \cdot \frac{1}{6} = \frac{1}{126}$.

Method 2:
Since we have 10 people, the total number of ways of sitting is 10!.
We tie 5 men as one unit and 5 women as one unit. We have 2! ways to seat them.
In each unit, we have 5! Ways to order them. So the number of favorable ways is 2! × 5! × 5!.

The probability is $\frac{2 \times 5! \times 5!}{10!} = \frac{1}{126}$.

Example 10: Suppose you toss two fair six-sided dice. Which probability is the greatest?
A. P(one die is even and the other is odd)
B. P(at least one die is prime and the sum is odd).
C. P(the sum is even and the product is a multiple of 5).
D. P("doubles" and the product is odd)
E. P(the sum is greater than 9 and one die is less than 4).

Solution: A.
One must compute each probability:

A. $p(\text{even \& odd}) = \frac{3 \cdot 3 + 3 \cdot 3}{36} = \frac{1}{2}$

B. The success cases are (2, 1), (2, 3), (2, 5), (3, 4), (3, 6), (5, 4), (5, 6), and (1, 2), (3, 2), (5, 2), (4, 3), (6, 3), (4, 5), (6, 5). So $p = \frac{14}{36} = \frac{7}{18}$.

C. The success pairs are (1, 5), (3, 5), (5, 5), (5, 1), and (5, 3), so $p = \frac{5}{36}$.

D. Since only (1, 1) (3, 3) and (5, 5) are success cases, $p = \frac{3}{36} = \frac{1}{12}$.

E. $p = 0$.

6. TOTAL PROBABILITY

Law of total probability (Marginal Probability)
Let $A_1, A_2, A_3, \ldots,$ and A_n be mutually exclusive and exhaustive events. Then for any other event B,

$P(B) = P(A_1 B_1) + P(A_2 B_2) + P(A_3 B_3) + \ldots + P(A_n B_n)$
$= P(A_1)P(B_1|A_1) + P(A_2)P(B_2|A_2) + P(A_3)P(B_3|A_3) + \ldots + P(A_n)P(B_n|A_n)$

Note: Mutually exclusive means that $P(A_i B_j) = 0$
Exhaustive rule: $P(A_1 \text{ or } A_2 \text{ or } A_3 \text{ or} \ldots A_n) = 1$

Then since $A_1, A_2, A_3, \ldots,$ and A_n exhaustive, if B occurs, it must be in conjunction with exactly one of A_i's. That is $B = (A_1$ and $B_1)$ or $(A_2$ and $B_2)$ or $\ldots (A_n$ and $B_n)$.

Example 11: Two bags of marbles are pictured below. Five marbles are randomly selected from Bag A and placed into Bag B. One marble is then randomly selected from Bag B. What is the probability that the marble selected from Bag B is black? Express your answer as a common fraction.

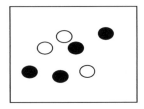
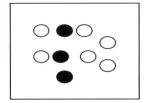

Bag A   Bag B

Solution: $\dfrac{41}{98}$.

Let B be the event that the marble selected from Bag B is black
A_1 be the event selecting 4 black marbles and 1 white marble from Bag A
B_1 be the event that the marble selected from Bag B is black after A_1

A_2 be the event selecting 3 black marbles and 2 white marble from Bag A
B_2 be the event that the marble selected from Bag B is black after A_2

A_3 be the event selecting 2 black marbles and 3 white marble from Bag A
B_3 be the event that the marble selected from Bag B is black after A_3
$B = A_1B_1 + A_2B_2 + A_3B_3$. Then

$$P(B) = P(A_1B_1) + P(A_2B_2) + P(A_3B_3) = P(A_1)P(B_1|A_1) + P(A_2)P(B_2|A_2) + P(A_3)P(B_3|A_3)$$

$$P(A_1) = \dfrac{\binom{3}{1}\binom{4}{4}}{\binom{7}{5}} = \dfrac{1}{7}. \quad P(A_2) = \dfrac{\binom{4}{3}\binom{3}{2}}{\binom{7}{5}} = \dfrac{4}{7} \quad P(A_3) = \dfrac{\binom{4}{2}\binom{2}{2}}{\binom{7}{5}} = \dfrac{2}{7}$$

$$P(B_1|A_1) = \dfrac{3+4}{14} = \dfrac{1}{2} \quad P(B_2|A_2) = \dfrac{3+3}{14} = \dfrac{3}{7} \quad P(B_3|A_3) = \dfrac{3+2}{14} = \dfrac{5}{14}$$

$$P(B) = \frac{1}{7} \times \frac{1}{2} + \frac{4}{7} \times \frac{3}{7} + \frac{2}{7} \times \frac{5}{14} = \frac{41}{98}.$$

Example 12: The numbers $1 \le a, b, c, d, e \le 2008$ are randomly chosen integers (repetition is allowed). What is the probability that $abc + de$ is even?
(A) 1/2 (B) 1/4 (C) 11/16 (D) 7/16 (E) 21/32

Solution: C.
Each of a, b, c, d, e is even or odd with probability 1/2. The expression $abc+de$ is even if either both of abc and de are odd, or both are even.
The probability that abc is odd is 1/8, the probability that de is odd is 1/4.
The probability that abc is even is 7/8, the probability that de is even is 3/4.
So the probability that $abc+de$ is even is $1/8 \times 1/4 + 7/8 \times 3/4 = 22/32 = 11/16$.

7. BINOMIAL PROBABILITY FORMULA:

$$P(r) = \binom{n}{r} p^r q^{n-r},$$

where P is the probability of exactly r successes. p is the probability of success in one trial and q is the probability of failure ($p + q = 1$). n is the number of independent trials. $0 \le r \le n$.

The negative binomial probability formula gives the probability of the total number of trials (n) required to obtain k successes, or in other word, the number of failures before the kth success occurs.

$$P = \binom{n-1}{k-1} p^k (1-p)^{n-k}$$

Example 13: An experiment consists of choosing with replacement an integer at random among the numbers from 1 to 9 inclusive. If we let M denote a number that is an integral multiple of 3 and N denote a number that is not an integral multiple of 3, which of the following sequences of results is least likely?
A. $M N N M N$ B. $N M M N$ C. $N M M N N$ D. $N N M N$ E. $M N M M$

Solution: C.

Since there are 3 multiples of 3 among the digits 1 to 9, the probability that a digit is of type M is 1/3 while the probability that it is of type N is 2/3.

The probability of each of the sequences occurring is

$$P(MNNMN) = \left(\frac{1}{3}\right)^2 \left(\frac{2}{3}\right)^3 = \frac{8}{243},$$

$$P(NMMN) = \left(\frac{1}{3}\right)^2 \left(\frac{2}{3}\right)^2 = \frac{4}{81},$$

$$P(NMMNM) = \left(\frac{1}{3}\right)^3 \left(\frac{2}{3}\right)^2 = \frac{4}{243},$$

$$P(NNMN) = \left(\frac{1}{3}\right) \left(\frac{2}{3}\right)^3 = \frac{8}{81},$$

$$P(MNMM) = \left(\frac{1}{3}\right)^3 \left(\frac{2}{3}\right) = \frac{2}{81}.$$

Algebra II Through Competitions   Chapter 19 Probability

PROBLEMS

Problem 1.   We randomly select 4 prime numbers without replacement from the first 10 prime numbers. What is the probability that the sum of the four selected numbers is odd?
A. 0.21   B. 0.30   C. 0.36   D. 0.40   E. 0.50

Problem 2.   A bag contains a number of marbles of which 80 are red, 24 are white, and the rest are blue. If the probability of randomly selecting a blue marble from this bag is 1 5, how many blue marbles are there in the bag?
A. 25   B. 26   C. 27   D. 28   E. 29

Problem 3.   Three standard dice are rolled, and someone tells you that the total of the top faces is 13. What is the probability that any one die has the number five on the top face?
A. 1/2   B. 4/7   C. 11/21   D. 91/216   E. 3/216.

Problem 4.   An integer is randomly selected from the set {100, 101, . . . , 999}. What is the probability that the sum of its digits is the same as the product of its digits?
A. 0   B. 1/900   C. 1/300   D. 1/150   E. 1/100

Problem 5. Fifteen numbers are picked from the set {1, 2, 3, . . . 20, 21}. Find the probability that at least three of those numbers are consecutive.
A. 0.1   B. 0.2   C. 0.4   D. 0.5   E. 1.0

Problem 6.   A random number generator on a computer selects three integers from 1 to 20. What is the probability that all three numbers are less than or equal to 5 ?
A. 1/64   B. 1/48   C. 1/36   D. 1/16   E. 1/4.

Problem 7.   What is the probability of obtaining an ace on both the first and second draws from an ordinary deck of 52 playing cards when the first card is not replaced before the second is drawn? There are four aces in such a deck.
A. 1/221   B. 4/221   C. 1/13   D. 1/17   E. 30/221.

Problem 8.   Three standard dice are rolled and the total of the top faces is 10. What is the probability that at least one die has the number "2" on its top face?
A. 4/9   B. 5/12   C. 12/23   D. 91/216   E. none of these

Problem 9.   A bag contains exactly 3 marbles –1 red, 1 white, and 1 blue. A girl draws a marble at random, replaces the marble, and continues to draw in this fashion. Find the probability that after 6 draws she has drawn exactly 2 marbles of each color. Express your answer as a common fraction reduced to lowest terms.

Problem 10. If three fair, standard cubical dice are thrown, find the probability that the sum of the numbers showing on the three uppermost faces is 8. Express your answer as a common fraction reduced to lowest terms.

Problem 11. If five cards are drawn at random without replacement from a standard deck of 52 cards, what is the probability that exactly two aces will be drawn?
A. 2,162/541,450   B. 2,162/54,145   C. 23,782/541,450   D. 10/221
E. none of the above.

Problem 12. Assume you start with $16. A fair coin is flipped, if it comes up heads $8 are added to that amount, otherwise half the money is lost. Find the probability that you will have less than $16 after the fourth flip.
A. 3/8   B. 7/16   C. 1/2   D. 9/16   E. 5/8.

Problem 13. An 8-ft-by-8-ft area has been tiled with 1-ft-by1-ft tiles. Two of the tiles were defective. What is the probability that the two defective tiles share an edge?
A. 1/9.   B. 1/16.   C. 1/18   D. 1/24   E. 1/32

Problem 14. Three fair 6-sided dice, each labeled 1, 2, $\cdots$, 6 are tossed. One is colored red, one is colored blue and one is colored yellow. What is the probability that the numbers at the top of the dice satisfy the inequality: red<blue<yellow?
A. 1/6   B. 5/12   C. 7/25   D. 4/45   E. 5/54

Problem 15. What is the probability that a woman with three children has more than one girl given that she has at least one girl?
A. 1/3   B. 1/2   C. 3/5   D. 4/7   E. 5/8

Problem 16. A boy has blue, green, and yellow marbles in a bag. There are three times as many green marbles as yellow marbles, and four times as many blue marbles as green marbles. If the boy selects a marble from the bag at random, what is the probability of selecting a yellow marble?
A. 1/16   B. 1/14   C. 1/12   D. 1/10   E. 1/8

Problem 17. A drawer contains 64 balls. Each ball is one of 8 colors, and there are 8 balls of each color. If the balls in the drawer are thoroughly mixed and you randomly choose two of them, what is the possibility that these two balls will have the same color?
A. 1/7   B. 1/8   C. 7/64   D. 9/64   E. 1/9

Problem 18. Mike and Dave play a game in which each independently throws a dart at a target. Mike hits the target with probability 0.6, while Dave hits the target with probability 0.3. Mike wins the game if he hits the target and Dave misses. Dave wins the game if he hits the target and Mike misses. Otherwise the game is a tie. What is the probability that the game is a tie?
A. 0.45 B. 0.46 C. 0.47 D. 0.48 E. 0.49

Problem 19.) If 7 distinct fair 6-sided dice are rolled at the same time, what is the probability that the sum will be 10?
A. $\dfrac{7}{279936}$ B. $\dfrac{7}{23328}$ C. $\dfrac{1}{139968}$ D. $\dfrac{1}{11664}$ E. none of these

Problem 20. If the letters a, A, b, B, c, and C are arranged at random in a row, what is the probability that the lower case letters appear in increasing alphabetical order?
a. $\dfrac{1}{6}$ b. $\dfrac{1}{2}$ c. $\dfrac{1}{720}$ d. $\dfrac{1}{36}$ e. $\dfrac{1}{30}$

Problem 21. A coin that comes up heads with probability $p > 0$ and tails with probability $1-p > 0$ independently on each flip is flipped eight times. Suppose the probability of three heads and five tails is equal to 1/25 of the probability of five heads and three tails. Let $p = m/n$, where m and n are relatively prime positive integers. Find $m + n$.

Algebra II Through Competitions   **Chapter 19 Probability**

SOLUTIONS

Problem 1. Solution: D.
The event happens precisely when the number 2 is one of the primes selected. This occurs with probability $\dfrac{\binom{9}{3}}{\binom{10}{4}} = \dfrac{9 \cdot 8 \cdot 7}{3 \cdot 2 \cdot 1} \cdot \dfrac{4 \cdot 3 \cdot 2 \cdot 1}{10 \cdot 9 \cdot 8 \cdot 7} = 0.40$.

Problem 2. Solution: B.
Let x be the number of blue marbles.
$\dfrac{x}{x+80+24} = \dfrac{1}{5} \Rightarrow 5x = x + 104 \Rightarrow x = 26$.

Problem 3. Solution: B.
Given three standard die. They are rolled and someone tells you that the total of the top faces is 13. What is the probability that any one die has the number five on the top face?
Outcomes: {(6, 6, 1)...(1, 6, 6),(6, 5, 2)...(2, 5, 6), (6, 4, 3)...(3, 4 ,6), (5, 5, 3)...(3, 5, 5), (5, 4, 4)...(4, 4, 5)}.
Thus there are 21 possible different outcomes 12 have a five. Answer 4/7.

Problem 4. Solution: D.
There are just six such numbers, 123, 132, 213, 231, 312, 321, so the probability is 6/900 = 1/150.

Problem 5. Solution: E.
Imagine putting the 15 numbers into seven boxes labeled 123, 456, 789, etc. Each number is put into the box that it helps to label. After all 15 numbers have been distributed among the boxes, some box must have three balls, by the Pigeon-Hole Principle. Thus the probability that some box has three consecutive numbers is 1.

Problem 6. Solution: A.
$\dfrac{5}{20} \times \dfrac{5}{20} \times \dfrac{5}{20} = \dfrac{1}{4} \times \dfrac{1}{4} \times \dfrac{1}{4} = \dfrac{1}{64}$.

Problem 7. Solution: A.
The probability of obtaining an ace on the first draw is 4/52 = 1/13. If the first card drawn is an ace there are 3 aces remaining in the deck, which now consists of 51 cards. Thus,

the probability of getting an ace on the second draw is 3/51 = 1/17. The required probability is the product of the two, which is 1/221.

Problem 8. Solution: A.
We have the following cases where we obtain a sum of 10:
(6, 3, 1), (6, 2, 2), (5, 4, 1), (5, 3, 2), (4, 4, 2), (4, 3, 3).
With ordering, we have 6 + 3 + 6 + 6 + 4 + 3 = 27 total ways.
We have the following cases with at least one "2": (6, 2, 2), (5, 3, 2), (4, 4, 2).
With ordering, we have 3 + 6 + 3 = 12 favorable total ways.
The probability is $P = \dfrac{12}{27} = \dfrac{4}{9}$.

Problem 9. Solution: 10/81.
The total number of ways is $3 \times 3 \times 3 \times 3 \times 3 \times 3 = 3^6$.
The number of favorable ways is $\dfrac{6!}{2! \times 2! \times 2!} = 90$. The probability is $P = \dfrac{90}{3^6} = \dfrac{10}{81}$.

Problem 10. Solution: 7/72
The number of ways to get a sum of 8 is the positive solution to the equation:
$x_1 + x_2 + x_3 = 8$, Which is $\binom{n-1}{r-1} = \binom{8-1}{3-1} = \binom{7}{2} = \dfrac{7 \times 6}{2 \times 1} = 21$.
The probability is $P = \dfrac{21}{6^3} = \dfrac{7}{72}$.

Problem 11. Solution: B.
Method 1:
The probability that the first card is an ace is 4/52. After the first card is drawn and is an ace, the probability that the next card is also an ace is 3/51. Then the probability that the next 3 cards are not aces is (48/50)(47/49)(46/48).
There are 5!/(2!3!) = 10 combinations of drawing the 2 aces within the 5 card set. Thus the probability is 10 times greater, or $10(\dfrac{4}{52} \cdot \dfrac{3}{51} \cdot \dfrac{48}{50} \cdot \dfrac{47}{49} \cdot \dfrac{46}{48}) = 2162/54145$.

Method 2:
The total number of outcomes will be the number of ways to draw 5 cards from a standard deck of 52 cards: $\binom{52}{5}$.

The number of favorable outcomes is the number of ways to draw 2 aces from 4 aces and to draw 3 cards from the rest (52 - 4 = 48 cards: $\binom{4}{2} \times \binom{48}{3}$.

The probability is $\dfrac{\binom{4}{2} \times \binom{48}{3}}{\binom{52}{5}} = \dfrac{2162}{54145}$.

Problem 12. Solution: D.
The sequences of winning or losing are shown here:
WWWW: 24-32-40-48 WWWL:24-32-40-20 WWLW:24-32-16-24
WWLL:24-32-16-8 WLWW:24-12-20-28 WLWL:24-12-20-10
WLLW:24-12-6-14 WLLL:24-12-6-3 LWWW:8-16-24-32
LWWL:8-16-24-12 LWLW:8-16-8-16 LWLL:8-16-8-4
LLWW:8-4-12-20 LLWL:8-4-12-6 LLLW:8-4-2-10
LLLL:8-4-2-1
As you can see, 9 of the 16 possibilities result in a final amount less than 16.

Problem 13. Solution: C.
There are $\binom{8^2}{2} = \dfrac{8^2(8^2-1)}{2} = \dfrac{64 \times 63}{2}$ ways of choosing two arbitrary squares for the defective tiles.

If the two defective tiles share and edge, then two cases must be considered.
Case1: One of the tiles was placed in any of the top 7 rows (8 × 7 ways), and the other was placed in the square below.
Case 2: One tile was placed in any of the left 7 columns (8 × 7 ways), and the other was placed the square to its right. So the probability is $\dfrac{2 \times 8 \times 7}{\left(\dfrac{64 \times 63}{2}\right)} = \dfrac{1}{18}$.

Problem 14. Solution: E.
We have the following 9 cases:

Y	B	R		
6	5	4, 3, 2, 1	$P_1 = \dfrac{1}{6} \times \dfrac{1}{6} \times \dfrac{4}{6}$	$= \dfrac{4}{216}$
6	4	3, 2, 1	$P_2 = \dfrac{1}{6} \times \dfrac{1}{6} \times \dfrac{3}{6}$	$= \dfrac{3}{216}$

6	3	2, 1	$P_3 = \frac{1}{6} \times \frac{1}{6} \times \frac{2}{6} = \frac{2}{216}$
6	2	1	$P_4 = \frac{1}{6} \times \frac{1}{6} \times \frac{1}{6} = \frac{1}{216}$
5	4	3, 2, 1	$P_5 = \frac{1}{6} \times \frac{1}{6} \times \frac{3}{6} = \frac{3}{216}$
5	3	2, 1	$P_6 = \frac{1}{6} \times \frac{1}{6} \times \frac{2}{6} = \frac{2}{216}$
5	2	1	$P_7 = \frac{1}{6} \times \frac{1}{6} \times \frac{1}{6} = \frac{1}{216}$
4	3	2, 1	$P_8 = \frac{1}{6} \times \frac{1}{6} \times \frac{2}{6} = \frac{2}{216}$
4	2	1	$P_9 = \frac{1}{6} \times \frac{1}{6} \times \frac{1}{6} = \frac{1}{216}$
3	2	1	$P_9 = \frac{1}{6} \times \frac{1}{6} \times \frac{1}{6} = \frac{1}{216}$

The total probability is $P = P_1 + P_2 + \cdots P_9 = \frac{4+3+2+1+3+2+1+2+1+1}{216} = \frac{20}{216} = \frac{5}{54}$.

Problem 15. Solution: D.
The possible combinations are:

G G G
G G B, G B G, B G G
G B B, B G B, B B G.

The total number of combinations is 7 and we have 4 cases where she has more than one girl. The probability is 4/7.

Problem 16. Solution: A.
We have the following equations: $3y = g$ and $4g = b$.

The probability is $P = \frac{y}{y+g+b} = \frac{\frac{g}{3}}{\frac{g}{3}+g+4g} = \frac{g}{16g} = \frac{1}{16}$

Problem 17. Solution: E.

First we choose one color, say, red. We have $\binom{8}{1} = 8$ ways to do so.

Then we select two marbles of the same color.

The probability to is $P(A) = 8 \times \dfrac{\binom{8}{2}}{\binom{64}{2}} = \dfrac{8 \times 4 \times 7}{32 \times 63} = \dfrac{1}{9}$

Problem 18. Solution: B.
We have two cases of a tie:
Mike hits the target and Dave also hits the target: $P_1 = 0.6 \times 0.3 = 0.18$
Mike misses the target and Dave also misses the target: $P_2 = 0.4 \times 0.7 = 0.28$
The answer is $P = P_1 + P_2 = 0.18 + 0.28 = 0.46$.

Problem 19. Solution: B.
Method 1 (official solution):
{1,1,1,1,1,1,4} can occur 7 ways; {1,1,1,1,1,2,3} can occur $7 \times 6 = 42$ ways
{1,1,1,1,2,2,2} can occur $\binom{7}{3} = 35$ ways.
So the total probability is $\dfrac{7 + 42 + 35}{6^7}$.

Method 2:
The number of ways to get a sum of 10 is the positive solution to the equation:
$x_1 + x_2 + x_3 + x_4 + x_5 + x_6 + x_7 = 10$, which is $\binom{n-1}{r-1} = \binom{10-1}{7-1} = \binom{9}{6} = \binom{9}{3} = \dfrac{9 \times 8 \times 7}{3 \times 2 \times 1} = 84$

The probability is $P = \dfrac{84}{6^7} = \dfrac{7}{23328}$.

Problem 20. Solution: A.
Method 1:
There are 6 possible arrangements of a, b, c and only one of them is in alphabetical order.

Method 2:
We place a, b, and c in alphabetical order as shown below.
  a b c

Then we put one of the three letters A, B, or C. We have 4 ways to put any one letter, say, letter A.

$$\begin{array}{cccc} a & b & c & \\ \uparrow & \uparrow & \uparrow & \uparrow \end{array}$$

After we put A, we have 5 ways to put another letter, say, B.

We see we have 6 ways to put letter C.

$$\begin{array}{cccccc} a & b & c & A & B & \\ \uparrow & \uparrow & \uparrow & \uparrow & \uparrow & \uparrow \end{array}$$

The number of ways to arrange these six letter in a row is 6!.

The desired probability is $P = \dfrac{4 \times 5 \times 6}{6!} = \dfrac{1}{6}$.

Method 3:

We place a, b, and c in alphabetical order as shown below.

$$\begin{array}{ccc} a & b & c \end{array}$$

Then we have 4 places to put other 3 letters.

$$\begin{array}{cccc} a & b & c & \\ \uparrow & \uparrow & \uparrow & \uparrow \end{array}$$

Case I: Three letters *ABC* are tied together. We have $\binom{4}{1} = 4$ ways to put them. After they are in place, we have 3! ways to order them. $4 \times 6 = 24$.

Case II: Two letters are tied together and one letter is alone. We have 3 ways to do so. After this, we have $\binom{4}{2} = 6$ ways to put them. After they are in place, we have 4 ways to order them. $3 \times 6 \times 4 = 72$.

Case III: Each letter is on its own. We have $\binom{4}{3} = 4$ ways to put them. After they are in place, we have 3! = 6 ways to order them. $4 \times 6 = 24$.

The desired probability is $P = \dfrac{24 + 72 + 24}{6!} = \dfrac{1}{6}$.

Problem 21. Solution: 011.

The conditions of the problem imply that $\binom{8}{3} p^3 (1-p)^5 = \dfrac{1}{25} \binom{8}{3} p^5 (1-p)^3$, and hence $(1-p)^2 = \dfrac{1}{25} p^2$, so that $1 - p = \dfrac{1}{5} p$. Thus $p = \dfrac{5}{6}$, and $m + n = 11$.

Algebra II Through Competitions — Chapter 20 Trigonometry

1. BASIC CONCEPTS

1.1. Definition of six functions in a circle

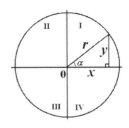

$$\sin\alpha = \frac{y}{r};\quad \cos\alpha = \frac{x}{r};\quad \tan\alpha = \frac{y}{x};\quad \cot\alpha = \frac{x}{y};$$

$$\sec\alpha = \frac{r}{x};\quad \csc\alpha = \frac{r}{y}.$$

1.2. Definition of six functions in a unit circle ($r = 1$)

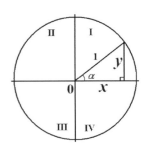

$$\sin\alpha = y;\quad \cos\alpha = x;\quad \tan\alpha = \frac{y}{x};\quad \cot\alpha = \frac{x}{y};$$

$$\sec\alpha = \frac{1}{x};\quad \csc\alpha = \frac{1}{y}.$$

1.3. Graphs of the sine and Cosine Functions

The properties of the graphs of $y = a\sin(bx - c)$ and $y = a\cos(bx - c)$:

amplitude $= |a|$.

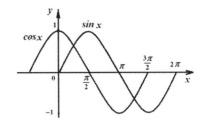

period $= 2\pi/b$

The **phase shift** and the resulting interval for one cycle are solutions to the equations $bx - c = 0$ and $bx - c = 2\pi$.

For $y = \sin x$, **amplitude** $= |1|$, **period** $= 2\pi$, and the **phase shift** is 0.

Example 1. If $\sin\alpha = 3\cos\alpha$ then what is $\sin\alpha\cos\alpha$?

(A) $\frac{1}{6}$ (B) $\frac{1}{5}$ (C) $\frac{2}{9}$ (D) $\frac{1}{4}$ (E) $\frac{3}{10}$

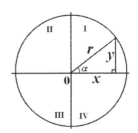

Solution: (E).
We know that in a unit circle $\sin\alpha = y$, and $\cos\alpha = x$.
We are given that $y = 3x$ \hfill (1)

292

We also know that $x^2 + y^2 = 1$ (2)

Substituting (1) into (2): $x^2 + 9x^2 = 1$ $\Rightarrow$ $x^2 = \dfrac{1}{10}$.

Therefore $\sin\alpha\cos\alpha = y \cdot x = 3x^2 = 3 \cdot \dfrac{1}{10} = \dfrac{3}{10}$.

Example 2. If the measure of one angle of a rhombus is $60°$, then the ratio of the length of its longer diagonal to the length of its shorter diagonal is

A. $2:1$ B. $\sqrt{3}:1$ C. $\sqrt{2}:1$ D. $\sqrt{3}:2$ E. $\sqrt{2}:2$

Solution: B.

The ratio is $\dfrac{\sqrt{3}a}{a} = \sqrt{3}:1$.

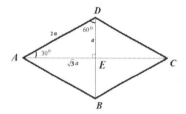

Example 3. Triangle ABC has a right angle at C. If $\sin A = \dfrac{2}{3}$, then $\tan B$ is

(A) $\dfrac{3}{5}$ (B) $\dfrac{\sqrt{5}}{3}$ (C) $\dfrac{2}{\sqrt{5}}$ (D) $\dfrac{\sqrt{5}}{2}$ (E) $\dfrac{5}{3}$

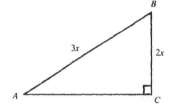

Solution: (D).

In the figure, $\sin A = \dfrac{BC}{AB} = \dfrac{2}{3}$. So for some $x > 0$, $BC = 2x$, $AB = 3x$ and $AC = \sqrt{(AB)^2 - (BC)^2} = \sqrt{5}x$. Thus $\tan B = \dfrac{AC}{BC} = \dfrac{\sqrt{5}}{2}$.

1.4. Signs of six functions

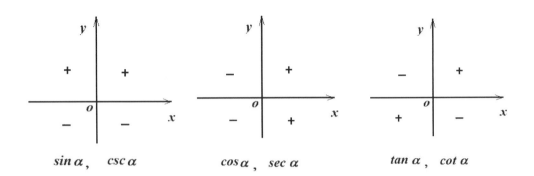

Example 4. If α is an obtuse angle, which one of the following is positive?
A. cos α − sin α   B. sin α cos α   C. tan α + cot α   D. sin α − cot α

Solution: D.
Since α is an obtuse angle, we know that sinα > 0, cosα < 0, tanα < 0, cotα < 0.
A, B, C are all negative. So the answer is D.

Example 5. For which value of θ listed below is it true that $0.5^{\sin\theta} < 1$ and $0.5^{\cos\theta} > 1$?
A. 5°   B. 55°   C. 125°   D. 205°   E. 285°

Solution: C.

$0.5^{\sin\theta} < 1 \quad \Rightarrow \quad (\frac{1}{2})^{\sin\theta} < 1 \quad \Rightarrow \quad \frac{1}{2^{\sin\theta}} < 1 \quad \Rightarrow \quad 2^{\sin\theta} > 1$

Take logs on both sides of $2^{\sin\theta} > 1$, $\sin\theta > 0$

$0.5^{\cos\theta} > 1 \quad \Rightarrow \quad (\frac{1}{2})^{\cos\theta} > 1 \quad \Rightarrow \quad \frac{1}{2^{\cos\theta}} > 1 \quad \Rightarrow \quad 2^{\cos\theta} < 1$

Take logs on both sides of $2^{\cos\theta} < 1$, $\cos\theta < 0$.

We see that in the region $\frac{\pi}{2} < \theta < \pi$, or
$90° < \theta < 180°$, both $\sin\theta > 0$ and $\cos\theta < 0$.
The answer is C.

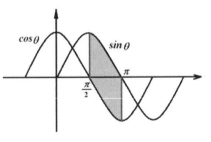

1.5. Cofunction

Sine and Cosine, Tangent and Cotangent, Secant and Cosecant, are cofunctions of each other.

1.6. Values of special angles

Angle	$0 = 0°$	$\pi/6 = 30°$	$\pi/4 = 45°$	$\pi/3 = 60°$	$\pi/2 = 90°$	$\pi = 180°$
sin	0	1/2	$\sqrt{2}/2$	$\sqrt{3}/2$	1	0
cos	1	$\sqrt{3}/2$	$\sqrt{2}/2$	1/2	0	-1
tan	0	$\sqrt{3}/3$	1	$\sqrt{3}$	∞	0
csc	∞	2	$\sqrt{2}$	$2\sqrt{3}/3$	1	∞
sec	1	$2\sqrt{3}/3$	$\sqrt{2}$	2	∞	-1
cot	∞	$\sqrt{3}$	1	$\sqrt{3}/3$	0	∞

1.7. Terminal side of an angle

In trigonometry, we can visualize an angle as being formed by rotating one of the sides about the vertex while keeping the other side fixed.
For example, $\angle AOB$ (α) in the figure is formed by rotating the side OB while OA is fixed.

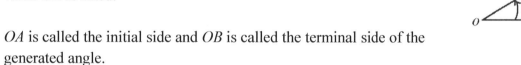

OA is called the initial side and OB is called the terminal side of the generated angle.

All angles having the same terminal side as angle α (including α) can be expressed as $360k + \alpha$ (k is any integer) or $2k\pi + \alpha$.

An angle is called a first quadrant angle if its terminal side is in the first quadrant.
An angle is called a second quadrant angle if its terminal side is in the second quadrant, and so on for the other two quadrants. Coterminal angles have equal functional values.

Example 6. If α is an angle in first quadrant, what are the quadrants of 2α and $\dfrac{1}{2}\alpha$ in?

Solution:
Since α is an angle in first quadrant, $2n\pi < \alpha < 2n\pi + \dfrac{\pi}{2}$. n is integer.

Therefore $4n\pi < 2\alpha < 4n\pi + \pi$; $n\pi < \dfrac{\alpha}{2} < n\pi + \dfrac{\pi}{4}$.

2α is in the first or second quadrant or the terminal side is on positive y-axis. $\frac{1}{2}\alpha$ is in first or third quadrant.

1.8. The reference angle

The reference angle is the acute angle made by the terminal side of the given angle and the x-axis.

Example 7. The reference angle for an angle $A = 3\pi/5$ (radians) is
A. π B. $\pi/5$ C. $2\pi/5$ D. $3\pi/5$
  E. $4\pi/5$

Solution: C.
The reference angle is the acute angle made by the terminal side of the given angle and the x-axis. $\alpha = \pi - \frac{3\pi}{5} = \frac{2\pi}{5}$.

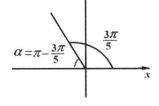

1.9. Reduction formulas

A trigonometric reduction formula simplifies an expression in terms of more easily calculated or readily available values. Any trigonometric function whose argument is $\pm\alpha + n \times 90°$ can be written simply in terms of α. For examples: $\sin(360° + \alpha) = \sin\alpha$; $\sin 390° = \sin(360° + 30°) = \sin 30° = \frac{1}{2}$.

If you know the following method, you do not need to remember all the reduction formulas.
$f(\pm\alpha + n \times 90°) = \pm g(\alpha)$, where
n may be any integer, positive, negative, or zero.
f is any one of the six trigonometric functions.
α may be any real angle measure.
If n is even, then g is the same function as f.
If n is odd, then g is the cofunction as f.
(Remember that sine and cosine, tangent and cotangent, secant and cosecant, are cofunctions of each other).

The second ± sign is determined by the sign of the reference angle of the original function.

Example 8. Reduce: sin (360° + α):

Solution:
(1) Determine the final function: 360 = 90 × 4. $n = 4$ which is even. The final function is the same as the original function. sin (360° + α) = ? sin α.

(2) Determine the sign: The reference angle is on the first quadrant. We know that sin α is positive in first quadrant, so the sign is "+". sin (360° + α) = + sin α.

Example 9. Reduce: cos (90° + α):

Solution:
(1) Determine the final function: 90 = 90 × 1, so $n = 1$ which is odd. The final function is the cofunction of the original function. cos (90° + α) = ? sin α.

(2) Determine the sign: The reference angle is on the second quadrant. We know that cos α is negative in second quadrant, so the sign is "−". cos (90° + α) = − sin α.

Example 10. Evaluate $\cos(\frac{\pi}{2} - x)\csc x$.

A. 0   B. 2   C. 3   D. $\frac{\pi}{2}$ E. 1

Solution: E.
$$\cos(\frac{\pi}{2} - x)\csc x = \sin x \cdot \csc x = \sin x \cdot \frac{1}{\sin x} = 1$$

2. TRIGONOMETRIC HEXAGON

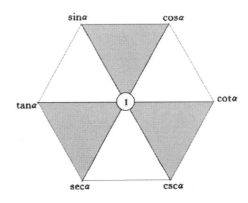

2.1. Reciprocal relations (Diagonals):

$\sin \alpha \csc \alpha = 1$; $\cos \alpha \sec \alpha = 1$; $\tan \alpha \cot \alpha = 1$.

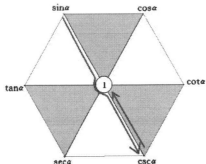

2.2. Vertices of the shaded equilateral triangles (Pythagorean relations)

$\sin^2 \alpha + \cos^2 \alpha = 1$; $\tan^2 \alpha + 1 = \sec^2 \alpha$; $1 + \cot^2 \alpha = \csc^2 \alpha$.

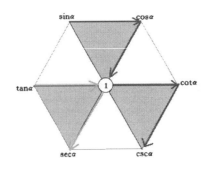

Example 11. Simplify $\dfrac{\cos x}{1-\sin x}+\dfrac{\cos x}{1+\sin x}$.

A. $\dfrac{2}{\cos x}$ B. $\sin x - \cos x$ C. $\sin x + \tan x$ D. $2\cos x$ E. $2\tan x$

Solution: A.
$$\dfrac{\cos x}{1-\sin x}+\dfrac{\cos x}{1+\sin x}=\dfrac{\cos x[(1+\sin x)+(1-\sin x)]}{(1-\sin x)(1+\sin x)}=\dfrac{2\cos x}{1-\sin^2 x}=\dfrac{2\cos x}{\cos^2 x}=\dfrac{2}{\cos x}.$$

Example 12. Express the rational number
$s=\sin^2(10°)+\sin^2(20°)+\sin^2(30°)+\cdots+\sin^2(80°)$ in lowest terms.
A. 1   B. 2   C. 3   D. 4   E. 5

Solution: D.
$s=\sin^2(10°)+\sin^2(20°)+\sin^2(30°)+\cdots+\sin^2(80°)$
$=[\sin^2(10°)+\sin^2(80°)]+[\sin^2(20°)+\sin^2(70°)]+\cdots+[\sin^2(40°)+\sin^2(50°)]$
$=[\sin^2(10°)+\cos^2(10°)]+[\sin^2(20°)+\cos^2(20°)]+\cdots+[\sin^2(40°)+\cos^2(40°)]=4$.

Example 13. If the system of equations
$y = 7\sin x + 3\cos x$
$y = 7\cos x + 3\sin x$
is solved simultaneously for $0 \le x \le \pi$, the value of y must be:
A. $4\sqrt{2}$ B. $2\sqrt{5}$ C. 2   D. $5\sqrt{2}$ E. -2

Solution: D.
By substitution, $7\cos x + 3\sin x = 7\sin x + 3\cos x \Rightarrow \sin x = \cos x \Rightarrow x = \pi/4$
$\Rightarrow y = 7\cos \pi/4 + 3\cos \pi/4 = 7\cdot\sqrt{2}/2 + 3\cdot\sqrt{2}/2 = 5\sqrt{2}$

2.3. Quotient relations (Vertices of the isosceles triangles):

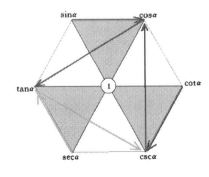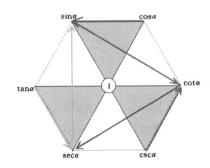

$$\tan\alpha = \frac{\sin\alpha}{\cos\alpha}; \quad \cot\alpha = \frac{\cos\alpha}{\sin\alpha}.$$

Example 14. Which of the following is equal to $\sec^5 x \cot x - \sec^3 x \cot x$?

A. $\dfrac{\cos x}{\sin^2 x}$ B. $\dfrac{\cos x}{\sin^4 x}$ C. $\dfrac{\sin x}{\cos^2 x}$ D. $\dfrac{\sin x}{\cos^4 x}$ E. None of these

Solution: D.
$\sec^5 x \cot x - \sec^3 x \cot x = \sec^3 x \cot x (\sec^2 x - 1) = \sec^3 x \cdot \cot x \cdot \tan^2 x = \sec^3 x \cdot \tan x$
$= \dfrac{1}{\cos^3 x} \cdot \dfrac{\sin x}{\cos x} = \dfrac{\sin x}{\cos^4 x}.$

Example 15. If $\sin x - \cos x = \dfrac{1}{2}$, then what is the value of $\sin^3 x - \cos^3 x$?

A. $\dfrac{1}{2}$ B. $\dfrac{3}{4}$ C. $\dfrac{9}{16}$ D. $\dfrac{5}{8}$ E. $\dfrac{11}{16}$.

Solution: E.
From $\sin x - \cos x = \dfrac{1}{2}$, we have by squaring both sides: $(\sin x - \cos x)^2 = \dfrac{1}{4}$ $\Rightarrow$
$\sin^2 x + \cos^2 x - 2\sin x \cdot \cos x = \dfrac{1}{4}$ $\Rightarrow$ $\sin x \cdot \cos x = \dfrac{3}{8}.$

$$\sin^3 x - \cos^3 x = (\sin x - \cos x)(\sin^2 x + \sin x \cdot \cos x + \cos^2 x)$$
$$= \frac{1}{2}(1 + \sin x \cdot \cos x) = \frac{1}{2}(1 + \frac{3}{8}) = \frac{11}{16}.$$

Example 16. Given $0 < x < \pi$, $\sin x + \cos x = \frac{1}{5}$, find $\tan x$.

A. $-\frac{4}{3}$ B. $-\frac{3}{4}$ C. $\frac{4}{3}$ D. $\frac{3}{4}$ E. $\frac{3}{5}$

Solution: A.
Method 1:
When $0 < x < \pi/2$, $\sin x > 0$, $\cos x > 0$.
$(\sin x + \cos x)^2 = \sin^2 x + \cos^2 x + 2\sin x \cos x = 1 + 2\sin x \cos x > 1 \Rightarrow \sin x + \cos x > 1$.
When $x = \pi$, $\sin x + \cos x = 1$.
So we conclude that x is the angle in quadrant II ($\pi/2 < x < \pi$) where $\sin x > 0$, $\cos x < 0$.
So we also know that $\sin x - \cos x > 0$.
$(\sin x - \cos x)^2 = \sin^2 x + \cos^2 x - 2\sin x \cos x = 2(\sin^2 x + \cos^2 x) - (\sin x + \cos x)^2$
$= 2 - (\frac{1}{5})^2 = \frac{49}{25}$. Therefore $\sin x - \cos x = \frac{7}{5}$.

$$\begin{cases} \sin\alpha - \cos\alpha = \frac{7}{5} \\ \sin\alpha + \cos\alpha = \frac{1}{5} \end{cases} \Rightarrow \begin{cases} \cos\alpha = -\frac{3}{5} \\ \sin\alpha = \frac{4}{5} \end{cases} \Rightarrow \tan x = \frac{\sin x}{\cos x} = -\frac{4}{3}.$$

Method 2:
Squaring both sides of equation $\sin x + \cos x = \frac{1}{5}$: $(\sin x + \cos x)^2 = (\frac{1}{5})^2 \Rightarrow$
$\sin^2 x + 2\sin x \cos x + \cos^2 x = \frac{1}{25} \Rightarrow 2\sin x \cos x = \sin 2x = -\frac{24}{25}$.

We know that $\sin 2\alpha = \frac{2\tan\alpha}{1 + \tan^2\alpha}$. So we have $\frac{2\tan x}{1 + \tan^2 x} = -\frac{24}{25}$ (1)

Since $1 + \tan^2 x > 0$, $\tan x < 0$. We know that $0 < x < \pi$, so we conclude that x is the angle in quadrant II ($\pi/2 < x < \pi$) where $\sin x > 0$, $\cos x < 0$.

(1) becomes $50\tan x = -24(1+\tan^2 x)$ $\Rightarrow \Rightarrow 25\tan x = -12(1+\tan^2 x)$

$\Rightarrow 12\tan^2 x + 25\tan x + 12 = 0$ $\Rightarrow (3\tan x + 4)(4\tan x + 3) = 0$.

So $3\tan x + 4 = 0$ $\Rightarrow \tan x = -\dfrac{4}{3}$. or $4\tan x + 3 = 0$ $\Rightarrow \tan x = -\dfrac{3}{4}$.

When $\tan x = -\dfrac{3}{4}$, $\sin x = \dfrac{3}{5}$, $\cos x = -\dfrac{4}{5}$, $\sin x + \cos x = -\dfrac{1}{5}$.

When $\tan x = -\dfrac{4}{3}$, $\sin x = \dfrac{4}{5}$, $\cos x = -\dfrac{3}{5}$, $\sin x + \cos x = \dfrac{1}{5}$.

So the answer is $\tan x = -\dfrac{4}{3}$.

3. TRIGONOMETRIC IDENTITIES

3.1. Angle – sum relationships (1)

$\sin(\alpha + \beta) = \sin\alpha\cos\beta + \cos\alpha\sin\beta$ \hfill (3.1)

Proof:

 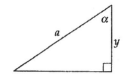 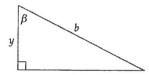

By definition, we have: $y = a\cos\alpha = b\cos\beta$.

By the area formula, we have:
$\dfrac{1}{2}ab\sin(\alpha + \beta) = \dfrac{1}{2}ay\sin\alpha + \dfrac{1}{2}by\sin\beta$.

Substituting the value of y into the above equation we get:
$\dfrac{1}{2}ab\sin(\alpha + \beta) = \dfrac{1}{2}ab\cos\beta\sin\alpha + \dfrac{1}{2}ba\cos\alpha\sin\beta$.

Since $ab \neq 0$, we can divide both sides by ab: $\sin(\alpha + \beta) = \sin\alpha\cos\beta + \cos\alpha\sin\beta$.

3.2. Angle – sum relationships (2)

Let $\alpha = \frac{\pi}{2} + \gamma$ in (1).

$\sin(\alpha + \beta) = \sin\alpha\cos\beta + \cos\alpha\sin\beta$ then becomes:

$\sin(\frac{\pi}{2} + \gamma + \beta) = \sin(\frac{\pi}{2} + \gamma)\cos\beta + \cos(\frac{\pi}{2} + \gamma)\sin\beta$. $\sin(\frac{\pi}{2} + \gamma + \beta) = \cos(\gamma + \beta)$ and

$\sin(\frac{\pi}{2} + \gamma)\cos\beta + \cos(\frac{\pi}{2} + \gamma)\sin\beta = \cos\gamma\cos\beta - \sin\gamma\sin\beta$.

Therefore, $\cos(\gamma + \beta) = \cos\gamma\cos\beta - \sin\gamma\sin\beta$ \hfill (3.2)

Example 17. $\sin(x + 30°) + \cos(x + 60°)$
A. $\sin(x) + \cos(x)$ B. $\sin(x) - \cos(x)$ C. $\sin(x)$ D. $\cos(x)$ E. none of these.

Solution: D.
$\sin(x + 30°) + \cos(x + 60°) = \sin x \cos 30° + \cos x \sin 30° + \cos x \cos 60° - \sin x \sin 60°$
$= \frac{\sqrt{3}}{2}\sin x + \frac{1}{2}\cos x + \frac{1}{2}\cos x - \frac{\sqrt{3}}{2}\sin x = \cos x$.

Example 18. Find the value of $\sin\frac{5\pi}{12}$.

Solution:

$\sin\frac{5\pi}{12} = \sin(\frac{\pi}{4} + \frac{\pi}{6}) = \sin\frac{\pi}{4}\cos\frac{\pi}{6} + \cos\frac{\pi}{4}\sin\frac{\pi}{6} = \frac{\sqrt{2}+\sqrt{6}}{4}$.

3.3. Angle – sum relationships (3)

Now we will use the two formulas (1) and (2) to derive all of the trigonometry identities.
(1) ÷ (2):

$$\frac{\sin(\alpha+\beta)}{\cos(\alpha+\beta)} = \frac{\sin\alpha\cos\beta + \cos\alpha\sin\beta}{\cos\alpha\cos\beta - \sin\alpha\sin\beta} = \frac{(\cos\alpha\cdot\cos\beta)(\frac{\sin\alpha}{\cos\alpha} + \frac{\sin\beta}{\cos\beta})}{(\cos\alpha\cdot\cos\beta)(1 - \frac{\sin\alpha\sin\beta}{\cos\alpha\cos\beta})}$$

So $\tan(\alpha + \beta) = \frac{\tan\alpha + \tan\beta}{1 - \tan\alpha\tan\beta}$ \hfill (3.3)

Example 19. Find $\tan(\alpha + \beta)$ if $\sin 2\alpha = \dfrac{3}{5}, (\dfrac{\pi}{4} < \alpha < \dfrac{\pi}{2})$ and $\tan(\alpha - \beta) = \dfrac{1}{2}$.

A. -2 B. -1 C. $-\dfrac{10}{11}$ D. $-\dfrac{2}{11}$

Solution: (A).

We know that $\dfrac{\pi}{4} < \alpha < \dfrac{\pi}{2}, \dfrac{\pi}{2} < 2\alpha < \pi$. $\sin 2\alpha = \dfrac{3}{5}$, $\cos 2\alpha = -\dfrac{4}{5}$, and $\tan 2\alpha = -\dfrac{3}{4}$.

Therefore $\tan(\alpha + \beta) = \tan[2\alpha - (\alpha - \beta)] = \dfrac{\tan 2\alpha - \tan(\alpha - \beta)}{1 + \tan 2\alpha \cdot \tan(\alpha - \beta)} = \dfrac{-\dfrac{3}{4} - \dfrac{1}{2}}{1 + (-\dfrac{3}{4}) \times \dfrac{1}{2}} = -2.$

Example 20. If $\tan x + \tan y = 25$ and $\cot x + \cot y = 30$, what is $\tan(x + y)$?

Solution: 150.

Since $\cot x + \cot y = 30$, $\dfrac{1}{\tan x} + \dfrac{1}{\tan y} = 30$, or $\tan x + \tan y = 30 \tan x \cdot \tan y$.

$\tan x \cdot \tan y = \dfrac{\tan x + \tan y}{30} = \dfrac{25}{30} = \dfrac{5}{6}$.

$\tan(x + y) = \dfrac{\tan x + \tan y}{1 - \tan x \cdot \tan y} = \dfrac{25}{1 - \dfrac{5}{6}} = 150$.

3.4. Angle – difference relationships (1)

Let $\beta = -\beta$ in (1), (2), and (3). We will get:
$\sin(\alpha - \beta) = \sin \alpha \cos \beta - \cos \alpha \sin \beta$ \hfill (3.4)
$\cos(\alpha - \beta) = \cos \alpha \cos \beta + \sin \alpha \sin \beta$ \hfill (3.5)
$\tan(\alpha - \beta) = \dfrac{\tan \alpha - \tan \beta}{1 + \tan \alpha \tan \beta}$ \hfill (3.6)

Example 21. Find the value of $\cos 15°$.

Solution:

$\cos 15° = \cos(45° - 30°) = \cos 45° \cos 30° + \sin 45° \sin 30° = \dfrac{\sqrt{2}}{2} \cdot \dfrac{\sqrt{3}}{2} + \dfrac{\sqrt{2}}{2} \cdot \dfrac{1}{2} = \dfrac{\sqrt{6} + \sqrt{2}}{4}$.

Example 22. What is the exact value of $\sin\left(x - \dfrac{\pi}{2}\right) \cdot \tan x \cdot \csc x$?

A. 2   B. 1   C. 0   D. –1   E. –2

Solution: D.

By the formula $\sin(\alpha - \beta) = \sin\alpha\cos\beta - \cos\alpha\sin\beta$, $\sin(x - \dfrac{\pi}{2}) = \sin x \cos\dfrac{\pi}{2} - \cos x \sin\dfrac{\pi}{2}$, with that $\sin\dfrac{\pi}{2} = 1$, $\cos\dfrac{\pi}{2} = 0$, so $\sin(x - \dfrac{\pi}{2}) = -\cos\alpha$.

$\sin\left(x - \dfrac{\pi}{2}\right) \cdot \tan x \cdot \csc x = -\cos x \cdot \dfrac{\sin x}{\cos x} \cdot \dfrac{1}{\sin x} = -1$.

3.5. Double-angle relationships (1)

Let $\beta = \alpha$ in (1), (2), and (3), we have:
$\sin 2\alpha = 2\sin\alpha\cos\alpha$ (3.7)
$\cos 2\alpha = \cos^2\alpha - \sin^2\alpha = 2\cos^2\alpha - 1 = 1 - 2\sin^2\alpha$ (3.8)
$\tan(2\alpha) = \dfrac{2\tan\alpha}{1 - \tan^2\alpha}$ (3.9)

Example 23. Evaluate $\cos 20° \cdot \cos 40° \cdot \cos 80°$.

A. 1   B. $\dfrac{2}{3}$ C. $\dfrac{1}{2}$ D. $\dfrac{1}{8}$ E. none of these

Solution: D.

$\cos 20° \cdot \cos 40° \cdot \cos 80° = \dfrac{2\sin 20°\cos 20°\cos 40°\cos 80°}{2\sin 20°}$

$= \dfrac{\sin 40°\cos 40°\cos 80°}{2\sin 20°} = \dfrac{2\sin 40°\cos 40°\cos 80°}{4\sin 20°}$

$= \dfrac{\sin 80°\cos 80°}{4\sin 20°} = \dfrac{2\sin 80°\cos 80°}{8\sin 20°} = \dfrac{\sin 160°}{8\sin 20°} = \dfrac{\sin(180-20)°}{8\sin 20°} = \dfrac{\sin 20°}{8\sin 20°} = \dfrac{1}{8}$.

Example 24. Find $\cos 72°$ and $\cos 144°$.

Solution: We use an acute 36°-72°-72° isosceles triangle.

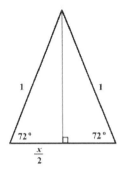 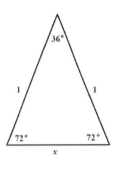 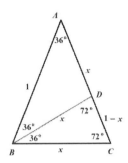

$$\cos 72° = \frac{\frac{x}{2}}{1} \quad \Rightarrow \quad x = 2\cos 72°$$

Draw the angle bisector of $\angle B$ to meet AC at D.

By the angle bisector theorem, we have $\dfrac{1}{x} = \dfrac{x}{1-x} \quad \Rightarrow \quad x^2 + x - 1 = 0$

$x_1 = \dfrac{-1+\sqrt{5}}{2}$ (the golden ratio). Thus $\cos 72° = \dfrac{-1+\sqrt{5}}{4}$.

By (3.8), $\cos(144°) = \cos(2\times 72°) = 2\cos^2(72°) - 1 = \left[2\left(\dfrac{-1+\sqrt{5}}{4}\right)^2 - 1\right]$

$= \dfrac{3-\sqrt{5}}{4} - 1 = \dfrac{-1-\sqrt{5}}{4}$. So $\cos(144°) = \dfrac{-1-\sqrt{5}}{4}$.

3.6. Power-reduction formula

$$\cos 2\alpha = 1 - 2\sin^2 \alpha \quad \Rightarrow \quad \sin^2 \alpha = \frac{1}{2}(1 - \cos 2\alpha) \tag{3.10}$$

$$\cos 2\alpha = 2\cos^2 \alpha - 1 \quad \Rightarrow \quad \cos^2 \alpha = \frac{1}{2}(1 + \cos 2\alpha) \tag{3.11}$$

Example 25. Find the exact value of $\sin\dfrac{7\pi}{12}$.

A. $\dfrac{\sqrt{3}}{4}$ B. $\dfrac{\sqrt{3}-\sqrt{6}}{3}$ C. $\dfrac{\sqrt{2-\sqrt{3}}}{2}$ D. $\dfrac{\sqrt{2+\sqrt{3}}}{2}$ E. $\dfrac{\sqrt{6}}{3}$

Solution: D.

By the formula $\sin^2\alpha = \frac{1}{2}(1-\cos 2\alpha)$, $\sin^2\frac{7\pi}{12} = \frac{1}{2}(1-\cos\frac{2\times 7\pi}{12}) = \frac{1}{2}[1-\cos(\pi+\frac{\pi}{6})]$

$= \frac{1}{2}(1+\cos\frac{\pi}{6}) = \frac{1}{2}(1+\frac{\sqrt{3}}{2}) = \frac{1}{4}(2+\sqrt{3})$. So $\sin\frac{7\pi}{12} = \sqrt{\frac{1}{4}(2+\sqrt{3})} = \frac{\sqrt{2+\sqrt{3}}}{2}$.

3.7. Half-angle relationships (1)

Let $\alpha = \alpha/2$ in (10) and (11). We get:

$$\sin\frac{\alpha}{2} = \pm\sqrt{\frac{1-\cos\alpha}{2}} \qquad (3.12)$$

$$\cos\frac{\alpha}{2} = \pm\sqrt{\frac{1+\cos\alpha}{2}} \qquad (3.13)$$

(12) ÷ (13): $\tan\frac{\alpha}{2} = \pm\sqrt{\frac{1-\cos\alpha}{1+\cos\alpha}} = \frac{\sin\alpha}{1+\cos\alpha} = \frac{1-\cos\alpha}{\sin\alpha}$ (3.14)

3.8. Double-angle relationships (2)

From (8), we have $\cos 2\alpha = \cos^2\alpha - \sin^2\alpha = \frac{\cos^2\alpha - \sin^2\alpha}{1} = \frac{\cos^2\alpha - \sin^2\alpha}{\cos^2\alpha + \sin^2\alpha}$

$= \frac{(\cos^2\alpha)(1-\frac{\sin^2\alpha}{\cos^2\alpha})}{(\cos^2\alpha)(1+\frac{\sin^2\alpha}{\cos^2\alpha})}$. So $\cos 2\alpha = \frac{1-\tan^2\alpha}{1+\tan^2\alpha}$ (3.15)

From (9), we have $\tan 2\alpha = \frac{\sin 2\alpha}{\cos 2\alpha} = \frac{2\tan\alpha}{1-\tan^2\alpha}$.

$\sin 2\alpha = \frac{2\tan\alpha}{1-\tan^2\alpha}\cdot\cos 2\alpha = \frac{2\tan\alpha}{1-\tan^2\alpha}\cdot\frac{1-\tan^2\alpha}{1+\tan^2\alpha} = \frac{2\tan\alpha}{1+\tan^2\alpha}$ (3.16)

Example 26. If $\sin x = 2\cos x$, what is the value of $\cos 2x$?

A. $\frac{1}{3}$ B. $-\frac{2}{5}$ C. $\frac{1}{4}$ D. $\frac{1}{2}$ E. $-\frac{3}{5}$

Solution: E.

$\sin x = 2\cos x \implies \dfrac{\sin x}{\cos x} = \tan x = 2$. $\cos 2\alpha = \dfrac{1-\tan^2 \alpha}{1+\tan^2 \alpha} = \dfrac{1-2^2}{1+2^2} = -\dfrac{3}{5}$.

Example 27. If $\tan\theta = -\dfrac{1}{3}$ and θ is in Quadrant II, find $\cos\theta$.

A. $-\dfrac{3}{\sqrt{10}}$ B. $-\dfrac{1}{\sqrt{10}}$ C. $-\dfrac{\sqrt{2}}{3}$ D. $\dfrac{1}{9}$ E. $\dfrac{1}{\sqrt{3}}$

Solution: A.

We know that $\cos 2\alpha = 2\cos^2 \alpha - 1$ and $\cos 2\alpha = \dfrac{1-\tan^2 \alpha}{1+\tan^2 \alpha}$.

So $2\cos^2 \theta - 1 = \dfrac{1-\tan^2 \theta}{1+\tan^2 \theta} \implies 2\cos^2 \theta - 1 = \dfrac{1-(-\frac{1}{3})^2}{1+(-\frac{1}{3})^2} = \dfrac{4}{5} \implies \cos^2 \theta = \dfrac{9}{10}$.

Since θ is in Quadrant II, $\cos\theta = -\sqrt{\dfrac{9}{10}} = -\dfrac{3}{\sqrt{10}}$.

3.9. Half-angle relationships (2)

Letting $2\alpha = \alpha$ in (9), (15), and (16), we get:

$$\tan(\alpha) = \dfrac{2\tan\dfrac{\alpha}{2}}{1-\tan^2\dfrac{\alpha}{2}} \tag{3.17}$$

$$\cos\alpha = \dfrac{1-\tan^2\dfrac{\alpha}{2}}{1+\tan^2\dfrac{\alpha}{2}} \tag{3.18}$$

$$\sin\alpha = \dfrac{2\tan\dfrac{\alpha}{2}}{1+\tan^2\dfrac{\alpha}{2}} \tag{3.19}$$

3.10. Function-product relationships

(1) + (4): $\sin(\alpha + \beta) + \sin(\alpha - \beta) = 2\sin\alpha\cos\beta$ or

$$\sin\alpha\cos\beta = \frac{1}{2}[\sin(\alpha+\beta) + \sin(\alpha-\beta)] \tag{3.20}$$

(1) − (4): $\sin(\alpha + \beta) - \sin(\alpha - \beta) = 2\cos\alpha\cos\beta$ or

$$\cos\alpha\sin\beta = \frac{1}{2}[\sin(\alpha+\beta) - \sin(\alpha-\beta)] \tag{3.21}$$

(2) + (5): $\cos(\alpha + \beta) + \cos(\alpha - \beta) = 2\cos\alpha\cos\beta$ or

$$\cos\alpha\cos\beta = \frac{1}{2}[\cos(\alpha+\beta) + \cos(\alpha-\beta)] \tag{3.22}$$

(2) − (5): $\cos(\alpha + \beta) - \cos(\alpha - \beta) = -2\sin\alpha\sin\beta$ or

$$\sin\alpha\sin\beta = -\frac{1}{2}[\cos(\alpha+\beta) - \cos(\alpha-\beta)] \tag{3.23}$$

Example 28. What is the value of the product $\sin\frac{\pi}{32}\cos\frac{\pi}{32}\cos\frac{\pi}{16}\cos\frac{\pi}{8}\cos\frac{\pi}{4}$?

A. $\frac{1}{2}$ B. $\frac{1}{4}$ C. $\frac{1}{8}$ D. $\frac{1}{16}$ E. $\frac{1}{32}$

Solution: D.

By the formula $\sin\alpha\cos\beta = \frac{1}{2}[\sin(\alpha+\beta) + \sin(\alpha-\beta)]$, $\sin\frac{\pi}{32}\cos\frac{\pi}{32} = \frac{1}{2}\sin\frac{\pi}{16}$.

So $\sin\frac{\pi}{32}\cos\frac{\pi}{32}\cos\frac{\pi}{16}\cos\frac{\pi}{8}\cos\frac{\pi}{4} = \frac{1}{2}\sin\frac{\pi}{16}\cos\frac{\pi}{16}\cos\frac{\pi}{8}\cos\frac{\pi}{4}$

$= \frac{1}{4}\sin\frac{\pi}{8}\cos\frac{\pi}{8}\cos\frac{\pi}{4} = \frac{1}{8}\sin\frac{\pi}{4}\cos\frac{\pi}{4} = \frac{1}{16}\sin\frac{\pi}{2} = \frac{1}{16}$.

3.11. Function-sum and function-difference relationships

Let $\alpha = \frac{x+y}{2}$ and $\beta = \frac{x-y}{2}$.

Then $\alpha + \beta = \frac{x+y}{2} + \frac{x-y}{2} = x$ and $\alpha - \beta = \frac{x+y}{2} - \frac{x-y}{2} = y$.

(20) becomes: $\sin\frac{x+y}{2}\cos\frac{x-y}{2} = \frac{1}{2}(\sin x + \sin y)$ or

$$\sin x + \sin y = 2\sin\frac{x+y}{2}\cos\frac{x-y}{2} \tag{3.24}$$

(21) becomes: $\cos\frac{x+y}{2}\sin\frac{x-y}{2} = \frac{1}{2}(\sin x - \sin y)$ or

$$\sin x - \sin y = 2\cos\frac{x+y}{2}\sin\frac{x-y}{2} \tag{3.25}$$

(22) becomes: $\cos\frac{x+y}{2}\cos\frac{x-y}{2} = \frac{1}{2}(\cos x + \cos y)$ or

$$\cos x + \cos y = 2\cos\frac{x+y}{2}\cos\frac{x-y}{2} \tag{3.26}$$

(23) becomes: $\sin\frac{x+y}{2}\sin\frac{x-y}{2} = -\frac{1}{2}(\cos x - \cos y)$ or

$$\cos x - \cos y = -2\sin\frac{x+y}{2}\sin\frac{x-y}{2} \tag{3.27}$$

Example 29. $\dfrac{\sin 10° + \sin 20°}{\cos 10° + \cos 20°}$ equals

A. $\tan 10° + \tan 20°$ B. $\tan 30°$ C. $\dfrac{1}{2}(\tan 10° + \tan 20°)$ D. $\tan 15°$ E. $\dfrac{1}{4}\tan 60°$

Solution: (D).

$\sin x + \sin y = 2\sin\dfrac{x+y}{2}\cos\dfrac{x-y}{2}$. $\cos x + \cos y = 2\cos\dfrac{x+y}{2}\cos\dfrac{x-y}{2}$

$\dfrac{\sin 10° + \sin 20°}{\cos 10° + \cos 20°} = \dfrac{\sin 15°}{\cos 15°} = \tan 15°$.

Example 30. Evaluate $\cos\dfrac{\pi}{5} + \cos\dfrac{2\pi}{5} + \cos\dfrac{3\pi}{5} + \cos\dfrac{4\pi}{5}$.

A. $-\dfrac{1}{2}$ B. 0 C. $\dfrac{1}{2}$ D. 1 E. $\dfrac{\sqrt{3}}{2}$

Solution: B.

By the formula $\cos x + \cos y = 2\cos\dfrac{x+y}{2}\cos\dfrac{x-y}{2}$,

$$\cos\frac{\pi}{5} + \cos\frac{2\pi}{5} + \cos\frac{3\pi}{5} + \cos\frac{4\pi}{5} = (\cos\frac{5\pi}{5} + \cos\frac{\pi}{5}) + (\cos\frac{3\pi}{5} + \cos\frac{2\pi}{5})$$

$$= 2\cos\frac{\frac{4\pi}{5}+\frac{\pi}{5}}{2}\cos\frac{\frac{4\pi}{5}-\frac{\pi}{5}}{2} + 2\cos\frac{\frac{3\pi}{5}+\frac{3\pi}{5}}{2}\cos\frac{\frac{3\pi}{5}-\frac{2\pi}{5}}{2}$$

$$= 2\cos\frac{\pi}{2}\cos\frac{3\pi}{10} + 2\cos\frac{\pi}{2}\cos\frac{\pi}{10} = 0.$$

4. SOLVING SIMPLE TRIGONOMETRIC EQUATIONS

Equations containing trigonometric functions are called trigonometric equations. Only very simple trigonometric equations can be solved by the method we learn from high school math.

We know that one value of a trigonometric function is corresponding to infinite many angles. Thus there are generally infinite many solutions for a trigonometric equation. The form of the solutions is usually different when a different method is mused. So it is difficult to check if a solution is correct or it is extraneous.
The best way to solve a trigonometric equation is to memorize the solution forms of the simplest trigonometric equations.

4.1. Simplest trigonometric equations

$\sin x = a$, $\cos x = a$, $\tan x = a$, and $\cot x = a$.

4.2. Solutions of the of the simplest trigonometric equations.

Equation		Solution
$\sin x = a$	$|a| > 1$	$\varnothing$
	$a = 1$	$x = 2k\pi + \frac{\pi}{2}$, k is integer.
	$a = -1$	$x = 2k\pi + \frac{3\pi}{2}$, k is integer.
	$|a| < 1$	$x = k\pi + (-1)^k \sin^{-1} a$, k is integer.
$\cos x = a$	$|a| > 1$	$\varnothing$

	$a = 1$	$x = 2k\pi, k$ is integer.
	$a = -1$	$x = (2k+1)\pi, k$ is integer.
	$|a| < 1$	$x = 2k\pi \pm \cos^{-1} a, k$ is integer.
$\tan x = a$		$x = k\pi + \tan^{-1} a, k$ is integer.
$\cot x = a$		$x = k\pi + \cot^{-1} a, k$ is integer.

Example 31. If $\sin 2x \sin 3x = \cos 2x \cos 3x$, then one value for x is
(A) 18° (B) 30° (C) 36° (D) 45° (E) 60°

Solution: (A).
The following statements are equivalent:
$$\sin 2x \sin 3x = \cos 2x \cos 3x,$$
$$\cos 2x \cos 3x - \sin 2x \sin 3x = 0,$$
$$\cos(2x + 3x) = 0$$
$$5x = 90° + 180°k, \quad k = 0, \pm 1, \pm 2, \ldots,$$
$$x = 18° + 36°k, \quad k = 0, \pm 1, \pm 2, \ldots.$$

Example 32. How many solutions does the trigonometric equation $\dfrac{1-\cos x}{\sin x} = \dfrac{\sqrt{3}}{3}$ have in the interval $[0, 2\pi]$?
A. 1 B. 2 C. 3 D. 4 E. 5

Solution: A.

By the formula $\tan\dfrac{\alpha}{2} = \dfrac{1-\cos\alpha}{\sin\alpha}$, we have $\dfrac{1-\cos x}{\sin x} = \tan\dfrac{x}{2} = \dfrac{\sqrt{3}}{3}$.

$\dfrac{x}{2} = \dfrac{\pi}{6} \Rightarrow x = \dfrac{\pi}{3}$. Or $\dfrac{x}{2} = \dfrac{7\pi}{6} \Rightarrow x = \dfrac{7\pi}{3}$.

Since our range is $[0, 2\pi]$, we know we have one solution $x = \dfrac{\pi}{3}$.

Example 33. How many solutions do the trigonometric equation $\dfrac{\cos x}{1-\sin x} = 2\cos x$ has in the interval $[0, 2\pi]$?

A. 5   B. 4   C. 3   D. 2   E. 1

Solution: C.

$\dfrac{\cos x}{1-\sin x} = 2\cos x \quad \Rightarrow \quad \cos x = 2\cos x(1-\sin x) \Rightarrow \quad \cos x = 2\cos x - 2\sin x \cos x$

$\Rightarrow \quad 2\sin x \cos x = \cos x$

If $\cos x = 0$, then $x = \pi/2$ and $3\pi/2$. Note that $x = \pi/2$ is extraneous since $\sin\dfrac{\pi}{2} = 1$ which leads the denominator $1 - \sin x = 0$.

If $\cos x \ne 0$, $2\sin x \cos x = \cos x \quad \Rightarrow \quad \sin x = \dfrac{1}{2}$.

Then $x = \pi/6$ and $15\pi/6$. There are 3 solutions.

Algebra II Through Competitions   **Chapter 20 Trigonometry**

PROBLEMS

Problem 1. If $\sin x + \sin^2 x = 1$, find the value of $\cos^2 x + \cos^4 x$.

A. 1   B. $\dfrac{1}{3\sqrt{5}}$ C. $\dfrac{1}{2}(3\sqrt{5} - 5)$ D. -1 E. $\dfrac{1}{2}$

Problem 2. The two shortest sides of a right triangle have lengths $\sqrt{3}$ and 2. Let α be the smallest interior angle of this triangle, what is the value of $\sin\alpha$?

A. $\sqrt{\dfrac{3}{7}}$ B. $\sqrt{\dfrac{4}{7}}$ C. $\sqrt{\dfrac{3}{5}}$ D. $\sqrt{\dfrac{3}{4}}$ E. $\sqrt{\dfrac{4}{5}}$

Problem 3. If the roots of $x^2 - bx + c = 0$ are $\sin\dfrac{\pi}{7}$ and $\cos\dfrac{\pi}{7}$, then $b^2 =$

A. c B. $1 + 2c$ C. $1 + c$ D. $1 - c$ E. $1 + c^2$

Problem 4. Which of the following is equal to $\sin^{\frac{1}{2}} x \cos x - \sin^{\frac{5}{2}} x \cos x$?

A. $\sec^5 x \tan^3 x$ B. $\cos^3 x \sqrt{\sin x}$ C. $\cos^2 x \sin^{\frac{1}{2}} x$ D. $\cos^3 x \sin^2 x$ E. None of these

Problem 5. If $\sin x = 2\cos x$, what is the value of $\sin 2x$?

A. $\dfrac{1}{3}$ B. $\dfrac{2}{5}$ C. $\dfrac{1}{4}$ D. $\dfrac{1}{2}$ E. $\dfrac{4}{5}$.

Problem 6. Given $\tan\alpha = 2$, find $\cos^2\alpha - \sin^2\alpha$.

A. $-\dfrac{4}{3}$ B. $-\dfrac{3}{5}$ C. $\dfrac{5}{3}$ D. $\dfrac{3}{4}$ E. $\dfrac{5}{4}$

Problem 7. If $\tan\theta = -\dfrac{1}{3}$ and θ is in Quadrant IV, find $\cos\theta$.

A. $\dfrac{3}{\sqrt{10}}$ B. $-\dfrac{1}{\sqrt{10}}$ C. $\dfrac{1}{\sqrt{10}}$ D. $-\dfrac{3}{\sqrt{10}}$ E. $\dfrac{1}{\sqrt{3}}$

Problem 8. If $\tan\theta = \dfrac{3}{4}$ and $0 < \theta < \dfrac{\pi}{2}$, find $\cos(\theta + \dfrac{\pi}{4})$.

A. $\dfrac{6+5\sqrt{2}}{10}$ B. $\dfrac{7\sqrt{2}}{10}$ C. $-\dfrac{\sqrt{2}}{10}$ D. $\dfrac{7\sqrt{2}}{10}$ E. $\dfrac{\sqrt{2}}{10}$.

Problem 9. For which value of θ listed below is it true that $2^{\cos\theta} > 1$ and $3^{\sin\theta} < 1$?
A. $10°$ B. $80°$ C. $150°$ D. $220°$ E. $290°$

Problem 10. Which value of θ listed below leads to
$$2^{\sin\theta} > 1 \quad \text{and} \quad 3^{\cos\theta} < 1?$$
A. $70°$ B. $140°$ C. $210°$ D. $280°$ E. $350°$

Problem 11. Express the rational number
$s = \cos^2(10°) + \cos^2(20°) + \cos^2(30°) + \cdots + \cos^2(80°)$ in lowest terms.
A. 1   B. 2   C. 3   D. 4   E. 5

Problem 12. The period of the function $\cos^4 x - \sin^4 x$ is
A. $\dfrac{\pi}{4}$ B. $\dfrac{\pi}{2}$ C. $\dfrac{3\pi}{4}$ D. π E. $\dfrac{3\pi}{2}$

Problem 13. Suppose x is a complex number for which $x + \dfrac{1}{x} = 2\cos 12°$. What is the value of $x^5 + \dfrac{1}{x^5}$?
A. 1   B. 2   C. 3   D. 4   E. 5

Problem 14. What is the value of $\cos^6 15° - \sin^6 15°$?
A. $\dfrac{3\sqrt{3}}{8}$ B. $\dfrac{7\sqrt{3}}{16}$ C. $\dfrac{15\sqrt{3}}{32}$ D. $\dfrac{31\sqrt{3}}{64}$ E. None of these

Problem 15. Let $f(x)$ be a function such that, for every real number x, $f(-x) + 3 f(x) = \sin x$. what is the value of $f\left(\dfrac{\pi}{2}\right)$?

A. -1 B. $-\dfrac{1}{2}$ C. 0   D. $\dfrac{1}{2}$ E. 1

Algebra II Through Competitions   **Chapter 20 Trigonometry**

Problem 16. Evaluate $\sin 67°\sin 22° + \sin 23°\sin 68°$.

A. $\dfrac{1}{\sqrt{2}}$ B. $\sin 89°$ C. 0   D. $\sqrt{2}$ E. $\dfrac{1}{2}$

Problem 17. If θ is an acute angle and $\sin\theta = 4/5$ then what is $\cos(2\theta)$?
A. $-16/25$ B. $-11/25$ C. $-9/25$ D. $-7/25$ E. $-4/25$

Problem 18. Given $0 < \alpha, \beta < 90°$, $\cos\alpha = \dfrac{3}{5}$ and $\cos(\alpha + \beta) = -\dfrac{12}{13}$, find $\sin\beta$.

A. $\dfrac{56}{65}$ B. $\dfrac{11}{12}$ C. $\dfrac{63}{65}$ D. $\dfrac{4}{5}$ E. none of these

Example 19: Find the value of $\tan\alpha$ if $\sin\alpha - \cos\alpha = \dfrac{1}{5}$, $0° < \alpha < 180°$.

(A) $\dfrac{3}{4}$ (B) $-\dfrac{3}{4}$ (C) $\dfrac{4}{3}$ (D) $-\dfrac{4}{3}$

Problem 20. If $0° < A < 180°$ and $\sin A + \cos A = \dfrac{7}{12}$, then the angle A satisfies

A. $0° < A < 45°$ B. $A = 45°$ C. $45° < A < 90°$ D. $90° < A < 180°$ E. none of these

Problem 21. If $\sin A = -12/13$, $\cos B = 3/5$, and A and B are in the same quadrant, then $\tan(A + B)$ is equal to
A. $56/33$ B. $-16/53$ C. $-56/33$ D. $56/63$ E. $16/53$

Problem 22. (2003 Tennessee Algebra II) The period of the sinusoidal function $f(x) = 5\sin(4x + \pi/4) + 10$ is equal to
A. 5   B. $\pi/2$ C. $\pi/4$ D. 2π E. $2\pi + 10$

Problem 23. How many solutions do the trigonometric equation $\dfrac{\sin x}{1+\cos x} = 1$ has in the interval $[0, 2\pi]$.
A. 1   B. 2   C. 3   D. 4   E. 5

Problem 24. Solve the equation $6\cos(t) - 4 = 0$, where $0 \le t \le 2\pi$. The sum of the solutions is:

A. π B. approx. 0.85107   C. 2π D. $\sqrt{3}\,\pi$ E. none of these.

Problem 25. Find the value of x where (x, y) is the solution to the system of equations:
$2(\cos\phi)\,x - 5(\sin\phi)\,y = \sec\phi$
$2(\sin\phi)\,x + 5(\cos\phi)\,y = 3\csc\phi$
A. 2   B. 1   C. $4\cos\phi$ D. $\sin\phi$ E. none of these

Problem 26. If $\log_{\sin x}(\cos x) = \dfrac{1}{2}$ and $0 < x < \dfrac{\pi}{2}$, find the value of $\sin x$.

A. $\dfrac{1}{3}\sqrt{3}$ B. $\dfrac{1}{2}(\sqrt{5}-1)$ C. $\dfrac{1}{2}(\sqrt{5}+1)$ D. $\dfrac{2}{3}\sqrt{3}$ E. $\sqrt{3}-\dfrac{1}{2}$

Problem 27: (2000 NC Math Contest) If $\begin{cases} x\sec\theta + y\tan\theta = 2\cos\theta \\ x\tan\theta + y\sec\theta = \cot\theta \end{cases}$, find what y equals.

A. $\dfrac{\cos 2\theta}{\sin\theta}$ B. $\sin\theta$ C. $\cos\theta$ D. $\sin 2\theta$ E. none of these

Problem 28. If $\tan^{-1} x - \cot^{-1} y = -45°$ then y is

A. $1 - x^2$ B. $1 + x^2$ C. $\dfrac{1+x}{1-x}$ D. $\dfrac{1-x}{1+x}$ E. None of the these

Problem 29. Given $-\dfrac{\pi}{2} < \sin^{-1} x < \dfrac{\pi}{2}$, then $\tan(\sin^{-1} x)$ must equal to:

A. $\dfrac{x}{1-x^2}$ B. $\dfrac{x}{x^2-1}$ C. $\dfrac{x}{\sqrt{1-x^2}}$ D. $\dfrac{x}{\sqrt{x^2-1}}$ E. none of these

Problem 30. Evaluate $\sin\left(\arcsin\dfrac{3}{5} + \arcsin\dfrac{8}{17}\right)$.

A. $\dfrac{4}{5}$ B. $\dfrac{77}{85}$ C. $\dfrac{84}{85}$ D. $\dfrac{91}{85}$ E. $\dfrac{15}{17}$

Problem 31. The minimum value of the function $2 + 3\sin x + 4\cos x$ is
A. -5 B. -3 C. -1 D. 1 E. 3

SOLUTIONS

Problem 1. Solution: A.
$\sin x + \sin^2 x = 1 \implies \sin^2 x = 1 - \sin x$.
So $\cos^2 x = 1 - \sin^2 x = 1 - (1 - \sin x) = \sin x$.
Thus $\cos^2 x + \cos^4 x = \cos^2 x (1 + \cos^2 x) = \sin x (1 + \sin x) = \sin x + \sin^2 x = 1$.

Problem 2. Solution: A.
The right triangle is shown in the figure. The smallest angle should be opposite to the shortest side, $\sqrt{3}$, in this case.

By Pythagorean Theorem, $AB = \sqrt{(\sqrt{3})^2 + 2^2} = \sqrt{7}$

Thus $\sin \alpha = \dfrac{\sqrt{3}}{\sqrt{7}} = \sqrt{\dfrac{3}{7}}$.

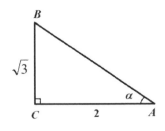

Problem 3. Solution: B.
By the Vieta's Theorem, the product of the roots is $\sin \dfrac{\pi}{7} \times \cos \dfrac{\pi}{7} = c$.

The sum of the two roots is $b = \sin \dfrac{\pi}{7} + \cos \dfrac{\pi}{7}$.

So $b^2 = (\sin \dfrac{\pi}{7} + \cos \dfrac{\pi}{7})^2 = \sin^2 \dfrac{\pi}{7} + 2\sin \dfrac{\pi}{7} \cos \dfrac{\pi}{7} + \cos^2 \dfrac{\pi}{7} = 1 + 2\sin \dfrac{\pi}{7} \cos \dfrac{\pi}{7} = 1 + 2c$.

Problem 4. Solution: B.
$\sin^{\frac{1}{2}} x \cos x - \sin^{\frac{5}{2}} x \cos x = \sin^{\frac{1}{2}} x \cos x (1 - \sin^2 x) = \sin^{\frac{1}{2}} x \cos x (1 - \sin^2 x)$
$= \sin^{\frac{1}{2}} x \cos x \cos^2 x = \sin^{\frac{1}{2}} x \cos^3 x = \sqrt{\sin x} \cos^3 x$.

Problem 5. Solution: E.
$\sin x = 2\cos x \implies \dfrac{\sin x}{\cos x} = \tan x = 2$.

By the formula $\sin 2\alpha = \dfrac{2\tan \alpha}{1 + \tan^2 \alpha}$, $\sin 2x = \dfrac{2\tan x}{1 + \tan^2 x} = \dfrac{2 \times 2}{1 + 2^2} = \dfrac{4}{5}$.

Problem 6. Solution: B.

$\cos^2\alpha - \sin^2\alpha = \cos 2\alpha = \cos 2\alpha = \dfrac{1-\tan^2\alpha}{1+\tan^2\alpha} = \dfrac{1-2^2}{1+2^2} = -\dfrac{3}{5}$.

Problem 7. Solution: A.

We know that $\cos 2\alpha = 2\cos^2\alpha - 1$ and $\cos 2\alpha = \dfrac{1-\tan^2\alpha}{1+\tan^2\alpha}$.

So $2\cos^2\theta - 1 = \dfrac{1-\tan^2\theta}{1+\tan^2\theta} \Rightarrow 2\cos^2\theta - 1 = \dfrac{1-(-\frac{1}{3})^2}{1+(-\frac{1}{3})^2} = \dfrac{4}{5} \Rightarrow \cos^2\theta = \dfrac{9}{10}$.

Since θ is in Quadrant II, $\cos\theta = \sqrt{\dfrac{9}{10}} = \dfrac{3}{\sqrt{10}}$.

Problem 8. Solution: E.

By the formula, $\cos(\gamma + \beta) = \cos\gamma\cos\beta - \sin\gamma\sin\beta$, we have

$\cos(\theta + \dfrac{\pi}{4}) = \cos\theta\cos\dfrac{\pi}{4} - \sin\theta\sin\dfrac{\pi}{4} = \dfrac{\sqrt{2}}{2}(\cos\theta - \sin\theta) = \dfrac{\sqrt{2}}{2}\cos\theta(1-\tan\theta)$

$= \dfrac{\sqrt{2}}{2}\cos\theta(1-\dfrac{3}{4}) = \dfrac{\sqrt{2}}{8}\cos\theta$.

We know that $\cos 2\alpha = 2\cos^2\alpha - 1$ and $\cos 2\alpha = \dfrac{1-\tan^2\alpha}{1+\tan^2\alpha}$.

So $2\cos^2\theta - 1 = \dfrac{1-\tan^2\theta}{1+\tan^2\theta} \Rightarrow 2\cos^2\theta - 1 = \dfrac{1-(\frac{3}{4})^2}{1+(\frac{3}{4})^2} = \dfrac{7}{25} \Rightarrow \cos^2\theta = \dfrac{16}{25}$.

Since θ is in Quadrant II, $\cos\theta = \sqrt{\dfrac{16}{25}} = \dfrac{4}{5}$. So $\cos(\theta + \dfrac{\pi}{4}) = \dfrac{\sqrt{2}}{8}\cos\theta = \dfrac{\sqrt{2}}{8} \times \dfrac{4}{5} = \dfrac{\sqrt{2}}{10}$.

Problem 9. Solution: E.

Take logs on both sides of $2^{\cos\theta} > 1$, $\cos\theta > 0$

Take logs on both sides of $3^{\sin\theta} < 1$, $\sin\theta < 0$.

We see that in the region $\dfrac{3\pi}{2} < \theta < 2\pi$, or

$270° < \theta < 360°$, both $\cos\theta > 0$ and $\sin\theta < 0$.

The answer is E.

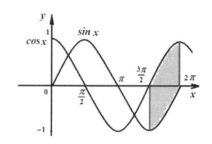

Problem 10. Solution: B.

Take logs on both sides of $3^{\cos\theta} < 1$, $\cos\theta < 0$

Take logs on both sides of $2^{\sin\theta} > 1$, $\sin\theta > 0$.

We see that in the region $\dfrac{\pi}{2} < \theta < \pi$, or $90° < \theta < 180°$, both $\sin\theta > 0$ and $\cos\theta < 0$. The answer is B.

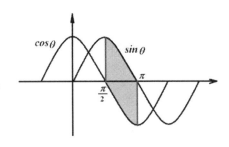

Problem 11. Solution: D.

$s = \cos^2(10°) + \cos^2(20°) + \cos^2(30°) + \cdots + \cos^2(80°)$
$= [\cos^2(10°) + \cos^2(80°)] + [\cos^2(20°) + \cos^2(70°)] + \cdots [\cos^2(40°) + \cos^2(50°)]$
$= [\cos^2(10°) + \sin^2(10°)] + [\cos^2(20°) + \sin^2(20°)] + \cdots [\cos^2(40°) + \sin^2(40°)] = 4$.

Problem 12. Solution: D.

$\cos^4 x - \sin^4 x = (\cos^2 x - \sin^2 x)(\cos^2 x + \sin^2 x) = \cos^2 x - \sin^2 x = \cos 2\alpha$. The period is then $\dfrac{2\pi}{2} = \pi$.

Problem 13. Solution: A.

$x + \dfrac{1}{x} = 2\cos 12° \quad \Rightarrow \quad x^2 - 2x\cos 12° + 1 = 0 \Rightarrow$

$x = \dfrac{2\cos 12° \pm \sqrt{4\cos^2 12° - 4}}{2} = \cos 12° \pm i\sin 12°$.

By DeMoivre's Theorem: $(\cos\theta + i\sin\theta)^n = \cos(n\theta) + i\sin(n\theta)$, we get
$x^5 = (\cos 12° \pm i\sin 12°)^5 = \cos 60° \pm i\sin 60°$

$\dfrac{1}{x^5} = x^{-5} = (\cos 12° \pm i\sin 12°)^{-5} = \cos(-60°) \pm i\sin(-60°) = \cos 60° \mp i\sin 60°$.

So $x^5 + \dfrac{1}{x^5} = \cos 60° \pm i\sin 60° + \cos 60° \mp i\sin 60° = 2\cos 60° = 2 \times \dfrac{1}{2} = 1$.

Problem 14. Solution: C.

$\cos^6 15° - \sin^6 15° = (\cos^2 15°)^3 - (\sin^2 15°)^3$

By the formula $x^3 - y^3 = (x - y)(x^2 + xy + y^2)$,

$(\cos^2 15°)^3 - (\sin^2 15°)^3 = (\cos^2 15° - \sin^2 15°)(\cos^4 15° + \cos^2 15°\sin^2 15° + \sin^4 15°)$

Since $\cos 2\alpha = \cos^2 \alpha - \sin^2 \alpha$, $(\cos^2 15° - \sin^2 15°)(\cos^4 15° + \cos^2 15° \sin^2 15° + \sin^4 15°)$
$= \cos 30°(\cos^4 15° + \cos^2 15° \sin^2 15° + \sin^4 15°)$
$= \cos 30°(\cos^4 15° + 2\cos^2 15° \sin^2 15° + \sin^4 15° - \cos^2 15° \sin^2 15°)$
$= \cos 30°[(\cos^2 15° + \sin^2 15°)^2 - \cos^2 15° \sin^2 15°] = \cos 30°(1 - \cos^2 15° \sin^2 15°)$
$= \cos 30°(1 - \frac{4}{4}\cos^2 15° \sin^2 15°) = \cos 30°(1 - \frac{1}{4}\sin^2 30°) = \frac{\sqrt{3}}{2}(1 - \frac{1}{4} \times \frac{1}{4}) = \frac{15\sqrt{3}}{32}$.

Problem 15. Solution: D.

Let $x = \frac{\pi}{2}$, $f(-x) + 3f(x) = \sin x$ $\Rightarrow$ $f(-\frac{\pi}{2}) + 3f(\frac{\pi}{2}) = \sin\frac{\pi}{2} = 1$ (1)

Let $x = -\frac{\pi}{2}$, $f(-x) + 3f(x) = \sin x$ $\Rightarrow$ $f(\frac{\pi}{2}) + 3f(-\frac{\pi}{2}) = \sin(-\frac{\pi}{2}) = -1$ (2)

(1) × 3: $3f(-\frac{\pi}{2}) + 9f(\frac{\pi}{2}) = 3$ (3)

(3) − (2): $8f(\frac{\pi}{2}) = 4 \Rightarrow f(\frac{\pi}{2}) = \frac{4}{8} = \frac{1}{2}$.

Problem 16. Solution: A.
$\sin 67° \sin 22° + \sin 23° \sin 68° = \sin 67° \sin(90 - 68)° + \sin(90 - 67)° \sin 68°$
$= \sin 67° \cos 68° + \cos 67° \sin 68° = \sin(67 + 68)° = \sin 135° = \sin(180 - 45)°$
$= \sin 45° = \frac{\sqrt{2}}{2} = \frac{1}{\sqrt{2}}$.

Problem 17. Solution: D.
By the formula $\cos 2\alpha = 1 - 2\sin^2 \alpha$, $\cos(2\theta) = 1 - 2\sin^2\theta = 1 - 2 \times (\frac{4}{5})^2 = 1 - \frac{32}{25} = -\frac{7}{25}$.

Problem 18. Solution: C.
We know that $\cos\alpha = \frac{3}{5}$. So $\sin a = \sqrt{1 - \cos^2\alpha} = \sqrt{1 - \frac{9}{25}} = \frac{4}{5}$.

We know that $\cos(\alpha + \beta) = -\frac{12}{13}$. So $\sin(a + \beta) = \sqrt{1 - \cos^2(\alpha + \beta)} = \sqrt{1 - (-\frac{12}{13})^2} = \frac{5}{13}$.

By the formula $\cos(\gamma + \beta) = \cos\gamma\cos\beta - \sin\gamma\sin\beta$,

$\cos(\alpha + \beta) = \cos\alpha\cos\beta - \sin\alpha\sin\beta = -\dfrac{12}{13}$, or $\dfrac{3}{5}\cos\beta - \dfrac{4}{5}\sin\beta = -\dfrac{12}{13}$.

$3\cos\beta - 4\sin\beta = -\dfrac{60}{13} \quad \Rightarrow \quad 3\cos\beta - 4\sin\beta = -\dfrac{60}{13}$ \hfill (1)

By the formula $\sin(\alpha + \beta) = \sin\alpha\cos\beta + \cos\alpha\sin\beta$,

$\sin\alpha\cos\beta + \cos\alpha\sin\beta = \dfrac{5}{13} \Rightarrow \dfrac{4}{5}\cos\beta + \dfrac{3}{5}\sin\beta = \dfrac{5}{13} \Rightarrow 4\cos\beta + 3\sin\beta = \dfrac{25}{13}$ \hfill (2)

(1) × 4: $12\cos\beta - 16\sin\beta = -\dfrac{240}{13}$ \hfill (3)

(2) × 3: $12\cos\beta + 9\sin\beta = \dfrac{75}{13}$ \hfill (4)

(4) − (3): $25\sin\beta = \dfrac{75}{13} - (-\dfrac{240}{13}) = \dfrac{315}{13} \Rightarrow \sin\beta = \dfrac{315}{13 \times 25} = \dfrac{63}{65}$.

Problem 19: Solution: C.

$(\sin\alpha + \cos\alpha)^2 = \sin^2\alpha + \cos^2\alpha + 2\sin\alpha\cos\alpha = 2(\sin^2\alpha + \cos^2\alpha) - (\sin\alpha - \cos\alpha)^2$

$= 2 - (\dfrac{1}{5})^2 = \dfrac{49}{25}$.

Therefore $\sin\alpha + \cos\alpha = \pm\dfrac{7}{5}$.

$\begin{cases} \sin\alpha - \cos\alpha = \dfrac{1}{5} \\ \sin\alpha + \cos\alpha = \dfrac{7}{5} \end{cases} \Rightarrow \begin{cases} \cos\alpha = \dfrac{3}{5} \\ \sin\alpha = \dfrac{4}{5} \end{cases} \Rightarrow \tan\alpha = \dfrac{\sin\alpha}{\cos\alpha} = \dfrac{4}{3}$

From $\begin{cases} \sin\alpha - \cos\alpha = \dfrac{1}{5} \\ \sin\alpha + \cos\alpha = -\dfrac{7}{5} \end{cases}$, we have $\begin{cases} \sin\alpha = -\dfrac{3}{5} \\ \cos\alpha = -\dfrac{4}{5} \end{cases} \Rightarrow \tan\alpha = \dfrac{3}{4}$.

Which is impossible since (sin x) is always positive when $0 < \alpha < 180°$ (in Quadrants I or II). So we can conclude: $\tan\alpha = \dfrac{4}{3}$.

Problem 20. Solution: D.

When $0 < x < 90°$, $\sin x > 0$, $\cos x > 0$.
$(\sin x + \cos x)^2 = \sin^2 x + \cos^2 x + 2\sin x \cos x = 1 + 2\sin x \cos x > 1 \Rightarrow \sin x + \cos x > 1$.
When $x = \pi$, $\sin x + \cos x = 1$.
So we conclude that x is the angle in quadrant II ($90° < A < 180°$).

Problem 21. Solution: A.

$\sin A = -12/13 \Rightarrow \cos A = \sqrt{1 - \sin^2 A} = \sqrt{1 - (-\frac{12}{13})^2} = \frac{5}{13}$.

$\cos B = 3/5 \Rightarrow \sin B = \sqrt{1 - \cos^2 B} = \sqrt{1 - (\frac{3}{5})^2} = \frac{4}{5}$.

By the formula $\tan(\alpha + \beta) = \dfrac{\tan \alpha + \tan \beta}{1 - \tan \alpha \tan \beta}$,

$\tan(A + B) = \dfrac{\sin(A+B)}{\cos(A+B)} = \dfrac{\sin A \cos B + \cos A \sin B}{\cos A \cos B - \sin A \sin B} = \dfrac{-\frac{12}{13} \times \frac{3}{5} + \frac{5}{13} \times \frac{4}{5}}{\frac{5}{13} \times \frac{3}{5} - (-\frac{12}{13}) \times \frac{4}{5}} = \dfrac{-\frac{16}{13 \times 5}}{\frac{63}{13 \times 5}} = -\dfrac{16}{63}$

Note: The answer key is A.

Problem 22. Solution: B.

The period is $\dfrac{2\pi}{4} = \dfrac{\pi}{2}$.

Problem 23. Solution: A.

By the formula $\tan\dfrac{\alpha}{2} = \dfrac{\sin \alpha}{1 + \cos \alpha}$, $\dfrac{\sin x}{1 + \cos x} = 1 \Rightarrow \tan\dfrac{x}{2} = 1$.

So $\dfrac{x}{2} = \dfrac{\pi}{4} \Rightarrow x = \dfrac{\pi}{2}$.

or $\dfrac{x}{2} = \dfrac{5\pi}{4} \Rightarrow x = \dfrac{5\pi}{2}$ (ignored since it is outside the region).

Problem 24. Solution: C.

$\cos(t) = \dfrac{2}{3}$.

For any value t, $\cos(t) = \cos(2\pi - t)$. Since $0 \leq t \leq 2\pi$, then $0 \leq 2\pi - t \leq 2\pi$. The fact that $-1 < \cos(t) < 1$ from the equation above only confirms the fact that a solution exists. There are only two solutions, and the sum of these two solution is $t + (2\pi - t) = 2\pi$.

Problem 25. Solution: A.
Multiplying both sides of equation $2(\cos\phi) x - 5(\sin\phi) y = \sec\phi$ by $(\cos\phi)$:
$$2x\cos^2\phi - 5y\sin\phi\cos\phi = 1 \tag{1}$$
Multiplying both sides of equation $2(\sin\phi) x + 5(\cos\phi) y = 3\csc\phi$ by $(\sin\phi)$:
$$2x\sin^2\phi + 5y\sin\phi\cos\phi = 3 \tag{2}$$
(1) + (2): $2x(\sin^2\phi + \cos^2\phi) = 4 \quad \Rightarrow \quad x = 2$.

Problem 26. Solution: B.
$$\log_{\sin x}(\cos x) = \frac{1}{2} \quad \Rightarrow \quad (\sin x)^{\frac{1}{2}} = \cos x.$$
Squaring both sides: $\sin x = \cos^2 x \quad \Rightarrow \quad \sin x = 1 - \sin^2 x \quad \Rightarrow \sin^2 x + \sin x - 1 = 0$
$$\Rightarrow \quad \sin x = \frac{-1 \pm \sqrt{1^2 - 4 \times 1 \times (-1)}}{2} = \frac{-1 \pm \sqrt{5}}{2}.$$
Since $0 < x < \frac{\pi}{2}$, $\sin x = \frac{-1 + \sqrt{5}}{2}$.

Problem 27: Solution: A.
We are given that
$$\begin{cases} x\sec\theta + y\tan\theta = 2\cos\theta \\ x\tan\theta + y\sec\theta = \cot\theta \end{cases}$$

By Cramer's Rule:
$$y = \frac{\begin{vmatrix} \sec\theta & 2\cos\theta \\ \tan\theta & \cot\theta \end{vmatrix}}{\begin{vmatrix} \sec\theta & \tan\theta \\ \tan\theta & \sec\theta \end{vmatrix}} = \frac{\sec\theta\cot\theta - 2\tan\theta\cos\theta}{\sec^2\theta - \tan^2\theta} = \frac{\frac{1}{\sin\theta} - 2\sin\theta}{1} = \frac{1 - 2\sin^2\theta}{\sin\theta} = \frac{\cos 2\theta}{\sin\theta}.$$

Problem 28. Solution: D.
Let $\tan^{-1} x = \alpha$, $\cot^{-1} y = \beta$.

$\tan^{-1} x - \cot^{-1} y = -45°$ $\Rightarrow$ $\alpha - \beta = -45°$ (1)

Take the tangent of both sides of equation (1): $\tan(\alpha - \beta) = -1$.

We know that $\tan(\alpha - \beta) = \dfrac{\tan\alpha - \tan\beta}{1 + \tan\alpha \tan\beta}$, so $-1 = \dfrac{\tan\alpha - \dfrac{1}{\cot\beta}}{1 + \tan\alpha \times \dfrac{1}{\cot\beta}} = \dfrac{x - \dfrac{1}{y}}{1 + x \times \dfrac{1}{y}} \Rightarrow$

$-1(1 + \dfrac{x}{y}) = x - \dfrac{1}{y}$ $\Rightarrow -1 - \dfrac{x}{y} = x - \dfrac{1}{y}$ $\Rightarrow$ $\dfrac{1}{y} - \dfrac{x}{y} = x + 1$

$\Rightarrow \dfrac{1-x}{y} = x + 1 \Rightarrow y = \dfrac{1-x}{x+1}$.

Problem 29. Solution: C.

Let $\sin^{-1} x = \alpha$ $\Rightarrow$ $\sin\alpha = x$.

Then $\cos\alpha = \sqrt{1 - \sin^2\alpha} = \sqrt{1 - x^2}$. $\tan(\sin^{-1} x) = \tan\alpha = \dfrac{\sin\alpha}{\cos\alpha} = \dfrac{x}{\sqrt{1 - x^2}}$.

Problem 30. Solution: B.

Let $\arcsin\dfrac{3}{5} = \alpha$ and $\arcsin\dfrac{8}{17} = \beta$. $\sin\alpha = \dfrac{3}{5}$ $\Rightarrow$ $\cos\alpha = \sqrt{1 - \sin^2\alpha} = \sqrt{1 - (\dfrac{3}{5})^2}$

$= \sqrt{\dfrac{16}{25}} = \dfrac{4}{5}$. $\sin\beta = \dfrac{8}{17}$ $\Rightarrow$ $\cos\beta = \sqrt{1 - \sin^2\beta} = \sqrt{1 - (\dfrac{8}{17})^2} = \dfrac{15}{17}$.

$\sin(\alpha + \beta) = \sin\alpha\cos\beta + \cos\alpha\sin\beta = \dfrac{3}{5} \times \dfrac{15}{17} + \dfrac{4}{5} \times \dfrac{8}{17} = \dfrac{77}{85}$.

Problem 31. Solution: B.

Note that in general, $a \cos x + b \sin x = R \cos(x - \alpha)$, where $R = \sqrt{a^2 + b^2}$ and α is determined by $\tan\alpha = b/a$.

$2 + 3 \sin x + 4 \cos x = 2 + 5 \cos(x - \alpha)$.

The smallest value of $\cos(x - \alpha)$ is -1. Thus the minimum value of the function $2 + 3 \sin x + 4 \cos x$ is $2 - 5 = -3$.

Algebra II Through Competitions **Chapter 21 Analytic Geometry Circles**

1. EQUATIONS OF CIRCLES

Standard form: $(x-a)^2+(y-b)^2=r^2$, center: (a, b), radius r ($r>0$).

General form: $x^2+y^2+Dx+Ey+F=0$.

When completing the square: $(x+\dfrac{D}{2})^2+(y+\dfrac{E}{2})^2=\dfrac{D^2+E^2-4F}{4}$.

(a) When $D^2+E^2-4F>0$, the equation stands for a circle with the center $(-\dfrac{D}{2}, -\dfrac{E}{2})$, and radius $r=\dfrac{1}{2}\sqrt{D^2+E^2-4F}$;

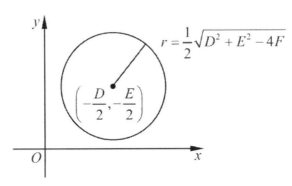

(b) When $D^2+E^2-4F=0$, the equation stands for a point $(-\dfrac{D}{2}, -\dfrac{E}{2})$.

(c) When $D^2+E^2-4F<0$, the equation does not stands for any curve.

The equation of a circle with two ends of the diameter at (x_1, y_1) and (x_2, y_2):
$(x-x_1)(x-x_2)+(y-y_1)(y-y_2)=0$.

The parameter equation of a circle with the center (a, b) and radius r $\begin{cases} x=a+r\cos\theta, \\ y=b+r\sin\theta. \end{cases}$

(θ is the parameter).

Example 1. The radius of the circle $x^2-4x+y^2+2y=4$ is:
A. 1   B. 2   C. 3   D. 4   E. 5

Solution: C.

326

Method 1:
We see that $D^2 + E^2 - 4F = (-4)^2 + 2^2 - 4 \times (-4) = 4 > 0$.
The radius is $r = \frac{1}{2}\sqrt{D^2 + E^2 - 4F} = \frac{1}{2}\sqrt{36} = 3$.

Method 2:
$x^2 - 4x + y^2 + 2y = (x^2 - 4x + 4 - 4) + (y^2 + 2y + 1 - 1) = (x-2)^2 + (y+1)^2 - 5$.
So $(x-2)^2 + (y+1)^2 - 5 = 4 \Rightarrow (x-2)^2 + (y+1)^2 = 3^2$.
The radius is 3.

Example 2. Find the center and r, the radius, of $x^2 - 12x + y^2 + 6y + 29 = 0$.
A. center $(6, -3)$; $r = 16$ B. center $(6, -3)$; $r = 4$ C. center $(-6, 3)$; $r = 16$
D. center $(-6, 3)$; $r = 4$ E. none of these

Solution: B.
Method 1:
We see that $D^2 + E^2 - 4F = (-12)^2 + 6^2 - 4 \times 29 = 64 > 0$, the equation stands for a circle with the center $(-\frac{-12}{2}, -\frac{6}{2})$, or $(6, -3)$, and radius is $r = \frac{1}{2}\sqrt{D^2 + E^2 - 4F} = \frac{1}{2}\sqrt{64} = 4$.

Method 2:
The equation of the circle can be found by completing the square twice:
$x^2 - 12x + y^2 + 6y = -29 \Rightarrow (x^2 - 12x + 36) + (y^2 + 6x + 9) = -29 + 36 + 9 \Rightarrow$
$(x - 6)^2 + (y + 3)^2 = 42$. It is now apparent that the center is at $(6, -3)$ and the radius is 4.

Example 3. The points $(4, 7)$ and $(-2, -1)$ lie on opposite ends of a diameter of a circle. What is the distance between the points where the circle intersects the x-axis?
A. 5   B. 6   C. 7   D. 8   E. 9

Solution: D.
We know that the equation of a circle with two ends of the diameter at (x_1, y_1) and (x_2, y_2): $(x - x_1)(x - x_2) + (y - y_1)(y - y_2) = 0 \Rightarrow (x - 4)(x + 2) + (y - 7)(y + 1) = 0$
When the circle intersects the x-axis, $y = 0$. So we have
$(x - 4)(x + 2) - 7 = 0 \Rightarrow (x - 4)(x + 2) - 7 = 0 \Rightarrow x^2 - 2x - 15 = 0$

We know that $|x_2 - x_1| = \frac{\sqrt{b^2 - 4ac}}{|a|} = \frac{\sqrt{(-2)^2 - 4 \times 1 \times (-15)}}{1} = \sqrt{64} = 8$.

2. RELATIONSHIP OF A CIRCLE AND A POINT

The distance from a point $P(x_0, y_0)$ to a circle $(x-a)^2 + (y-b)^2 = r^2$ is

$$d = \sqrt{(a-x_0)^2 + (b-y_0)^2} \tag{2.1}$$

If
(1) $d > r \Leftrightarrow P$ is outside of the circle.
(2) $d = r \Leftrightarrow P$ is on the circle.
(3) $d < r \Leftrightarrow P$ is inside of the circle.

Example 4. How far is the furthest point on $x^2 + y^2 + 8x - 6y = 0$ away from the origin?
A. 10 units   B. 14 units   C. 5 units   D. 7 units   E. Cannot be determined.

Solution: A.
$x^2 + y^2 + 8x - 6y = 0 \Rightarrow (x+4)^2 + (y-3)^2 = 5^2$.
The center of the circle is at $(-4, 3)$ and its radius is 5.
The furthest point on the circle is 10 units away from the origin.

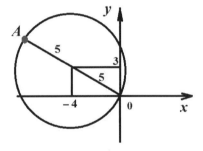

Example 5. A circle of radius 4 is centered at the origin; every second, its radius increases by 3 units. A second circle of radius 12 is centered at (30, 0); every second, its radius decreases by 1 unit. This process continues until the circles meet. At that time, the point (27, 4) lies
A. inside the first circle.   B. on the first circle.   C. inside the second circle.
D. on the second circle.   E. between the circles.

Solution: D.
The equation of the first circle: $x^2 + y^2 = 4^2$.
The equation of the first circle: $(x - 30)^2 + y^2 = 12^2$.
Let n be the number of seconds needed for two circles to meet.
We have $4 + 3n = 18 + n \Rightarrow n = 7$.
The equation of the first circle becomes: $x^2 + y^2 = 25^2$.

The equation of the first circle: circle $(x-30)^2 + y^2 = 5^2$.
The distance from a point $P(x_0, y_0)$ to a circle $(x-a)^2 + (y-b)^2 = r^2$ is
$d = \sqrt{(a-x_0)^2 + (b-y_0)^2} = \sqrt{(27)^2 + (4)^2} = \sqrt{745} > \sqrt{625} = 25 = r_1$ for the first circle or
$d = \sqrt{(a-x_0)^2 + (b-y_0)^2} = \sqrt{(30-27)^2 + (0-4)^2} = 5 = r_2$ for the second circle
So the point (27, 4) lies outside the first circle but on the second circle.

3. RELATIONSHIP OF A CIRCLE AND A LINE

Two methods are commonly used to check the positions of a line and a circle:
(1) Discriminant method
Solve the system of equations of the line and the circle to eliminate one variable. Let the discriminant of the roots of the quadratic equation obtained be Δ.

$\Delta > 0$ $\Leftrightarrow$ Line and circle intersect.
$\Delta = 0$ $\Leftrightarrow$ Line and the circle are tangent.
$\Delta < 0$ $\Leftrightarrow$ Line and the circle apart.

(2). Use d, the distance from the center of the circle with radius r to the line.
$d < r$ $\Leftrightarrow$ Line and circle intersect.
$d = r$ $\Leftrightarrow$ Line and the circle are tangent.
$d > r$ $\Leftrightarrow$ line and the circle are apart.

The distance from a line $Ax + By + C = 0$ to a circle $(x-a)^2 + (y-b)^2 = r^2$ with center (a, b) is d and $d = \dfrac{|Aa + Bb + C|}{\sqrt{A^2 + B^2}}$ (3.1)

Example 6. Find all points (x, y) that have an x-coordinate twice the y-coordinate and that lie on the circle of radius of 5 with center at (2, 6).
A. (6, 3) only B. (2, 1) only C. (4, 2) and (6, 3) D. (2, 1) and (0, 0) E. (6, 3) and (2, 1).

Solution: E.
The two conditions are $(x-2)^2 + (y-6)^2 = 25$ and $x = 2y$.
Substituting and solving we have: $(2y-2)^2 + (y-6)^2 = 25$, $y^2 - 4y + 3 = 0$,
$(y-3)(y-1) = 0$ so $y = 3$ and $y = 1$. The corresponding x values are 6 and 2.

Example 7. For which of the following values of a does the line $y = a(x - 3)$ and the circle $(x - 3)^2 + y^2 = 25$ have two points of intersection, one in the 1st quadrant and one in the 4th quadrant?
A. -1 B. 0 C. 1 D. 2 E. None of A, B, C, and D.

Solution: D.
Method 1:
We substitute the equation $y = a(x - 3)$ into $(x - 3)^2 + y^2 = 25$: $(x-3)^2 + [a(x-3)]^2 = 25$

$\Rightarrow (x-3)^2(1+a^2) = 25 \Rightarrow (x-3)^2 = \dfrac{25}{1+a^2} \Rightarrow x = 3 \pm \sqrt{\dfrac{25}{1+a^2}}$

Since the two points of intersection are in the 1st quadrant and the 4th quadrant, respectively, we know that $x > 0$. It is always true that $x = 3 + \sqrt{\dfrac{25}{1+a^2}} > 0$. So we must

have $3 - \sqrt{\dfrac{25}{1+a^2}} > 0 \Rightarrow 3 > \sqrt{\dfrac{25}{1+a^2}} \Rightarrow 1+a^2 > \dfrac{25}{9} \Rightarrow a^2 - \dfrac{16}{9} > 0$

$\Rightarrow (a - \dfrac{4}{3})(a + \dfrac{4}{3}) > 0$. The solution is $a > \dfrac{4}{3}$

or $a < -\dfrac{4}{3}$. So the answer is D.

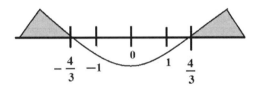

Method 2 (official solution):
The line goes through the center of the circle, $(3, 0)$ so it must intersect the circle twice. For $a = -1$ the intersection includes a point in the second quadrant. For $a = 2$, the line is $y = 2x - 6$, so $(x - 3)^2 + (2x - 6)^2 = 25$, which becomes $(x - 3)^2 = 5$, which has two positive solutions. For $a = 0$ the intersection includes only points on the x-axis. For $a = 1$, the intersection includes a point of the third quadrant.

4. EQUATION OF THE TANGENT LINE
(1). The equation of the tangent line passing through a point (x_0, y_0) on the circle $x^2 + y^2 = r^2$ is $xx_0 + yy_0 = r^2$ (4.1)

(2). The equation of the tangent line passing through a point (x_0, y_0) on the circle $(x-a)^2 + (y-b)^2 = r^2$ is $(x_0 - a)(x - a) + (y_0 - b)(y - b) = r^2$ (4.2)

Algebra II Through Competitions — Chapter 21 Analytic Geometry Circles

(3). The equation of the tangent line passing through a point (x_0, y_0) on the circle $x^2 + y^2 + Dx + Ey + F = 0$ is $xx_0 + yy_0 + \frac{D}{2}(x+x_0) + \frac{E}{2}(y+y_0) + F = 0$ \hfill (4.3)

(4). The equation of the tangent line with the slope k to the circle $x^2 + y^2 = r^2$ is
$$y = kx \pm r\sqrt{1+k^2}.$$ \hfill (4.4)

(5). From $P(x_0, y_0)$, a point outside the circle $x^2 + y^2 = r^2$, we draw two tangent lines. The equation of the line (chord) formed by connecting two points of tangent is
$xx_0 + yy_0 = r^2$ \hfill (4.5)

Example 8. Find the equation of a line that is tangent to circle $x^2 + 4x + y^2 + 6y = 12$ at point $(1, 1)$.
A. $3x + 4y = 7$ B. $4x + 6y = 12$ C. $4x - 3y = 1$ D. $3x - 4y = -1$ E. $3x - 2y = 1$.

Solution: A.
Method 1:
By (4.3), the equation of the tangent line passing through a point $(1, 1)$ on the circle $x^2 + 4x + y^2 + 6y = 12$ is $x \times 1 + y \times 1 + \frac{4}{2}(x+1) + \frac{6}{2}(y+1) - 12 = 0$, or $3x + 4y = 7$.

Method 2:
Circle rewritten as $(x + 2)^2 + (y + 3)^2 = 25$ thus its center is $(-2, -3)$. Slope of line from center to $(1, 1) = 4/3$, thus slope of tangent line is $-3/4$.
Equation of the tangent line is $3x + 4y = 1$.
Note: the official solution contains a typo and the solution should be $3x + 4y = 7$.

Example 9. Given a circle centered at the origin, and line tangent to this circle. Find the y-intercept of that line if the point of tangency is $(\sqrt{3}, 2)$.

A. $2 + \sqrt{3}$ B. $\dfrac{2+3\sqrt{3}}{2}$ C. $\dfrac{7}{2}$ D. $\dfrac{7+2\sqrt{3}}{3}$ E. none of these

Solution: C.
Method 1:
By (4.1), the equation of the tangent line passing through a point $(\sqrt{3}, 2)$ on the circle $x^2 + y^2 = r^2$ is $x \times \sqrt{3} + y \times 2 = r^2$, where $r^2 = (\sqrt{3})^2 + 2^2 = 7$. The y-intercept is obtained by letting $x = 0$ in $x \times \sqrt{3} + y \times 2 = 7 \Rightarrow y = 7/2$.

331

Method 2 (official solution):

The slope of the radius from (0, 0) to the point of tangency, ($\sqrt{3}$, 2) is $\frac{2-0}{\sqrt{3}-0} = \frac{2}{\sqrt{3}}$. The tangent line will have a slope which is the negative reciprocal of this, so its slope is $-\frac{\sqrt{3}}{2}$. The slope from the desired y-intercept would be

$-\frac{\sqrt{3}}{2} = \frac{2-b}{\sqrt{3}-0} = \frac{2-b}{\sqrt{3}} \Rightarrow 2(2-b) = -\sqrt{3}\cdot\sqrt{3} \Rightarrow 4-2b = -3 \Rightarrow 2b = 7$, so $b = 7/2$.

Example 10. Write an equation of the line that is tangent to the circle $x^2 + y^2 = 25$ at (−3, 4).

A. $3x + 4y = 25$ B. $3x − 4y = 25$ C. $−3x + 4y = 25$ D. $−3x − 4y = 25$
E. none of these

Solution: C.
Method 1:
By (4.1), the equation of the tangent line passing through a point (−3, 4) on the circle $x^2 + y^2 = 5^2$ is $x \times (-3) + y \times 4 = 25$, or $-3x + 4y = 25$.

Method 2:
The slope of the line containing the center and the point (−3, 4) is $k = \frac{4-0}{-3-0} = -\frac{4}{3}$.

The slope of the tangent line is $-\frac{1}{k} = \frac{3}{4}$.

The equation of the tangent line is $y - 4 = \frac{3}{4}[x - (-3)]$, or

$y - 4 = \frac{3}{4}x + \frac{9}{4} \quad \Rightarrow \quad y = \frac{3}{4}x + \frac{25}{4} \quad \Rightarrow \quad -3x + 4y = 25$.

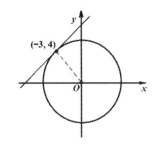

5. RELATIONSHIP OF TWO CIRCLES

Two circles with the center O_1, and O_2, the radius are R and r. $\odot O_1$: $(x - a_1)^2 + (y - b_1)^2 = r^2$ and $\odot O_2$: $(x - a_2)^2 + (y - b_2)^2 = R^2$ (5.1)

The distance between two centers $O_1(a_1, b_1)$ and $O_2(a_2, b_2)$ is d and
(1). If $d < |R - r|$, one circle is inside the other. Zero common tangent line.

(2). If $d = |R - r|$, one circle is internally tangent to the other. One common tangent line.

(3) If $|R - r| < d < r + R$, one circle intersects the other. Two common tangent lines.

(4). If $d = r + R$, one circle is externally tangent to the other. Three common tangent lines.

(5). If $d > r + R$, one circle is away from the other. Four common tangent lines.

(6). If $C_1 : x^2 + y^2 + D_1 x + E_1 y + F_1 = 0$ and $C_2 : x^2 + y^2 + D_2 x + E_2 y + F_2 = 0$ intersect, the equation of the line passing through the two points of intersection is
$$(D_1 - D_2)x + (E_1 - E_2)y + F_1 - F_2 = 0 \tag{5.2}$$

(7). The equation of the circles passing through two points (x_1, y_1) and (x_2, y_2) is
$$(x - x_1)(x - x_2) + (y - y_1)(y - y_2) + k[(x - x_1)(y_2 - y_1) - (y - y_1)(x_2 - x_1)] = 0 \tag{5.3}$$
Where k is a parameter.
Note: when $k = 0$, the equation is the equation of the circle whose diameter is the line segment connecting points (x_1, y_1) and (x_2, y_2).

(8). The equation of the circles passing through the point $A(a, b)$ is
$$(x - a)^2 + (y - b)^2 + k_1(x - a) + k_2(y - b) = 0 \tag{5.4}$$

Example 11. The graphs of $x^2 + y^2 = 24x + 10y - 120$ and $x^2 + y^2 = k^2$ intersect when k satisfies $0 \leq a \leq k \leq b$, and for no other positive values of k. Find $b - a$.
A. 10   B. 14   C. 26   D. 34   E. 144.

Solution: B.
The first circle $(x - 12)^2 + (y - 5)^2 = 49$ is centered at $(12, 5)$ and has radius 7, while the second is centered at $(0, 0)$ and has radius k.

We know that the two circle intersect when $|R - r| < d < r + R$, that is $|7 - k| < d < k + 7$. $d = \sqrt{(12 - 0)^2 + (5 - 0)^2} = \sqrt{13^2} = 13$. We also know that $k > 0$. So $|7 - k| < 13 < k + 7$.
From $13 < k + 7$, we get $k > 6$.
From $|7 - k| < 13$, we get $|7 - k| < 13 \Rightarrow \quad k < 20$. So $6 \leq k \leq 20$, so $b - a = 14$.

Example 12. Find the exact distance between the centers of the two circles whose respective equations are $x^2 + 4x + y^2 + 2y = -4$ and $(x+5)^2 + (y-1)^2 = 81$.

Solution: $\sqrt{13}$.
For the first circle $x^2 + 4x + y^2 + 2y = -4$, we see that $D^2 + E^2 - 4F = (4)^2 + 2^2 - 4 \times 4 = 4 > 0$. So the center is at $(-\frac{4}{2}, -\frac{2}{2})$, or $(-2, -1)$, and radius is $r = \frac{1}{2}\sqrt{D^2 + E^2 - 4F} = \frac{1}{2}\sqrt{4} = 1$. The second circle $(x+5)^2 + (y-1)^2 = 81$ is centered at $(-5, 1)$ and has radius 9. The exact distance between the centers of the two circles is $\sqrt{(-5+2)^2 + (1+1)^2} = \sqrt{13}$.

Example 13. What is the equation of the circle that passes through the points A(2, 2), B(5, 3), and C(3, –1)?

Solution:
Method 1:
By (5. 4), the equation of the circles passing through the point $A(2, 2)$ is
$(x-2)^2 + (y-2)^2 + k_1(x-2) + k_2(y-2) = 0$.
Since the circle also passes through points B and C, we have:
$9 + 1 + 3k_1 + k_2 = 0$ (1)
$1 + 19 + k_1 - 3k_2 = 0$ (2)
Solving the system of equations (1) and (2): $k_1 = -3$, and $k_2 = 2$.
The equation is then $x^2 + y^2 - 8x - 2y + 12 = 0$.

Method 2:
The equation of the line passing through points A and B is $x - 3y + 4 = 0$.
The equation of the circle with the diameter AB is $(x-2)(x-5) + (y-2)(y-3) = 0$.
The equation of the circles passing through points A and B is
$(x-2)(x-5) + (y-2)(y-3) + k(x-3y+4) = 0$.
Since the circle also passes through point C, we have $-2 + 12 + 10k = 0$ $\Rightarrow$ $k = -1$
The equation is then $x^2 + y^2 - 8x - 2y + 12 = 0$.

Method 3:

We take any point $P(x, y)$ on the circle other than points A, B, and C. We know that $\angle APB = \angle ACB$ (or $\angle APB = 180° - \angle ACB$). Thus $\dfrac{k_{AP} - k_{BP}}{1 + k_{AP} \cdot k_{BP}} = \dfrac{k_{AC} - k_{BC}}{1 + k_{AC} \cdot k_{BC}}$ or

$$\frac{\dfrac{y-2}{x-2} - \dfrac{y-3}{x-5}}{1 + \dfrac{y-2}{x-2} \times \dfrac{y-3}{x-5}} = \frac{\dfrac{2+1}{2-3} - \dfrac{3+1}{5-3}}{1 + \dfrac{2+1}{2-3} \times \dfrac{3+1}{5-3}}.$$

Simplifying, we get the equation: $x^2 + y^2 - 8x - 2y + 12 = 0$.

Algebra II Through Competitions — Chapter 21 Analytic Geometry Circles

PROBLEMS

Problem 1. The radius of the circle given by $x^2 - 6x + y^2 + 4y = 12$ is
A. 5　　B. 6　　C. 7　　D. 8　　E. 36

Problem 2. The radius of the circle given by $x^2 - 6x + y^2 + 4y = 36$ is
A. 5　　B. 6　　C. 7　　D. 8　　E. 36

Problem 3. A circle C contains the points (0, 6), (0, 10), and (8, 0). What is the second *x*-intercept?
A. 7.00　　B. 7.25　　C. 7.50　　D. 7.75　　E. 9.00

Problem 4. The radius of the circle with equation $2x^2 + 2y^2 + 8x - 12y + 22 = 0$ is
A. $\sqrt{2}$　　B. 2　　C. $\sqrt{22}$　　D. 22　　E. none of these

Problem 5. What is the radius of the circle that passes through the points (1, 2), (−3, −8), and (−9, 6)?
A. $4\sqrt{4}$　　B. $\sqrt{70}$　　C. $2\sqrt{15}$　　D. $\sqrt{58}$　　E. $3\sqrt{6}$

Problem 6. Which of the lines given below divides the following circle into two equal parts: $x^2 + 4x + y^2 + 6y = 10$
A. $y = mx + 5$　　B. $y = mx + 2m - 3$　　C. $y = mx - 3m + 2$
D. $y = mx + 10$　　E. none of these.

Problem 7. Find the center of the smaller circle that passes through the point located in (2, 16) and that is tangent to both axes.
A. (4, 4)　　B. (7,7)　　C. (10,10)　　D. (26, 26)　　E. ($35 - 8\sqrt{3}, 35 - 8\sqrt{3}$)

Problem 8. Find the distance between the origin and the center of the circle $x^2 + y^2 - 6x - 4y + 11 = 0$.
A. $\sqrt{13}$　　B. 13　　C. $2\sqrt{13}$　　D. $5\sqrt{2}$　　E. $2\sqrt{5}$

Problem 9. Let L denote the line which passes through the point (7, 1) and the center of the circle $x^2 + y^2 - 10x + 6y + 9 = 0$. Which of the following points is also on the line L?
A. (5, −7)　　B. (4, −10)　　C. (−3, 7)　　D. (8, 3)　　E. none of these

Algebra II Through Competitions   **Chapter 21 Analytic Geometry Circles**

Problem 10. Find an equation of the line tangent to the circle $x^2 + y^2 = 2$ at the point (1, 1).
A. $3x - y = 4$ B. $y = 2x - 1$ C. $y = -2x + 3$ D. $3x - 2y = 1$ E. $x + y = 2$.

Problem 11. The line $y = x$ intersects the circle $(x - 3)^2 + (y - 2)^2 = 1$ in two points. What is the sum of the x-coordinates of the two points?
A. 3   B. 4   C. 9/2   D. 5   E. 11/2.

Problem 12. Determine the equation of the line passing through the points of intersection of the circles $(x + 2)^2 + (y + 1)^2 = 9$ and $(x + 3)^2 + (y + 4)^2 = 16$.
A. $2x + 6y = -13$ B. $10x - 6y = -7$ C. $-2x + 6y = -7$
D. $-10x + 6y = 25$ E. $2x - 6y = 13$

Problem 13. A circle has the equation of $(x - 7)^2 + (y + 1)^2 = 10$. A line is tangent to the circle at the point (10, -2). This tangent line passes through the point (0, y). Find the value of y.

Problem 14. Find the equation of the tangent line to the circle defined by $(x - 2)^2 + (y - 1)^2 = 25$ at the point (5, -3).

Problem 15. A circle has an equation of $(x + 2)^2 + (y - 1)^2 = 100$. A line is tangent to the circle at the point (6, -5). Find the slope of this tangent. Express your answer as a common fraction to the lowest terms.

SOLUTIONS

Problem 1. Solution: A.
Complete the squares by adding 9 and 4 to both sides to get $x^2 - 6x + 9 + y^2 + 4y + 4 = (x - 3)^2 + (y + 2)^2 = 12 + 9 + 4 = 25 = 5^2$. So the radius is 5.

Problem 2. Solution: C.
Complete the squares by adding 9 and 4 to both sides to get $x^2 - 6x + 9 + y^2 + 4y + 4 = 36 + 9 + 4 = 49 = 7^2$. So the radius is 7.

Problem 3. Solution: D.
The center of C must lie on the line $y = 8$, because the center is the same distance from (0, 6) as it is from (0, 10). It must also lie on the line that perpendicularly bisects the segment from (0, 6) to (8, 0), an equation for which is $y - 3 = (4/3) \cdot (x - 4)$. Solving these two equations simultaneously gives $x = 7.75$.

Problem 4. Solution: A.
$2x^2 + 2y^2 + 8x - 12y + 22 = 0 \quad \Rightarrow \quad x^2 + y^2 + 4x - 6y + 11 = 0$
$\Rightarrow \quad (x^2 + 4x + 4) - 4 + (y^2 - 6y + 9) - 9 + 11 = 0 \Rightarrow (x+2)^2 + (y-3)^2 = 2$.
The radius is $\sqrt{2}$.

Problem 5. Solution: D.
Method 1:
By (5.3), we know that the equation of the circles passing thorugh two points (x_1, y_1) and (x_2, y_2) is $(x - x_1)(x - x_2) + (y - y_1)(y - y_2) + k[(x - x_1)(y_2 - y_1) - (y - y_1)(x_2 - x_1)] = 0$,
where k is a parameter.
With two points (1, 2), (−3, −8), we then have
$(x - 1)(x + 3) + (y - 2)(y + 8) + k[(x - 1)(-8 - 2) - (y - 2)(-3 - 1)] = 0 \quad \Rightarrow$
$x^2 + 2x + y^2 + 6y - 19 + k(-10x + 4y + 2) = 0$.
Since the circle also passes through the points (−9, 6), we get
$(-9)^2 + 2(-9) + 6^2 + 6 \times 6 - 19 + k(-10 \times (-9) + 4 \times 6 + 2) = 0 \quad \Rightarrow$
$\qquad 116 + 116k = 0 \quad \Rightarrow \quad k = -1$
The equation of the circle is $x^2 + 2x + y^2 + 6y - 19 + (-1)(-10x + 4y + 2) = 0 \Rightarrow$

$x^2+12x+y^2+2y-21=0$. The radius is $r=\frac{1}{2}\sqrt{D^2+E^2-4F}=\frac{1}{2}\sqrt{12^2+2^2-4\times(-21)}=\sqrt{58}$.

Method 2:
Let the center of the circle be (x, y).
We have $(x-1)^2+(y-2)^2=(x+9)^2+(y-6)^2 \Rightarrow 5x-2y=-28$ (1)
We have $(x-1)^2+(y-2)^2=(x+3)^2+(y+8)^2 \Rightarrow 2x+5y=-17$ (2)
(1) × 2 − (2) × 5: $29y=-29 \Rightarrow y=-1$ and $x=-6$.
The radius is then $\sqrt{(x-1)^2+(y-2)^2}=\sqrt{(-6-1)^2+(-1-2)^2}=\sqrt{49+9}=\sqrt{58}$.

Method 3:
The general form of a circle is $x^2+y^2+Dx+Ey+F=0$.
For the point (1, 2), we have
$1^2+2^2+D+2E+F=0 \Rightarrow D+2E+F=-5$ (1)
For the point (−3, −8), we have
$(-3)^2+(-8)^2-3D-8E+F=0 \Rightarrow 3D+8E-F=73$ (2)
For the point (−9, 6), we have
$(-9)^2+6^2-9D+6E+F=0 \Rightarrow 9D-6E-F=117$ (3)
(1) + (2): $4D+10E=68 \Rightarrow 2D+5E=34$ (4)
(1) + (3): $10D-4E=112$ (5)
(4) × 5 − (5): $29E=58 \Rightarrow E=2$. Thus $D=12$ and $F=-21$.
The equation of the circle is $x^2+12x+y^2+2y-21=0 \Rightarrow$
$x^2+12x+6^2-6^2+y^2+2y+1-1-21=0 \Rightarrow (x+6)^2+(y+1)^2=58$
The radius is $r=\sqrt{58}$.

Problem 6. Solution: B.
Any line which passes through the circle's center, (−3, −2), must contain a diameter of the circle. If the slope of the equation is m, then such a line can be described as
$y = m(x + 2) - 3 \Rightarrow y = mx + 2m - 3$.

Problem 7. Solution: C.
Since all radii are congruent, the radius extending from the center, (a, a) to (2, 16) and the radius extending from (a, a) to one of the axes are congruent.

$\sqrt{(a-2)^2+(a-16)^2}=a \Rightarrow a^2-36a+260=0$
$\Rightarrow (a-10)(a-26)=0$
There is more than one circle that is tangent to the axes and contains the point (2, 16), but the smaller of the 2 circles has a radius of 10 and is centered at (10, 10).

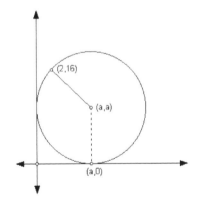

Problem 8. Solution: A.
Complete the squares to find the center of the circle.
$x^2-6x+9+y^2-4y+4=-11+9+4=2$, so $(x-3)^2+(y-2)^2=2$, putting the center at (3, 2). This point is $\sqrt{(3-0)^2+(2-0)^2}=\sqrt{13}$ from the origin.

Problem 9. Solution: D.
$x^2+y^2-10x+6y+9=0 \Rightarrow (x-5)^2+(y+3)^2=25$. So L passes through the center (5, −3) and (7, 1) and has slope $m=(-3-1)/(5-7)=2$ and the equation of L is $y=2x-13$. The only choice that lies on the line is (8,3), choice D.

Problem 10. Solution: E.
The center of our circle is (0, 0) and the line connecting the center of the circle with (1, 1) has slope 1. Thus the tangent line must have slope −1 and pass through (1, 1). The point-slope form of the equation of the tangent line is $y-1=-(x-1)$, which may be rearranged into $x+y=2$.

Problem 11. Solution: D.
Replacing y with x yields $(x-3)^2+(x-2)^2=1$ which reduces to $2x^2-10x+12=0$, the sum of whose zeros is $-(-10/2)=5$. Alternatively, solve the quadratic for $x=2$ and $x=3$. For yet another solution, notice that the center of the circle is the point (3, 2) and the radius is 1. Since $3-3=0$ and $3-2=1$, one of the points is (3, 3). The other is (2, 2) since $(2-3)^2=1$ and $2-2=0$. Thus the sum is 5.

Problem 12. Solution: A.
By (5.2), if $C_1: x^2+y^2+D_1x+E_1y+F_1=0$ and $C_2: x^2+y^2+D_2x+E_2y+F_2=0$ intersect, the equation of the line passing through the two points of intersection is $(D_1-D_2)x+(E_1-E_2)y+F_1-F_2=0$.

$(x+2)^2 + (y+1)^2 = 9$ $\Rightarrow$ $x^2 + y^2 + 4x + 2y - 4 = 0$

$(x+3)^2 + (y+4)^2 = 16$ $\Rightarrow$ $x^2 + y^2 + 6x + 8y + 9 = 0$

So the equation is $(D_1 - D_2)x + (E_1 - E_2)y + F_1 - F_2 = 0$ or

$(6-4)x + (8-2)y + 9 - (-4) = 0$ $\Rightarrow$ $2x + 6y = -13$.

Problem 13. Solution: -32

By (4.2), we know that the equation of the tangent line passing through a point (x_0, y_0) on the circle $(x-a)^2 + (y-b)^2 = r^2$ is $(x_0 - a)(x-a) + (y_0 - b)(y-b) = r^2$.

So we have $(10-7)(x-7) + (-2+1)(y+1) = 10$ $\Rightarrow$ $3(x-7) - (y+1) = 10$ $\Rightarrow$

$3x - y = 32$

At the point $(0, y)$ we have $\Rightarrow$ $3 \times 0 - y = 32 \Rightarrow$ $y = -32$.

Problem 14. Solution: $y = \dfrac{3}{4}x - \dfrac{27}{4}$.

By (4.2), we know that the equation of the tangent line passing through a point $(5, -3)$ on the circle $(x-2)^2 + (y-1)^2 = 25$ is $(5-2)(x-2) + (-3-1)(y-1) = 25$.

So we have $3(x-2) - 4(y-1) = 25$ $\Rightarrow$ $3x - 27 = 4y$ $\Rightarrow$ $y = \dfrac{3}{4}x - \dfrac{27}{4}$.

Problem 15. Solution: 4/3.

As shown in the figure, the slope of the line containing points $(-2, 1)$ and $(6, -5)$ is

$k_1 = \dfrac{-5-1}{6-(-2)} = \dfrac{-6}{8} = -\dfrac{3}{4}$.

The slope of the tangent line is k_2 and then $k_1 \times k_2 = -1$. So

$k_2 = -\dfrac{1}{k_1} = \dfrac{4}{3}$.

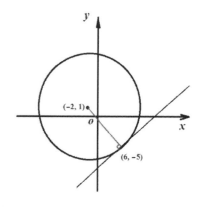

1. ELLIPSES

1.1. Definition:

Ellipse: An **ellipse** is the set of all points P in a plane such that the sum of the distances of P from two fixed points in the plane is constant ($d_1 + d_2 = 2a$).

Focus/foci: Each of the fixed points, F_1 and F_2, is called a **focus**, and together they are called **foci**.

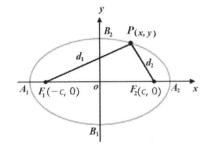

Major axis: The line segment A_1A_2 through the foci is the **major axis**.

Minor axis: The perpendicular bisector B_1B_2 of the major axis is the **minor axis**.

Vertex: Each end of the major axis, A_1 or A_2, is called a **vertex**.

Center: The midpoint of the line segment F_1F_2 is called the **center** of the ellipse.

_a, b, c:_ The length of the major axis is $2a$. The length of the minor axis is $2b$. The distance between the foci is $2c$.
$b^2 = a^2 - c^2$. (1.1)

Semimajor axis/semiminor axis: The number a is called the semimajor axis and the number b the semiminor axis.

Eccentricity: The **eccentricity** of an ellipse is the ratio c/a (horizontal major axis).

$$e = \frac{c}{a} = \sqrt{1 - \left(\frac{b}{a}\right)^2} \qquad 0 < e < 1 \qquad (1.2)$$

Notes: An ellipse has two foci in addition to the center. The graph is symmetric with respect to the x axis, y axis, and origin.

1.2 Equation Of Ellipse

Standard form of equation with center at (0, 0):

By the distance formula, we have
$$\sqrt{(x+c)^2+(y-0)^2}+\sqrt{(x-c)^2+(y-0)^2}=2a$$
After eliminating radicals and simplifying:
$$(a^2-c^2)x^2+a^2y^2=a^2(a^2-c^2) \quad\Rightarrow\quad \frac{x^2}{a^2}+\frac{y^2}{a^2-c^2}=1$$

Let $b^2 = a^2 - c^2$. $b > 0$. We get $\dfrac{x^2}{a^2}+\dfrac{y^2}{b^2}=1$ \hfill (2.1)

The parametric form:
$$\begin{cases} x = a\cos\theta \\ y = =b\sin\theta \end{cases} \tag{2.2}$$

If we start with the foci on the y axis instead of on the x axis, we obtain
$$\frac{y^2}{a^2}+\frac{x^2}{b^2}=1 \qquad (a>b) \tag{2.3}$$
where the relationship among a, b, and c remains the same as before: $b^2 = a^2 - c^2$. $b > 0$.

The general equation of an ellipse when the major/minor axes are unknown:
$Ax^2 + By^2 = C$ (A, B, and C have the same sign) \hfill (2.4)

Standard form of equation with center (h, k):

The standard form of the equation of an ellipse with center (h, k) and major and minor axes of lengths $2a$ and $2b$, respectively, where $0 < b < a$, is

$$\frac{(x-h)^2}{a^2}+\frac{(y-k)^2}{b^2}=1 \qquad \text{Major axis is horizontal.} \tag{2.5}$$
Coordinates of the Foci: *(h + c, k)* and *(h - c, k)*,
Coordinates of the Vertices: *(h + a, k)* and *(h - a, k)*.

$$\frac{(x-h)^2}{b^2}+\frac{(y-k)^2}{a^2}=1. \qquad \text{Major axis is vertical.} \tag{2.6}$$
Coordinates of the Foci: *(h, k + c)* and *(h, k - c)*
Coordinates of the Vertices: *(h, k + a)* and *(h, k - a)*

Example 1. Find the length of the major axis of the ellipse whose equation is $\dfrac{x^2}{49}+\dfrac{y^2}{25}=1$.

Solution: 14.
The length of the major axis is $2a = 2 \times 7 = 14$.

Example 2. The center (C) and vertices (V) of the ellipse $4x^2 + 9y^2 - 16x - 54y + 61 = 0$ are

A. C(–2, 3); V(–2 ± 3,3)   B. C(2, 3); V(2 ± 3, 3)   C. C(–2, 3); V(–2 ± 3, –3)
D. C(2, 3); V(2 ± 3, 0)   E. C(0, 0); V(2 ± 3, 3).

Solution: B.
Grouping/associating must be done in order to reduce the left-hand side to the sum of 2 squared binomials.
$4(x^2 - 4x + 4) + 9(y^2 - 6x + 9) = -61 + 16 + 81$
$$\frac{(x-2)^2}{3^2} + \frac{(y-3)^2}{2^2} = 1$$
The $x - 2$ term and the $y - 3$ term tell us that the center is (2, 3). Vertices of the major axis must lie on the line $y = 3$, and they are $(2 \pm 3, 3)$.

Example 3. An ellipse with equation $\frac{x^2}{16} + \frac{y^2}{36} = 1$ has which point as a focus?

A. $(0, -2\sqrt{3})$ B. $(0, 2\sqrt{5})$ C. $(-2\sqrt{5}, 0)$ D. $(2\sqrt{13}, 0)$ E. $(0, 6)$

Solution: B.
The length of the major axis (y-axis) is $2a$ and $a = 6$. The length of the minor axis is $2b$. and $b = 4$. The distance between the foci is $2c$ and $b^2 = a^2 - c^2$.
So $c^2 = a^2 - b^2 = 20$. The coordinate of one focus is $(0, \sqrt{20})$ or $(0, 2\sqrt{5})$.

Example 4. The x-coordinates of the foci of the ellipse that has $x^2 - 2x + 2y^2 = 3$ as its equation are:

A. $\pm \sqrt{2}$ B. $1 \pm \sqrt{2}$ C. $\pm \sqrt{6}$ D. $1 \pm \sqrt{6}$ E. $-1 \pm \sqrt{6}$.

Solution: B.
$x^2 - 2x + 2y^2 = 3 \Rightarrow (x^2 - 2x + 1 - 1) + 2y^2 = 3 \Rightarrow (x-1)^2 + 2y^2 = 4$
$\Rightarrow \frac{(x-1)^2}{2^2} + \frac{y^2}{2} = 1$.
The length of the major axis (y – axis) is $2a$ and $a = 2$. The length of the minor axis is $2b$. and $b = \sqrt{2}$. The distance between the foci is $2c$ and $b^2 = a^2 - c^2$.
So $c^2 = a^2 - b^2 = 2$. The coordinates of foci are $(1, \sqrt{2})$ and $(1, -\sqrt{2})$.

Example 5. Find the foci of the ellipse $9x^2 - 18x + 4y^2 + 16y = 11$.
A. $(0, \sqrt{5})$ and $(0, -\sqrt{5})$ B. $(5, 0)$ and $(-5, 0)$ C. $(1, -2+\sqrt{5})$ and $(1, -2-\sqrt{5})$
D. $(1+\sqrt{5}, -2)$ and $(1-\sqrt{5}, -2)$ E. $(2, 3)$ and $(2, -3)$.

Solution: C.
$9x^2 - 18x + 4y^2 + 16y = 11 \Rightarrow 9(x^2 - 2x) + 4(y^2 + 4y) = 11 \Rightarrow$
$9(x^2 - 2x + 1 - 1) + 4(y^2 + 4y + 4 - 4) = 11 \Rightarrow 9(x-1)^2 - 9 + 4(y-1)^2 - 19 = 11 \Rightarrow 9(x-1)^2 + 4(y-1)^2 = 36$
$\Rightarrow \dfrac{(x-1)^2}{2^2} + \dfrac{(y+2)^2}{3^2} = 1$.

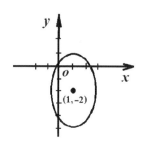

The length of the major axis (y-axis) is $2a$ and $a = 3$. The length of the minor axis is $2b$. and $b = 2$. The distance between the foci is $2c$ and $b^2 = a^2 - c^2$.
So $c^2 = a^2 - b^2 = 5$. The coordinates of foci are $(1, -2+\sqrt{5})$ and $(1, -2-\sqrt{5})$

Example 6. The graph of $25x^2 + 9y^2 - 100x - 54y = 44$ looks like:

A. 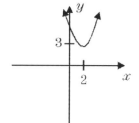   B. 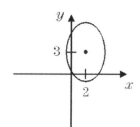   C.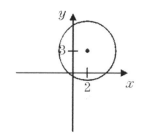

D. 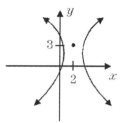   E.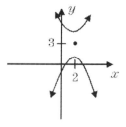

Solution: B.

Completing the squares of $25x^2 + 9y^2 - 100x - 54y = 44$:
$25(x^2 - 4x + 2^2 - 2^2) + 9(y^2 - 6y + 3^2 - 3^2) = 44 \Rightarrow 25(x-2)^2 + 9(y-3)^2 = 225$

$$\Rightarrow \quad \frac{(x-2)^2}{3^2} + \frac{(y-3)^2}{5^2} = 1.$$

This is an ellipse with the center (2, 3) and the vertical major axis.

Example 7. Draw a curve with the aid of two tacks and a piece of string. Tie the two ends of the string to the two tacks. Stick the two tacks into a piece of paper in such a way that they are closer together than the length of the string. Keeping the string taut with a pencil, trace out a curve (to make a complete loop, lift the string over one of the tacks when necessary). It can be proved that this curve is:
A. a circle   B. an ellipse   C. either a circle of an ellipse   D. a hyperbola   E. a parabola

Solution: B.
$d_1 + d_2$ is the length of the string, which is constant.
By the distance formula, we have

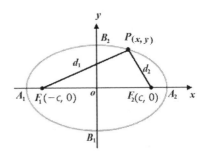

$$\sqrt{(x+c)^2 + (y-0)^2} + \sqrt{(x-c)^2 + (y-0)^2} = 2a$$

After eliminating radicals and simplifying:
$(a^2 - c^2)x^2 + a^2 y^2 = a^2(a^2 - c^2) \quad \Rightarrow$

$$\frac{x^2}{a^2} + \frac{y^2}{a^2 - c^2} = 1$$

Let $b^2 = a^2 - c^2$. $b > 0$. We get $\frac{x^2}{a^2} + \frac{y^2}{b^2} = 1$, exactly the form of (2.1), which represents an ellipse.

Example 8. $\frac{x^2}{25} + \frac{y^2}{16} = 1$ is the equation of an ellipse. What is the eccentricity of this ellipse?

A. 0   B. 3/5   C. 3/4   D. 4/5   E. none of these

Solution: B.
The eccentricity is the ratio c/a. For the given ellipses,

$$e = \frac{c}{a} = \sqrt{1 - \left(\frac{b}{a}\right)^2} = \sqrt{1 - \left(\frac{4}{5}\right)^2} = \sqrt{\left(\frac{9}{25}\right)^2} = \frac{3}{5}.$$

2. HYPERBOLAS

2.1. Terms And Equations Of Hyperbolas

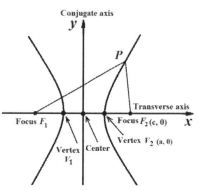

A **hyperbola** is the set of all points P in a plane such that the absolute value of the difference of the distances of P to two fixed points in the plane is a positive constant. Each of the fixed points, F_1 and F_2, is called a **focus**. The intersection points V_1 and V_2 of the line through the foci and the two branches of the hyperbola are called **vertices**, and each is called a **vertex**. The line segment V_1V_2 is called the **transverse axis**.

The midpoint of the transverse axis is the **center** of the hyperbola.

The graph is symmetric with respect to the x axis, y axis, and origin.

The perpendicular bisector of the transverse axis, extending from one side of the asymptote rectangle to the other, is called the **conjugate axis** of the hyperbola.

Transverse axis length = $2a$. Conjugate axis length = $2b$.

$$b^2 = c^2 - a^2 \quad b > 0 \tag{4.1}$$

$$\frac{x^2}{a^2} - \frac{y^2}{b^2} = 1 \tag{4.2}$$

If we start with the foci on the y axis at $F_1(0, -c)$ and $F_2(0, c)$ instead of on the x axis, we obtain

$$\frac{y^2}{a^2} - \frac{x^2}{b^2} = 1 \tag{4.3}$$

(4.2) and (4.3) can be written as:
$$Ax^2 - By^2 = c \tag{4.4}$$
A, B, and C have the same sign.

Note that (a) equations (4.2) and (4.3) always equal (=) 1; (b) Equations are always minus (−); (c) a^2 is always the first denominator, and (d) If x^2 is first then the hyperbola is horizontal. If y^2 is first then the hyperbola is vertical.

The polar form:

$$\begin{cases} x = a\sec\theta \\ y = b\tan\theta \end{cases} \qquad (4.5)$$

Eccentricity: The **eccentricity** of a hyperbola is the ratio c/a.

$$e = \frac{c}{a} = \sqrt{1 + \left(\frac{b}{a}\right)^2} \qquad (e > 1) \qquad (4.6)$$

Hyperbolas have a pair of asymptotes that pass through the center of the hyperbola. The **asymptotes** $y = \pm\frac{b}{a}x$ are for the graph of equation 4.2 ($\frac{x^2}{a^2} - \frac{y^2}{b^2} = 1$).

The **asymptotes** are $y = \pm\frac{a}{b}x$ for the graph of equation 4.3 ($\frac{y^2}{a^2} - \frac{x^2}{b^2} = 1$).

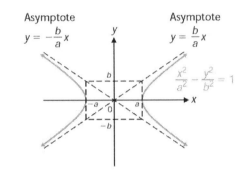

The slopes of the asymptotes are $\pm b/a$ for a horizontal hyperbolas and $\pm a/b$ for a vertical hyperbola.

The hyperbola approaches these lines as a point $P(x, y)$ on the hyperbola moves away from the origin. An easy way to draw the asymptotes is to first draw the rectangle as in the figure below, then extend the diagonals. We refer to this rectangle as the **asymptote rectangle**.

The rectangle used to graph the hyperbola is called the **fundamental rectangle**. From the above derivation, we can generalize to either of two forms.

Horizontal hyperbola with center (h, k): $\dfrac{(x-h)^2}{a^2} - \dfrac{(y-k)^2}{b^2} = 1$ \qquad (4.8)

Coordinates of the Foci: $(h + c, k)$ and $(h - c, k)$.
Coordinates of the Vertices: $(h + a, k)$ and $(h - a, k)$.
Equation of the Asymptotes: $y = k \pm \dfrac{b}{a}(x - h)$.

Vertical hyperbola with center (h, k) $\dfrac{(y-k)^2}{a^2} - \dfrac{(x-h)^2}{b^2} = 1$ \qquad (4.9)

Coordinates of the Foci: $(h, k + c)$ and $(h, k - c)$.
Coordinates of the Vertices: $(h, k + a)$ and $(h, k - a)$.

Equation of the Asymptotes: $y = k \pm \dfrac{a}{b}(x-h)$.

Example 9. Which hyperbola has asymptotes $y = \pm \dfrac{3}{4}x$?

A. $\dfrac{x^2}{9} - \dfrac{y^2}{16} = 1$ B. $\dfrac{y^2}{3} - \dfrac{x^2}{4} = 1$ C. $\dfrac{y^2}{16} - \dfrac{x^2}{9} = 1$ D. $\dfrac{x^2}{3} - \dfrac{y^2}{4} = 1$ E. $\dfrac{x^2}{16} - \dfrac{y^2}{9} = 1$

Solution: E.

We know that $y = \pm \dfrac{b}{a}x$ and $y = \dfrac{3}{4}x$. So we can have $a = 4$ and $b = 3$. The equation can be written as $\dfrac{x^2}{16} - \dfrac{y^2}{9} = 1$.

Example 10. Which of the following is an equation of the hyperbola with center (0, 0), vertex at (4, 0) and an asymptote of $y = \dfrac{3}{2}x$?

A. $\dfrac{x^2}{16} - \dfrac{y^2}{36} = 1$ B. $\dfrac{x^2}{16} - \dfrac{y^2}{24} = 1$ C. $\dfrac{x^2}{4} - \dfrac{y^2}{9} = 1$ D. $-\dfrac{x^2}{16} + \dfrac{y^2}{36} = 1$ E. $-\dfrac{x^2}{4} - \dfrac{y^2}{9} = 1$.

Solution: A.

Since one vertex is at (4, 0), $a = 4$.
We know that $y = \pm \dfrac{b}{a}x$ and $y = \dfrac{3}{2}x$. So $b = 3 \times 2 = 6$. Since the center is at (0, 0), the equation is can be written as $\dfrac{x^2}{4^2} - \dfrac{y^2}{6^2} = 1$, or $\dfrac{x^2}{16} - \dfrac{y^2}{36} = 1$.

Example 11. The hyperbola whose equation is $2x^2 - y^2 + 6y = 5$ has foci:
A. $(-\sqrt{6}, 3)$ and $(\sqrt{6}, 3)$.   B. $(\pm\sqrt{6}, 3)$.   C. $(0, -3)$ and $(0, 9)$.   D. $(0, 3 \pm \sqrt{6})$.   E. $(0, \pm\sqrt{6})$.

Solution: D.
$2x^2 - y^2 + 6y = 5 \Rightarrow 2x^2 - (y^2 - 6y + 3^2) - 3^2 = 5 \Rightarrow 2x^2 - (y+3)^2 = -4$
$\Rightarrow \dfrac{(y-3)^2}{4} - \dfrac{x^2}{2} = 1$.

So $a^2 = 4$ and $b^2 = 2$. Since $b^2 = c^2 - a^2$, $c = \sqrt{4+2} = \sqrt{6}$.

Foci is $\sqrt{6}$ units up & down from the center, $(0, 3 \pm \sqrt{6})$.

Example 12. Find the slopes of the asymptotes of a hyperbola with the equation $\dfrac{y^2}{a^2} - \dfrac{x^2}{b^2} = 1$.

A. a/b and $-a/b$. B. a and $-a$. C. b and $-b$. D. b/a and $-b/a$ E. $1/a$ and $1/b$.

Solution: A.

The asymptotes of a hyperbola with the equation $\dfrac{y^2}{a^2} - \dfrac{x^2}{b^2} = 1$ is $y = \pm \dfrac{a}{b} x$. So the slopes are a/b and $-a/b$.

Example 13. Find the foci of the hyperbola with the equation $9(y-3)^2 - 4(x+1)^2 = 36$

Solution:

$9(y-3)^2 - 4(x+1)^2 = 36 \quad \Rightarrow \quad \dfrac{(y-3)^2}{2^2} - \dfrac{(x+1)^2}{3^2} = 1$.

So $a^2 = 4$ and $b^2 = 9$. Since $b^2 = c^2 - a^2$, $c = \sqrt{4+9} = \sqrt{13}$.

Foci is $\sqrt{13}$ units up & down from the center, $(-1, 3+\sqrt{13}), (-1, 3-\sqrt{13})$.

3. GENERAL FORM OF A CONIC

$Ax^2 + Bxy + Cy^2 + Dx + Ey + F = 0$ (5.1)

is a conic. $A, B, C, D, E,$ and F are constants. $A, B,$ and C are not simultaneously 0.
Define $\Delta = B^2 - 4AC$.
(a). If $\Delta < 0$, (5.1) defines an ellipse.
Special case: (5.1) defines a circle, a point, nothing if $A = C, B = 0$.
(b). If $\Delta > 0$, (5.1) defines a hyperbola.
Special case: (5.1) defines two intersecting lines if left hand side can be factored.
(c). If $\Delta = 0$, (5.1) defines a parabola.

Example 14. $16x^2 - 9y^2 + 64x + 18y = 89$ is the equation of:
A. a circle B. an ellipse C. a hyperbola D. two intersecting lines E. none of the above

Solution: C.

$A = 16$, $B = 0$, $C = -9$.
$\Delta = B^2 - 4AC = 0 - 4 \times 16 \times (-9) > 0$. The given equation is the equation of a hyperbola.

Example 15. The graph of the equation $4x^2 + 11 = y^2 + 2y + 16x$ is:

A. a line B. a point C. a hyperbola D. an ellipse E. a parabola

Solution: C.
The given equation can be written as $4x^2 - y^2 - 16x - 2y + 11 = 0$.
$A = 4$, $B = 0$, $C = -1$.
$\Delta = B^2 - 4AC = 0 - 4 \times 4 \times (-1) > 0$. The given equation is the equation of a hyperbola.

Example 16. The graph of the equation $9x^2 + y^2 - 54x + 6y + 81 = 0$ is:
A. A point   B. An ellipse   C. A hyperbola   D. A circle   E. No graph

Solution: B.
$A = 9$, $B = 0$, $C = 1$.
$\Delta = B^2 - 4AC = 0 - 4 \times 9 \times 1 < 0$. The given equation is the equation of an ellipse.

4. MAXIMUM AND MINIMUM VALUES

Example 17: Find the smallest value of $y = \sqrt{x^2 - 4x + 7}$.

Solution:

We square both sides of $y = \sqrt{x^2 - 4x + 7}$: $y^2 = x^2 - 4x + 7 \Rightarrow \dfrac{y^2}{3} - \dfrac{(x-2)^2}{3} = 1$ ($y \geq 0$).

This graph is the upper branch of the hyperbola with center $(2, 0)$, the major axis at y-axis, and $a = \sqrt{3}$. When $x = 2$, the smallest value y is $\sqrt{3}$.

Example 18. Suppose real numbers x and y satisfy $x^2 + 9y^2 - 4x + 6y + 4 = 0$. What is the maximum value of $4x - 9y$?
A. 15   B. $9\sqrt{3}$ C. 16   D. $12\sqrt{2}$ E. 18

Solution: C.
Method 1:

Let $4x - 9y = m$ $\Rightarrow$ $4x = m + 9y$ (1)
$x^2 + 9y^2 - 4x + 6y + 4 = 0$ (2)
Since (x, y) is on the graph (2), (1) represents a line tangent to (2).

Substituting (1) into (2): $\dfrac{(m+9y)^2}{16} + 9y^2 - (m+9y) + 6 + 4 = 0$ (3)

(3) can be simplifying into: $225y^2 - y(18m - 48) + m^2 - 16m + 64 = 0$ ($)

Δ must be zero in (4): $m^2 - 22m + 96 = 0$ $\Rightarrow$ $(m-6)(m-16) = 0$.

$m = 6$ or $m = 16$. The smallest value of $4x - 9y$ is 6 and the maximum value is 16.

Method 2:

$x^2 + 9y^2 - 4x + 6y + 4 = 0$ $\Rightarrow$ $\dfrac{(x-2)^2}{1^2} + \dfrac{(y+\frac{1}{3})^2}{(\frac{1}{3})^2} = 1$ $\Rightarrow$ $\dfrac{(4x-8)^2}{4^2} + \dfrac{(-9y-3)^2}{9^2 \times (\frac{1}{3})^2} = 1$

by Cauchy inequality:

$1 = \dfrac{(4x-8)^2}{4^2} + \dfrac{(-9y-3)^2}{9^2 \times (\frac{1}{3})^2} \geq \dfrac{(4x-8-9y-3)^2}{16 + 9^2 \times (\frac{1}{3})^2} = \dfrac{(4x-9y-11)^2}{5^2}$ or

$\dfrac{(4x-9y-11)^2}{5^2} \leq 1$ $\Rightarrow$ $(4x-9y-11)^2 \leq 5^2$ $\Rightarrow$ $4x-9y-11 \leq 5$ $\Rightarrow$ $4x-9y \leq 16$.

The maximum value of $4x - 9y$ is 16, which is achieved when $\dfrac{x-2}{1^2} = \dfrac{(y+\frac{1}{3})^2}{(\frac{1}{3})^2}$ $\Rightarrow$

$x - 9y = 5$. We solve $x - 9y = 5$ and $4x - 9y = 16$ to get $x = \dfrac{11}{3}$ and $y = -\dfrac{4}{27}$.

Algebra II Through Competitions **Chapter 22 Ellipses and Hyperbolas**

PROBLEMS

Problem 1. The length of the minor axis of the ellipse having equation $x^2 + 4y^2 + 2x - 24y + 33 = 0$ is which of these?
A. 1 B. 2 C. 3 D. 4 E. 5

Problem 2. Find the locus of points whose distance from the point $(-1, 0)$ is twice the distance from the point $(2, 3)$. What is an equation of this locus of points?
A. $y = -x + 3$ B. $y = x^2 - 2x + 3$ C. $x^2 + y^2 - 6x - 8y + 17 = 0$
D. $x^2 + 4y^2 - 6x - 32y + 53 = 0$ E. None of these

Problem 3. Which one of the following is a focus of the ellipse $\dfrac{x^2}{16} + \dfrac{(y+2)^2}{36} = 1$?

A. $(2(1+\sqrt{5}), 0)$ B. $(0, -2(1+\sqrt{5}))$ C. $(2(1+\sqrt{5}), -2(1+\sqrt{5}))$
D. $(2\sqrt{5}, -2\sqrt{5})$ E. $(0, -2\sqrt{5})$

Problem 4. What conic does the equation $4x^2 - y^2 - 8x - 4y - 9 = 0$ represent?
A. Circle B. Ellipse C. Hyperbola D. Parabola E. none of these

Problem 5. The conic section with the smallest eccentricity is:
A. Circle B. Ellipse C. Hyperbola D. Parabola E. none of these

Problem 6. Which of the following is an equation of a hyperbola with horizontal transverse axis and asymptotes $y = \pm\dfrac{4}{3}x$?

A. $x^2 - y^2 = 12$ B. $\dfrac{x^2}{16} - \dfrac{y^2}{9} = 1$ C. $\dfrac{x^2}{9} - \dfrac{y^2}{16} = 1$ D. $\dfrac{y^2}{9} - \dfrac{x^2}{16} = 1$ E. $\dfrac{y^2}{16} - \dfrac{x^2}{9} = 1$

Problem 7. Find the slopes of the asymptotes of a hyperbola with the equation $\dfrac{x^2}{a^2} - \dfrac{y^2}{b^2} = 1$.
A. a/b and $-a/b$. B. a and $-a$. C. b and $-b$. D. b/a and $-b/a$ E. $1/a$ and $1/b$.

Problem 8. The graph of $4x^2 - 9y^2 + 16x - 144y - 560 = 0$ is
A. a circle B. an ellipse C. a parabola D. a hyperbola E. two intersecting lines

Problem 9. How many points of intersection do the graphs of the following equations have?
$4x^2 + 9y^2 - 16x + 54y + 61 = 0$ and $y^2 - x^2 + 6y + 4x + 4 = 0$
A. 0   B. 1   C. 2   D. 3   E. 4.

Problem 10. The graph of $\dfrac{y^2}{36} + \dfrac{x^2}{4} = 0$ is

A. an ellipse   B. a point   C. a hyperbola   D. two intersecting lines   E. none of these

Problem 11. $\dfrac{x^2}{16} + \dfrac{y^2}{25} = 1$ is the equation of an ellipse. What is the eccentricity of this ellipse?

A. 0   B. 3/5   C. 3/4   D. 4/5   E. none of these

Problem 12. The graph of $\dfrac{y^2}{4} - \dfrac{x^2}{9} = 0$ is

A. a hyperbola   B. two intersecting lines   C. two parallel lines   D. a point   E. none of these

Problem 13. What is the shape of the graph represented by $x^2 + 6x - 4y - 2y^2 = 10$?
A. circle   B. parabola   C. ellipse   D. hyperbola   E. none of these

Problem 14. What is the smallest possible value of $x^2 + 2y^2 + 2xy + 6x + 2y$?
A. –11   B. –12   C. –13   D. –14   E. –15.

Problem 15: Point (x, y) is on the ellipse $4(x - 2)^2 + y^2 = 4$. Find the smallest value of y/x.

Problem 16. The area of an ellipse is calculated using the formula $A = \pi ab$, where a is half the length of the major axis of the ellipse and b is half the length of the minor axis of the ellipse. What is the area enclosed by $9x^2 + 25y^2 + 72x - 250y + 544 = 0$?
A. 9π B. 15π C. 25π D. 60π E. 225π.

Algebra II Through Competitions **Chapter 22 Ellipses and Hyperbolas**

Problem 17. The equation of the hyperbola whose vertices are at (0, 4) and (0, −4) and whose eccentricity is 2.5 can be expressed in the form $\dfrac{y^2}{k} - \dfrac{x^2}{w} = 1$. Find the value of $(2k + w)$.

Problem 18. Give the length of the major axis of the ellipse whose equation is:
$$\dfrac{4(x-3\tfrac{1}{2})^2}{225} + \dfrac{(y-\tfrac{5}{6})^2}{16} = 1.$$

Problem 19. What is the shape of the graph represented by $x^2 + 6x - 4y - 2y^2 = 10$?
A. circle B. parabola C. ellipse D. hyperbola E. none of these

Problem 20. The graph of the equation $25y^2 - 9x^2 - 100y - 54x + 10 = 0$ is:
A. a line B. a parabola C. a circle D. an ellipse E. a hyperbola

Problem 21. What is the shape (circle, ellipse, etc.) of the graph of the relation $x^2 + 4x + 6y - y^2 = 8$?
A. circle B. ellipse C. hyperbola D. parabola E. None of these

Problem 22. The graph of $x^2 + 2y^2 + 2x - 20y = -43$ is:
A. An ellipse with center (1, 5). B. An ellipse with center (−1, 5).
C. An hyperbola with center (1, 5). D. An hyperbola with center (−1, 5).
E. An parabola with vertex (−1, 5).

Algebra II Through Competitions   **Chapter 22 Ellipses and Hyperbolas**

SOLUTIONS

Problem 1. Solution: B.
The equation of the ellipse must be simplified by completing the square.
$$x^2 + 4y^2 + 2x - 24y + 33 = 0 \Rightarrow (x+1)^2 + 4(y-3)^2 = 4 \Rightarrow \frac{(x+1)^2}{2^2} + \frac{(y-3)^2}{1^2} = 1.$$
The minor axis is twice the semiminor axis, making the minor axis $2 \times 1 = 2$.

Problem 2. Solution: C.
Let the coordinate of the point be (x, y). $\sqrt{(x+1)^2 + (y-0)^2} = 2\sqrt{(x-2)^2 + (y-3)^2}$.
Squaring both sides: $(x+1)^2 + y^2 = 4[(x-2)^2 + (y-3)^2] \Rightarrow$
$x^2 + y^2 + 2x + 1 = 4(x^2 - 4x + 4 + y^2 - 6y + 9) \Rightarrow$
$x^2 + y^2 + 2x + 1 = 4x^2 - 16x + 4y^2 - 24y + 52 \Rightarrow$
$3x^2 - 18x + 3y^2 - 24y + 51 = 0 \Rightarrow x^2 + y^2 - 6x - 8y + 17 = 0$.

Problem 3. Solution: B.
The length of the major axis (y – axis) is $2a$ and $a = 6$. The length of the minor axis is $2b$. and $b = 4$. The distance between the foci is $2c$ and $b^2 = a^2 - c^2$.
So $c^2 = a^2 - b^2 = 20$. $c = 2\sqrt{5}$.
The coordinates of foci are $(0, -2 + 2\sqrt{5})$ and $(0, -2 - 2\sqrt{5})$. We see that $(0, -2 - 2\sqrt{5})$ is the same as $(0, -2(1+\sqrt{5}))$. So the answer is B.

Problem 4. Solution: C.
$4x^2 - y^2 - 8x - 4y - 9 = 0 \Rightarrow 4(x-1)^2 - (y+2)^2 = 9$.
$A = 4, B = 0, C = -1$.
$\Delta = B^2 - 4AC = 0 - 4 \times 4 \times (-1) > 0$. The given equation is the equation of a hyperbola.

Problem 5. Solution: A.
The eccentricity is ratio c/a. For hyperbolas, $e > 1$. For ellipses, $e = \frac{c}{a} = \sqrt{1 - \left(\frac{b}{a}\right)^2}$, $0 < e < 1$. For parabolas, $e = 1$. If $b = a$, the ellipse becomes a circle with $e = 0$. So the answer is A.

Problem 6. Solution: (C).

If the transverse axis is horizontal, then the hyperbola must be of the form $\frac{x^2}{a^2} - \frac{y^2}{b^2} = 1$.

The asymptotes will be of the form $y = \pm \frac{b}{a} x$ and $a = 4$ and $b = 3$.

So the hyperbola is $\frac{x^2}{9} - \frac{y^2}{16} = 1$.

Problem 7. Solution: D.

The asymptotes of a hyperbola with the equation $\frac{x^2}{a^2} - \frac{y^2}{b^2} = 1$ is $y = \pm \frac{b}{a} x$. So the slopes are b/a and $-b/a$.

Problem 8. Solution: E.
$4x^2 - 9y^2 + 16x - 144y - 560 = 4(x^2 + 4x) - 9(y^2 + 16y) - 560 = 4(x^2 + 4x + 4 - 4) - 9(y^2 + 16y + 64 - 64) - 560 = 4(x - 4)^2 - 9(y - 8)^2 = [2(x - 4) + 3(y - 8)][2(x - 4) - 3(y - 8)]$.
So the graph is two intersecting lines.

Problem 9. Solution: E.

$4x^2 + 9y^2 - 16x + 54y + 61 = 0 \implies \frac{(x-2)^2}{3^2} + \frac{(y+3)^2}{2^2} = 1$.

$y^2 - x^2 + 6y + 4x + 4 = 0 \implies \frac{(y-3)^2}{1^2} - \frac{(x-2)^2}{1^2} = 1$.

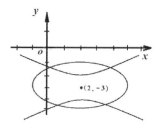

We plot the graphs and we get 4 points of intersection.

Problem 10. Solution: B.
The equation is true if and only if $y = x = 0$. So it is a point.

Problem 11. Solution: B.

The eccentricity is the ratio c/a. For the given ellipses,

$e = \frac{c}{a} = \sqrt{1 - \left(\frac{b}{a}\right)^2} = \sqrt{1 - \left(\frac{4}{5}\right)^2} = \sqrt{\left(\frac{9}{25}\right)^2} = \frac{3}{5}$.

Problem 12. Solution: B.

$$\frac{y^2}{4} - \frac{x^2}{9} = 0 \implies (\frac{y}{2} - \frac{x}{3})(\frac{y}{2} + \frac{x}{3}) = 0.$$

We get two lines: $2x - 3y = 0$ and $2x + 3y = 0$.

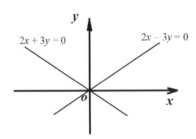

Problem 13. Solution: D.
The given equation can be written as $x^2 - 2y^2 + 6x - 2y - 10 = 0$ which can not be factored.
$A = 1$, $B = 0$, $C = -2$.
$\Delta = B^2 - 4AC = 0 - 4 \times 1 \times (-2) > 0$. The given equation is the equation of a hyperbola.

Problem 14. Solution: C.
Let the smallest possible value of $x^2 + 2y^2 + 2xy + 6x + 2y$ be m. Since m is real, both x and y are real.
The discriminant of $x^2 + x(2y + 6) + 2y^2 + 2y - m = 0$ should be greater or equal to zero.
Or $\Delta_x = (2y+6)^2 - 4(2y^2 + 2y - m) \geq 0 \implies y^2 - 4y - m - 9 \geq 0$ \hfill (1)
Since y is real, the discriminant of equation $y^2 - 4y - m - 9 = 0$ should be greater or equal to zero. So $\Delta_y = (-4)^2 - 4 \times 1 \times (-m - 9) \geq 0 \implies m \geq -13$.
The smallest value of m is 13.
This value is achieved when $x = 5$ and $y = 2$.
These values of x and y are obtained as follows:
$y^2 - 4y - m - 9 = 0 \implies y^2 - 4y - (-13) - 9 = 0$
$\implies y^2 - 4y + 4 = 0 \implies (y-2)^2 = 0 \implies y = 2$.
$x^2 + x(2y + 6) + 2y^2 + 2y - m = 0 \implies x^2 + x(2 \times 2 + 6) + 2 \times 2^2 + 2 \times 2 - (-13) = 0$
$\implies x^2 + 10x + 25 = 0 \implies (x+5)^2 = 0 \implies x = 5$.

Problem 15: Solution:
Method 1:
Let $y/x = k$. We have $y = kx$. This is a line that intersects with the ellipse at (x, y).
Substituting into the equation of the ellipse we get: $4(x - 2)^2 + k^2x^2 = 4$.
Simplifying into: $(4 + k^2)x^2 - 16x + 12 = 0$. Since they intersect, $\Delta \geq 0$.
$16^2 - 4(4 + k^2) \cdot 12 \geq 0 \implies 3k^2 \leq 4 \implies -\frac{2\sqrt{3}}{3} \leq k \leq \frac{2\sqrt{3}}{3}$.

The smallest value of k, or y/x is $-\dfrac{2\sqrt{3}}{3}$.

Note that the greatest value of k, or y/x is $\dfrac{2\sqrt{3}}{3}$.

Method 2:
Let $y/x = k$. the $y = kx$. The equation $4(x-2)^2 + y^2 = 4$ can be changed into the form of $\dfrac{(kx-2k)^2}{k^2} + \dfrac{y^2}{4} = 1$.

By Cauchy, we have $1 = \dfrac{(kx-2k)^2}{k^2} + \dfrac{y^2}{4} \geq \dfrac{(kx-2k+y)^2}{k^2+4} = \dfrac{4k^2}{k^2+4}$.

Thus $1 \geq \dfrac{4k^2}{k^2+4}$ $\Rightarrow$ $k^2 + 4 \geq 4k^2$ $\Rightarrow$ $k^2 \leq \dfrac{4}{3}$.

Solving we get $k \leq \dfrac{2\sqrt{3}}{3}$ or $k \geq -\dfrac{2\sqrt{3}}{3}$. The smallest value of y/x is $-\dfrac{2\sqrt{3}}{3}$. This value occurs when $\dfrac{kx-2k}{k^2} = \dfrac{y}{4}$, or $x = \mp\dfrac{3}{2}$ and $y = \pm\sqrt{3}$.

Problem 16. Solution: B.
$9x^2 + 25y^2 + 72x - 250y + 544 = 0$ $\Rightarrow$ $9(x+4)^2 + 25(y-5)^2 = 15^2$
$\Rightarrow$ $\dfrac{(x+4)^2}{5^2} + \dfrac{(y-5)^2}{3^2} = 1$.

So $a = 5$ and $b = 3$. The area is $A = \pi ab = 15\pi$.

Problem 17. Solution: 116.

The **eccentricity** of a hyperbola is the ratio c/a or $e = \dfrac{c}{a} = \sqrt{1 + \left(\dfrac{b}{a}\right)^2}$.

We know that $a = 4$. Thus $2.5 = \sqrt{1 + \left(\dfrac{b}{4}\right)^2}$ $\Rightarrow$ $6.25 = 1 + \left(\dfrac{b}{4}\right)^2$ $\Rightarrow$

$5.25 = \left(\dfrac{b}{4}\right)^2$ $\Rightarrow$ $b^2 = 84 \Rightarrow$ $b^2 = 84$

$\dfrac{y^2}{k} - \dfrac{x^2}{w} = 1$ $\Rightarrow$ $\dfrac{y^2}{16} - \dfrac{x^2}{84} = 1$. $2k + w = 2 \times 16 + 84 = 116$.

Problem 18. Solution: 15.

The length of the major axis is $2a = 2 \times (15/2) = 15$.

Problem 19. Solution: D.
The given equation can be written as $x^2 - 2y^2 + 6x - 4y - 10 = 0$.
$A = 1, B = 0, C = -2$.
$\Delta = B^2 - 4AC = 0 - 4 \times 1 \times (-2) > 0$. The given equation is the equation of a hyperbola.

Problem 20. Solution: E.
The given equation can be written as $-9x^2 + 25y^2 - 54x - 100y + 10 = 0$
$A = -9, B = 0, C = 25$.
$\Delta = B^2 - 4AC = 0 - 4 \times (-9) \times 25 > 0$. The given equation is the equation of a hyperbola.

Problem 21. Solution: C.
The given equation can be written as $x^2 - y^2 + 4x + 6y - 8 = 0$.
$A = 1, B = 0, C = -1$.
$\Delta = B^2 - 4AC = 0 - 4 \times 1 \times (-1) > 0$. The given equation is the equation of a hyperbola.

Problem 22. Solution: B.
$x^2 + 2y^2 + 2x - 20y = -43$ $\Rightarrow$ $(x+1)^2 + 2(y-5)^2 = 8$.
$\Rightarrow$ $\dfrac{(x+1)^2}{8} + \dfrac{(y-5)^2}{2^2} = 1$. The answer is B.

1. Divisibility Rules

Characteristic of number	Number divisible by:
Last digit is even	2
The sum of the digits is divisible by 3	3
The last two digits form a number divisible by 4	4
The last digit is 0 or 5	5
The number is divisible by both 2 and 3.	6
To find out if a number is divisible by seven, take the last digit, double it, and subtract it from the rest of the number. If you get an answer divisible by 7 (including zero), then the original number is divisible by seven. If you don't know the new number's divisibility, you can apply the rule again.	7
The last three digits form a number divisible by 8	8
The sum of the digits is divisible by 9	9
The numeral ends in 0	10
To find out if a number is divisible by eleven, add every other digit, and call that sum "x." Add together the remaining digits, and call that sum "y." Take the difference, $x - y$. If the difference is zero or a multiple of eleven, then the original number is a multiple of eleven.	11
The number is divisible by both 3 and 4.	12
Delete the last digit from the number, and then subtract 9 times the deleted digit from the remaining number. If what is left is divisible by 13, then so is the original number. Repeating the rule if necessary.	13
The last four digits form a number divisible by 16	16
A number is evenly divisible by 3×5 (relative prime)	15

Algebra II Through Competitions　　　　　　　**Chapter 23 Number Theory**

A number is evenly divisible by 2×9 (not 3×6) (relative prime)	18
A number is evenly divisible by 3×8 (neither 4×6 nor 2×12) (relative prime)	24
A number is evenly divisible by 4×9 (neither 6×6 nor 3×12)(relative prime)	36

TIP: If a number is divisible by two different prime numbers, then it is divisible by the products of those two numbers. Since 36, is divisible by both 2 and 3, it is also divisible by 6.

Example 1. The three digit number $2A3$ is added to the number 326 to give the three digit number $5B9$. If $5B9$ is divisible by 9, then $A + B =$
A. 5.　　　B. 6.　　　C. 7.　　　D. 8.　　　E. 9.

Solution: B.
$5B9$ is divisible by 9 when $B = 4$. Then $2A3 + 326 = 549 \Rightarrow 2A3 = 223$ and $A = 2$, So $A + B = 2 + 4 = 6$.

Example 2. How many integers between 199 and 301 are divisible by 4 or 10?
A. 26　　　B. 31　　　C. 35　　　D. 37　　　E. 39

Solution: B.
From 200 to 300, there are 26 multiples of 4 and 11 multiples of 10. However, this overcounts 6 numbers that are multiples of both 4 and 10. Thus, there are $26 + 11 - 6 = 31$ numbers from 200 to 300 that are divisible by 4 or 10.

Example 3. Let x be a two digit number such that 0 is not a digit of x and the digits of x are distinct. If the digits of x are reversed to form a new two digit number y, the sum $x + y$ must be divisible by
A. 2　　　B. 3　　　C. 5　　　D. 7　　　E. 11

Solution: E.
$x = 10a + b, a \neq 0, b \neq 0$
$y = 10b + a \Rightarrow x + y = 11a + 11b = 11(a + b)$
So the sum must be divisible by 11.

Example 4. Trevor started with a whole number that was a perfect square. When he added 100 to the number, he obtained a number that was one more than a perfect square. When he added 200 to his original number, he obtained another perfect square. His original number was evenly divisible by
A. 3　　　B. 5　　　C. 7　　　D. 11　　　E. none of these

362

Solution: C.
Let the original number be x. Two other square numbers be y^2 and z^2.

$$x^2 + 100 = y^2 + 1 \quad (1)$$
$$x^2 + 200 = z^2 \quad (2)$$

(2) − (1): $100 = z^2 - y^2 - 1 \Rightarrow z^2 - y^2 = 101 \Rightarrow (z - y)(z + y) = 101$.

We know that 101 is a prime number. So we have

$$\begin{cases} z + y = 101 \\ z - y = 1 \end{cases}$$

The solution is $z = 51$ and $y = 50$. So $x = 49$ which is divisible by 7.

2. Patterns of the Last Digit of a^n

The last digits of a^n have patterns shown in the table below.
For example, when $a = 2$,

$$2^1 = 2 \qquad 2^2 = 4, \qquad 2^3 = 8, \qquad 2^4 = 16,$$
$$2^5 = 32, \qquad 2^6 = 64, \qquad 2^7 = 128, \qquad 2^8 = 256,\ldots$$

The last digits of 2^n demonstrate a pattern: 2, 4, 8, 6, 2, 4, 8, 6, etc…

n	1	2	3	4	Period
1^n	1				1
2^n	2	4	8	6	4
3^n	3	9	7	1	4
4^n	4	6			2
5^n	5				1
6^n	6				1
7^n	7	9	3	1	4
8^n	8	4	2	6	4
9^n	9	1			2

Theorem 1. $a^{4k+r} \equiv a^r \pmod{10}$, where $r = 1, 2, 3,$ or 4 and k is a positive integer.

Example 5. Find the last digit of 3^{1999}.

Solution: 7.
The pattern for the last digits of 3^n is: 3, 9, 7, 1 (repeating every four numbers). When 1999 is divided by 4, the quotient is 499 and the remainder is 3. The last digit of 3^{1999} is the same as the last digit of 3^3.
So the last digit of 3^{1999} is 7.

Or $3^{1999} = 3^{4 \times 499 + 3} = 3^3 = 27$. The last digit is 7.

Example 6. The units digit of 7^{2010} is:
A . 1   B . 3   C . 5   D . 7   E . 9

Solution: E.
The last digits of 7^n have patterns: 7, 9, 3, 1.
$7^{2010} = 7^{502 \times 4 + 2}$.
The last digit of 7^{2010} is the same as the last digit of 7^2, which is 9.

Example 7. Find the remainder when 3^{98} is divided by 5.

A. 0   B. 1   C. 2   D. 3   E. 4

Solution: E.
The remainders when power of 3 are divided by five are as follows:
$3^0 \div 5 \to 1$; $3^1 \div 5 \to 3$; $3^2 \div 5 \to 4$; $3^3 \div 5 \to 2$. So you can see that these cycle through 1, 3, 4, 2, 1, 3, 4, 2.... Since 98 has remainder 2 when divided by 4, the remainder will be the same as 3^2, which is 4.

3. Divisors And Theorems

A divisor, also called a factor, of a number divides that number without leaving a remainder.
(1) an odd number is of the form $2k + 1$, for some integer k.
(2) an even number is of the form $2m$, for some integer m.

Theorem 2. There exists a unique pair (q, r) such that $a = qb + r$ or $\dfrac{a}{b} = q + \dfrac{r}{b}$, where a and b are integers, $b > 0$, q is the quotient, and r is the remainder with $0 \le r < b$.

If $r = 0$, $\dfrac{a}{b} = q$, which can be written as $a = qb$. This is equivalent to saying that a is divisible by b, $b|a$, and b divides a.

Theorem 3. The largest number which divides any two given numbers leaving the same remainder equals the difference of the two numbers.

Theorem 4. If $a|b$, $a|c$, then $a| (jb + kc)$, and j, k are any integers.

Example 8. The number 839 can be written as $19q + r$ where q and r are positive integers. What is the largest possible value of $q - r$?
A. 37 B. 39 C. 41 D. 45 E. 47

Solution: C.
Divide 839 by 19 to get $839 = 44 \cdot 19 + 3$. One pair of positive value is $q = 44$ and $r = 3$. If we try to make q any bigger, we are forced to make r negative.

Example 9. If we divide 344 by d the remainder is 3, and if we divide 715 by d the remainder is 2. Which of the following is true about d?
A. $10 \leq d \leq 19$ B. $20 \leq d \leq 29$ C. $30 \leq d \leq 39$ D. $40 \leq d \leq 49$ E. $50 \leq d \leq 59$

Solution: C.
The number d divides $341 = 11 \times 31$ and $713 = 23 \times 31$. The only common divisors are 1 and 31. Since we get nonzero remainders, $d = 31$.

Example 10. If n is divided by 90, the remainder is 1. What are the remainders, respectively, if $n + 179$ and $n + 182$ are divided by 90?
A. 89 and 3 B. 89 and 2 C. 0 and 3 D. 0 and 2 E. 2 and 3

Solution: C.
We know that n is divided by 90, the remainder is 1. So we have $n = q \times 90 + 1$ or $n - 1 = q \times 90$. That is, $(n - 1)$ is divisible by 90.
$n + 179 = (n - 1) + 180$ which is divisible by 90.
$n + 182 = (n - 1) + 180 + 3$ which has the remainder of 3 when divided by 90.
So the answer is C.

Example 11. For a certain integer n, $5n + 16$ and $8n + 29$ have a common factor larger than one. That common factor is

Algebra II Through Competitions   **Chapter 23 Number Theory**

A. 11   B. 13   C. 17   D. 19   E. 23

Solution: C.

Since $5n + 16$ and $8n + 29$ have a common factor larger than one, $\dfrac{8n+29}{5n+16}$ is an integer. Then there is some number p that divides both $8n + 29$ and $5n + 6$. That same p will also divide any linear combination of these terms; in particular, by Theorem 4, p divides $5(8n + 29) - 8(5n + 16) = 17$. Since 17 is a prime number, so the common factor is 17.

Example 12. Mickey and Minnie have x one dollar bills. Minnie noticed that if she stacked the dollar bills in stacks of 8, she had 7 bills left over. When Mickey made stacks of 6, he had 3 left over. Together they made stacks of 7 and had 4 left over. If they have less than $100, how many bills did they have left over when they made stacks of 5?
A. 4   B. 3   C. 2   D. 1   E. 0

Solution: A.
For stacks of 8 with 7 bills left over, $x = 8n + 7$, for some integer $n, n \geq 1$, so $x \in \{15, 23, 31, 39, \ldots, 95\}$. Similarly, $x = 6k + 3$, for some integer $k, k \geq 1$, so $x \in \{9, 15, 21, \ldots, 97\}$, and $x = 7m + 4$ for some integer $m, m \geq 1$, so $x \in \{11, 18, 25, \ldots, 95\}$. Since 39 is the only element of all 3 sets, we seek to find the remainder when 39 is divided by 5. That remainder is 4.

Theorem 5. (The Fundamental Theorem of Arithmetic)

Each composite natural number can be expressed in one and only one way as a product of primes.

For an integer n greater than 1, the unique prime factorization of n is $n = p_1^a p_2^b p_3^c \ldots p_k^m$, where $a, b, c, \ldots,$ and m are nonnegative integers, $p_1, p_2, \ldots, p_k$ are prime numbers.

Theorem 6. The **number of divisors** of n is: $d(n) = (a+1)(b+1)(c+1)\ldots(m+1)$

Theorem 7. The **sum of divisors** of n is:

Theorem 8. The **product of** $\sigma(n) = (\dfrac{p_1^{a+1}-1}{p_1 - 1})(\dfrac{p_2^{b+1}-1}{p_2 - 1})\ldots(\dfrac{p_k^{m+1}-1}{p_k - 1})$ **divisors** of n is:
$\tau(n) = (p_1^a p_2^b p_3^c)^{(a+1)(b+1)(c+1)/2}$.

366

Example 13. How many positive integer divisors does $N = 6^3 \cdot 150$ have?
A. 32 B. 75 C. 18 D. 24 E. 27

Solution: B.
The prime factorization of N is $2^3 \cdot 3^3 \cdot 2 \cdot 3 \cdot 5^2 = 2^4 3^4 5^2$. Thus N has $(4+1)(4+1)(2+1) = 75$ divisors.

Example 14. A certain integer N has exactly eight factors, counting itself and 1. The numbers 35 and 77 are two of the factors. What is the sum of the digits of N?
A. 9 B. 10 C. 16 D. 18 E. 20

Solution: C.
The divisors include 5, 7, and 11. But $5^1 \cdot 7^1 \cdot 11^1 = 385$ already has $(1+1)(1+1)(1+1) = 8$ divisors. So the sum of the digits is $3 + 8 + 5 = 16$.

Example 15. Let k be a positive integer such that $23 < k < 64$. Find the sum of all possible distinct values of k such that the product of all distinct positive integral divisors of k is k^2.

Solution: 561.
Since $2 \times 3 \times 5 \times 7 = 210 > 64$. So k has at most 3 different prime factors.
$k = p_1^a p_2^b p_3^c$.
We know that the product of all distinct positive integral divisors of k is
$(p_1^a p_2^b p_3^c)^{(a+1)(b+1)(c+1)/2}$.
So we have $(p_1^a p_2^b p_3^c)^{(a+1)(b+1)(c+1)/2} = (p_1^a p_2^b p_3^c)^2$.
Thus $(a+1)(b+1)(c+1) = 4$.

Case I: $(a+1)(b+1)(c+1) = 1 \times 1 \times 4$.
$a = 0$, $b = 0$, and $c = 3$.
$k = p_1^a p_2^b p_3^c = p_3^3$. We have $k = 3^3 = 27$.

Case II: $(a+1)(b+1)(c+1) = 1 \times 2 \times 2$.
$a = 0$, $b = 1$, and $c = 1$.
$k = p_1^a p_2^b p_3^c = p_2 p_3$.

367

We have $k = 2\times 13 = 26$, $k = 2\times 17 = 34$. $k = 2\times 19 = 38$. $k = 2\times 23 = 46$. $k = 2\times 29 = 58$. $k = 2\times 31 = 62$.
$k = 3\times 11 = 33$, $k = 3\times 13 = 39$, $k = 3\times 17 = 51$. $k = 3\times 19 = 57$.
$k = 5\times 7 = 35$, $k = 5\times 11 = 55$.
The sum is $27 + 26 + 34 + 38 + 46 + 58 + 62 + 33 + 39 + 51 + 57 + 35 + 55 = 561$.

Algebra II Through Competitions   **Chapter 23 Number Theory**

PROBLEMS

Problem 1. What is the sum of all the positive integers between 1 and 2011 that are divisible by five?
A. 81,003   B. 403,005   C. 404,215   D. 405,015   E. 810,030

Problem 2. How many integers between 100 and 1000 are multiples of 7?
A. 128   B. 130   C. 132   D. 134   E. 136

Problem 3. If x and y are two-digit positive integers with $xy = 555$, what is $x + y$?
(A) 116   (B) 188   (C) 52   (D) 45   (E) None of these

Problem 4. Find the sum of all distinct positive integers less than 199 that are integral multiples of 24 or integral multiples of 40.

Problem 5. Find the ones digit of 7^{7^7}.
A. 1   B. 3   C. 5   D. 7   E. 9

Problem 6. What is the remainder when you divide 3^{2008} by 13?
A. 0   B. 1   C. 3   D. 9   E. none of these

Problem 7. Let a, b, c, and d represent four distinct positive integers where $a^2 - b^2 = c^2 - d^2 = 9^2$. Find the value of $a + b + c + d$.
A. 27   B. 65   C. 98   D. 108   E. 111

Problem 8. The difference between the squares of two consecutive odd integers is 128. What is the product of the two integers?
A. 899   B. 1023   C. 1155   D. 783   E. 483

Problem 9. Compute the maximum value of $x + y$ for all positive integer pairs (x, y) that satisfy $15x + 55y = 2000$.
A. 104   B. 120   C. 128   D. 172   E. 196

Problem 10. N is a natural number with properties that $10 < N < 100$, if N is divided by 7 its remainder is 3, and if N is divided by 13 its remainder is also 3. Find the product of the digits of N.
A. 0   B. 9   C. 36   D. 42   E. 64

Problem 11. Which of the following primes is not a factor of $55^{100} + 55^{101} + 55^{102}$?
A. 3 B. 5 C. 7 D. 11 E. 13

Problem 12. How many positive integers n have the property that 62 is divisible by $n - 4$?
A. 0 B. 2 C. 4 D. 6 E. None of these

Problem 13. Exactly two of the divisors of $4^{12} - 1$ are between 60 and 80. Find the sum of these two divisors.
A. 126 B. 128 C. 129 D. 140 E. 142

Problem 14. If $3x$, $x/3$, and $15/x$ are integers, which of the following must also be an integer?
I. $x/3$ II. x III. $6x$
A. I only. B. II only. C. III only. D. I and III only. E. II and III only.

Problem 15. Two integers are said to be partners if both are divisible by the same set of prime numbers. The number of positive integers greater than 1 and less than 25 that have no partners in this set of integers is
A. 9. B. 10. C. 11. D. 12. E. 13.

Problem 16. Determine the largest prime number dividing $27! + 28!$.
A. 7 B. 23 C. 29 D. 53 E. None of these

Problem 17. For how many integers n between 1 and 1990 is the improper fraction $\dfrac{(n^2 + 7)}{(n^2 + 4)}$ NOT in lowest terms?
A. 0 B. 86 C. 90 D. 104 E. 105

Problem 18. The number of positive integers k for which the equation $kx + 12 = 3k$ has an integer solution for x is
A. 3 B. 4 C. 5 D. 6 E. 7

Problem 19. What is the product of all the even divisors of 1000?
(A) 32×10^{12} (B) 64×10^{14} (C) 128×10^{16} (D) 64×10^{18} (E) 10^{24}

Problem 20. For how many n in $\{1, 2, 3, \ldots, 100\}$ is the tens digit of n^2 odd?
A. 16 B. 17 C. 18 D. 19 E. 20

Problem 21. How many ordered pairs (m,n) of positive integers are solutions to $4/m + 2/n = 1$?
A. 1. B. 2. C. 3. D. 4. E. More than 4.

Problem 22. For integers n and m, let $gcd(n, m)$ denote the greatest common divisor of n and m. Compute $\sum_{n=1}^{30} gcd(n,30)$.
A. 232 B. 245 C. 465 D. 66 E. 135

SOLUTIONS:

Problem 1. Solution: D.
The first number is 5 and the last number is 2010. There are n such numbers in $2010 = 5 + (n-1) \times 5 \implies n = 402$. So The sum is $(5 + 2010) \times 5/2 = 405{,}105$.

Problem 2. Solution: A.
Method 1: The number of integers less than 1000 divisible by 7 is n and $994 = 7 + (n-1) \times 7 \implies n = 142$.
The number of integers less than 100 divisible by 7 is m and $98 = 7 + (m-1) \times 7 \implies m = 14$.
The answer is $142 - 14 = 128$.

Method 2: $\left\lfloor \dfrac{1000}{7} \right\rfloor - \left\lfloor \dfrac{100}{7} \right\rfloor = 142 - 14 = 128$.

Problem 3. Solution: C.
By prime factorization, we have $555 = 3 \cdot 5 \cdot 37$ and since x and y and are required to be two-digit integers, the only pair of integers that satisfy the problem conditions is 15 and 37.

Problem 4. Solution: 1144.
The sum of all distinct positive integers less than 199 that are integral multiples of 24:
$1 \times 24 + 2 \times 24 + 3 \times 24 + 4 \times 24 + 5 \times 24 + 6 \times 24 + 7 \times 24 + 8 \times 24 = 24 \times (1 + 2 + 3 + 4 + 5 + 6 + 7 + 8) = 24 \times (1 + 8) \times 8/2 = 864$.
The sum of all distinct positive integers less than 199 that are integral multiples of 40:
$1 \times 40 + 2 \times 40 + 3 \times 40 + 4 \times 40 = 40 \times (1 + 2 + 3 + 4) = 400$.
But we counted $lcm(24, 40) = 120$ twice. So the answer is $864 + 400 - 120 = 1144$.

Problem 5. Solution: D.
$7^4 = 2401$. So the last digit of 7^4 is 1. The last digit of 7^7 is the same as the last digit of 7^3, which is 3. So the last digit of 7^{7^7} is the same the last digit of 3^7. We know that $3^7 = (3^4)3^3 = 81 \times 27$. So the last digit of 3^7 is 7.

Problem 6. Solution: C.
$3^3 = 27$ and the remainder is 1 when 27 is divided by 13.

$(3^{2007}) = (3^3)^{669}$. So the remainder is 1 when 3^{2007} is divided by 13.
$3^{2008} = (3^{2007}) \times 3$.
So the remainder is 3 when 3^{2008} is divided by 13.

Problem 7. Solution: D.
$a^2 - b^2 = c^2 - d^2 = 9^2 \Rightarrow (a-b)(a+b) = (c-d)(c+d) = 81$. If a, b, c, d are distinct positive integers, then $1 \times 81 = 81$ or $3 \times 27 = 81$. Let $c - d = 1$ and $c + d = 81$. Then $c = 41$ and $d = 40$. Let $a - b = 3$ and $a + b = 27$. Then $a = 15$ and $b = 12$. So $a + b + c + d = 108$.

Problem 8. Solution: B.
Let the two consecutive odd integers be $2n - 1$ and $2n + 1$. The difference then is
$(2n + 1)^2 - (2n - 1)^2 = 8n = 128 \Rightarrow n = 16$. The product is
$(2n + 1)(2n - 1) = 4n^2 - 1 = 4(16^2) - 1 = 1023$.

Problem 9. Solution: C.
First notice that $15x + 55y = 2000 \Leftrightarrow 3x + 11y = 400$. Since 400 has remainder 1 when divided by 3, and $3x$ is always divisible by 3, we need a multiple of 11 what has remainder of 1 when divided by 3. This happens when $y = 2$, among others. So we have $3x + 11(2) = 400 \Rightarrow 3x = 400 - 22 = 378 \Rightarrow x = 126$. So the ordered pair $(126, 2)$ satisfies this equation. Each time we increase the y-values by 3, we must decrease the x-values by 11, so the set of ordered pairs $\{(126 - 11k, 2 + 3k), 0 \le k \le 11\}$ gives us all ordered pairs of positive integers which satisfy the equation. The maximum sum of the coordinates occurs with the ordered pair $(126, 2)$ and is 128.

Problem 10. Solution: C.
IF we consider the numbers between 10 and 100 which have remainder of 3 when divided by 13, we get 16, 29, 42, 55, 68, 81, and 94. For each of these the remainder is 2, 1, 0, 6, 5, 4, and 3, respectively when divided by 7. We are looking for the one with remainder 3, so the desired number is 94. The product of its digits is 36.

Problem 11. Solution: C.
$55^{100} + 55^{101} + 55^{102} = 55^{100} (1 + 55 + 55^2) = 5^{100} \times 11^{100} \times 3081$. 3081 is not divisible by 7.

Problem 12. Solution: D.
$62 = 2 \times 31$. There are 4 factors: 1, 2, 31, and 62.
So $\quad n - 4 = 1 \quad \Rightarrow \quad n = 5$.
$\quad\quad\, n - 4 = 2 \quad \Rightarrow \quad n = 6$.
$\quad\quad\, n - 4 = 31 \quad \Rightarrow \quad n = 35$.
$\quad\quad\, n - 4 = 62 \quad \Rightarrow \quad n = 66$.

We also have
$$n - 4 = -1 \Rightarrow n = 3.$$
$$n - 4 = -2 \Rightarrow n = 2.$$
$$n - 4 = -31 \Rightarrow n = -27 \text{ (ignored)}.$$
$$n - 4 = -62 \Rightarrow n = -58 \text{(ignored)}.$$
The answer is D.

Problem 13. Solution: B.
We know $4^{12} - 1 = (4^6)^2 - 1 = (4^6 - 1)(4^6 + 1) = [(4^3)^2 - 1](4^6 + 1)$
$= (4^3 - 1)(4^3 + 1)(4^6 + 1) = (64 - 1)(64 + 1) \times (4^6 + 1) = 63 \times 65 \times (4^6 + 1)$.
So the two divisors are 63 and 65. The sum is $65 + 63 = 128$.

Problem 14. Solution: C.
Look beyond integers 1 and 3 and consider non-integer rational numbers such as 1/3 for x. Then $x/3$ and x are not integers. Note $6x = 2 \cdot 3x$ so $6x$ must be an integer. Only III is an integer.

Problem 15. Solution: C.
The 11 such integers with no partners in the set are 5, 7, 11, 13, 14, 15, 17, 19, 21, 22, 23.

Problem 16. Solution: C.
$27! + 28! = 27! (1 + 28) = 27! \times 29$. So the answer is C.

Problem 17. Solution: A.
$$\frac{(n^2 + 7)}{(n^2 + 4)} = \frac{n^2 + 4 + 3}{n^2 + 4} = 1 + \frac{3}{n^2 + 4}.$$
No matter what the value of n from 1 to 1990 is, $n^2 + 4$ and 3 have no common factor. So the answer is A.

Problem 18. Solution: D.
We have $x = 3 - 12k$, so k must be a positive integer that divides 12 and each such choice will have an integer solution for x. So $k = 1, 2, 3, 4, 6, 12$ are the values of k.

Problem 19. Solution: D.
Since $1000 = 2^3 \cdot 5^3$ it has only four odd divisors, 1, 5, 25, and 125. The product of all 16 divisors is 1000^8 since the 16 divisors form pairs of numbers that have 1000 as their product. Thus the product P of the even divisors is $P = (2^3 \cdot 5^3)^8 \div (5 \cdot 5^2 \cdot 5^3) = 2^{24} \cdot 5^{24}$
$\div 5^6 = 2^{24} 5^{18} = 2^6 (2 \cdot 5)^{18} = 64 \cdot 10^{18}$.

Algebra II Through Competitions   **Chapter 23 Number Theory**

Problem 20. Solution: E.
There are 2 in each decile, $10a + 4$ and $10a + 6$. The tens digit of $(10a + 4)^2 = 100a^2 + 80a + 16$ is the units digit of $8a + 1$, while the tens digit of $(10a + 6)^2 = 100a^2 + 120a + 36$ is the units digit of $2a + 3$, both of which are odd for any integer a. All the other tens digits of perfect squares are even: $(10a + b)^2 = 100a^2 + 20a + b^2$, the tens digit of which is the tens digit of $2a + b^2$, which is even if the tens digit of b^2 is even. But the tens digit of b^2 is even if $b \ne 4$, $b \ne 6$.

Problem 21. Solution: D.
Since m and n must both be positive, it follows that $n > 2$ and $m > 4$. Because $4/m + 2/n = 1$ is equivalent to $(m - 4)(n - 2) = 8$, we need only find all ways of writing 8 as a product of positive integers. The four ways $1 \cdot 8, 2 \cdot 4, 4 \cdot 2$, and $8 \cdot 1$, correspond to the four solutions $(m,n) = (5, 10), (6, 6), (8, 4)$, and $(12, 3)$.

Problem 22. Solution: E.
Case I: 30 has 8 factors: 1, 2, 3, 5, 6, 10, 15, and 30. We know that $gcd(m, 30) = m$ if m is a factor of 30.
So we have $1 + 2 + 3 + 5 + 6 + 10 + 15 + 30 = 72$

Case II: There are 7 numbers that are relatively prime to 30 (7, 11, 13, 17, 19, 23, 29).
We know that $gcd(p, 30) = 1$ if p is relatively prime to 30. So we have $7 \times 1 = 7$.

Case III: We have more numbers as follows:
4, 8, 9, 12, 14, 16, 18, 20, 21, 22, 24, 25, 26, 27, and 28 with the gcd's:
2, 2, 3, 6, 2, 2, 6, 10, 3, 2, 6, 5, 2, 3, and 2.
The sum is $72 + 7 + (2 + 2 + 3 + 6 + 2 + 2 + 6 + 10 + 3 + 2 + 6 + 5 + 2 + 3 + 2) = 135$.

Chapter 24 Geometry

1. LINE SEGMENTS

Theorem 1: If $\triangle ABC \sim \triangle A_1B_1C_1$, then $\dfrac{a}{a_1} = \dfrac{b}{b_1} = \dfrac{c}{c_1}$ and the ratio of the areas is as follows:

$$\frac{S_{\triangle ABC}}{S_{\triangle A_1B_1C_1}} = \left(\frac{a}{a_1}\right)^2 = \left(\frac{b}{b_1}\right)^2 = \left(\frac{c}{c_1}\right)^2$$

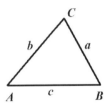

Theorem 2: If $DE // BC$, then $\triangle ABC \sim \triangle ADE$ and

$$\frac{AD}{AB} = \frac{AE}{AC} = \frac{DE}{BC}$$

$$\frac{AD}{DB} = \frac{AE}{EC} \quad \Rightarrow \quad \frac{AD}{AE} = \frac{DB}{EC}$$

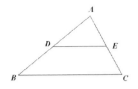

Theorem 3: If $\angle ACB = \angle ADC = 90°$, then

$$\frac{AC}{AD} = \frac{AB}{AC} \quad \Rightarrow \quad AC^2 = AB \times AD \tag{1}$$

$$\frac{BC}{AB} = \frac{BD}{BC} \quad \Rightarrow \quad BC^2 = AB \times BD \tag{2}$$

$$\frac{CD}{AD} = \frac{BD}{CD} \quad \Rightarrow \quad CD^2 = AD \times BD \tag{3}$$

$$\frac{1}{2} CD \times AB = \frac{1}{2} AC \times BC \quad \Rightarrow \quad CD \times AB = AC \times BC \tag{4}$$

$$\frac{1}{CD^2} = \frac{1}{BC^2} + \frac{1}{AC^2} \tag{5}$$

Theorem 4 (Angle Bisector Theorem):

The angle bisector of a triangle divides the opposite side into segments that are proportional to the adjacent sides.

$$\frac{AB}{AC} = \frac{BD}{CD} \quad \text{or} \quad \frac{AB}{BD} = \frac{AC}{CD}.$$

The length of the angle bisector is $AD^2 = AB \times AC - BD \times DC$.

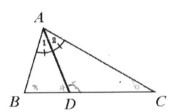

Theorem 5. Tangents to a circle from an outside point are congruent. $PA = PB$.

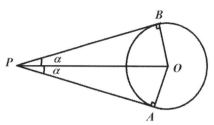

Example 1. An isosceles triangle ABC with point D on AB, has $AC = BC = BD$ and $AD = DC$. If $AB = 2$, find the length of CD.
A. 1   B. $\sqrt{2}$ C. $\sqrt{5}-1$
D. $\sqrt{10}-3$ E. $3-\sqrt{5}$

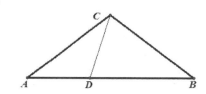

Solution: E.
Let $x = CD$.
Since $\triangle ABC \sim \triangle ABC \Rightarrow CD : AC = AC : AB \Rightarrow 2x = (AC)^2$.
$AD + BD = AB \Rightarrow x + AC = 2 \Rightarrow 2x = (2-x)^2 \Rightarrow x^2 - 6x + 4 = 0 \Rightarrow x = 3 - \sqrt{5}$.

Example 2. A small tree 5 feet from a lamp post casts a shadow 4 feet long. If the lamp post were 2 feet higher the shadow would only be 3 feet long. How tall is the tree?
A. 7/2 ft.   B. 24/5 ft.   C. 8/3 ft.   D. 14/3 ft.   E. 17/5 ft.

Solution: B.
$\triangle ABC \sim \triangle DEC$, then $\dfrac{AB}{BC} = \dfrac{DE}{EC}$ $\Rightarrow$ $\dfrac{a}{5+4} = \dfrac{b}{4}$

$\Rightarrow \quad a = \dfrac{9b}{4}$ (1)

Similarly $\dfrac{a+2}{5+3} = \dfrac{b}{3}$ (2)

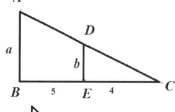

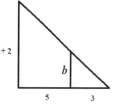

Substituting (1) into (2): $\dfrac{\dfrac{9b}{4}+2}{8} = \dfrac{b}{3} \Rightarrow b = \dfrac{24}{5}$.

Example 3: In Triangle ABC, $AB = 16$, $BC = 5$, and $AC = 19$. AD bisects $\angle CAB$, and CD bisects $\angle ACB$. Expressed in simplest radical form, $AD = k\sqrt{w}$, where k and w are positive integers. Find the value of $(k + w)$.
Solution: 59.

Applying the angle bisector theorem, we have

$$\frac{AB}{5-CE} = \frac{AC}{CE} \quad \Rightarrow \quad \frac{16}{5-CE} = \frac{19}{CE}$$

$CE = 19/7$.

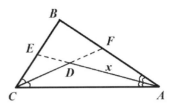

By the angle bisector length formula, we have $AE^2 = AB \times AC - CE \times EB \Rightarrow$

$$AE^2 = 16 \times 19 - \frac{19}{7} \times (5 - \frac{19}{7}) \Rightarrow \quad AE = \frac{16\sqrt{57}}{7}.$$

Applying the angle bisector theorem again, we have

$$\frac{CE}{ED} = \frac{AC}{AD} \quad \Rightarrow \quad \frac{\frac{19}{7}}{AE-x} = \frac{19}{x} \quad \Rightarrow \quad 7AE - 7x = x \Rightarrow$$

$$x = \frac{7AE}{8} = \frac{7}{8} \times \frac{16\sqrt{57}}{7} = 2\sqrt{57}.$$

$k = 2$ and $w = 57$. So $k + w = 2 + 57 = 59$.

Example 4. A circle is inscribed in a triangle with sides of lengths of 8, 17, and 19. Let the segments determined by the point of tangency on the side of length 8 be w and r, with $w < r$. Find the ratio of w to r.
A. 3:5   B. 5:7   C. 2:3   D. 1:5   E. 7:9

Solution: A.
$r + w = 8$ (1)
$19 - r = 17 - w$ (2)
Solving the system of equations: $r = 5$ and $w = 3$.
The ratio is 3:5.

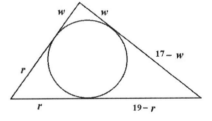

2. PYTHAGOREAN THEOREM

(For right triangles only): $a^2 + b^2 = c^2$ (a and b are two legs. c is the hypotenuse).

Proof:
In right triangle ABC, draw $CD \perp AB$. We know that $\triangle ABC \sim \triangle ACD$ $\sim \triangle CBD$ and:
$AC^2 = AB \times AD$ (1)
$BC^2 = AB \times BD$ (2)
(1) + (2): $AC^2 + BC^2 = AB \times AD + AB \times BD = AB(AD + BD) = AB \times AB = AB^2$

Example 5. For what positive value of x is there a right triangle with sides $2x + 2$, $6x$, and $6x + 2$?
A. 8   B. 6   C. 4   D. 5   E. 9

Solution: C.
The number x must satisfy $(2x + 2)^2 + (6x)^2 = (6x + 2)^2$ since $6x + 2$ is clearly the largest of the three numbers. Thus $4x^2 + 8x + 4 + 36x^2 - 36x^2 - 24x - 4 = 4x^2 - 16x = 0$ from which it follows that $x = 4$.

Example 6. When the base of a ladder is 16 *ft* from the base of a wall, 3 *ft* of the ladder projects beyond the top of the wall. When the base of the ladder is 9 *ft* from the base of the wall, 8 *ft* of the ladder projects beyond the top of the wall. How long is the ladder?
A. 15 *ft*   B 16.5 *ft*   C. 20 *ft*   D. 23 *ft*   E. 25.25 *ft*

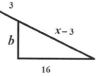

Solution: D.
Let the length be x.

$$b^2 + 16^2 = (x-3)^2 \quad (1)$$
$$b^2 + 9^2 = (x-8)^2 \quad (2)$$

(1) −(2): $16^2 - 9^2 = (x-3)^2 - (x-8)^2 \Rightarrow 175 = 10x - 55 \Rightarrow x = 23$.

3. QUADRATIC EQUATION/FUNCTION APPLICATION

Example 7. A rectangular garden measures 8 feet wide and 16 feet long. The gardener wants to put a strip of gravel of uniform width around the garden. There is enough gravel or 112 square feet. How wide will the strip be?
A. 2 feet   B. 1 foot   C. 6 inches   D. 4 feet   E. 3 feet.

Solution: A.
$(16 + 2x)(8 + 2x) - 8 \times 16 = 112 \Rightarrow 4x^2 + 48x - 112 = 0 \Rightarrow$
$x^2 + 12x - 28 = 0 \Rightarrow (x-2)(x+14) = 0$. So $x = 2$.

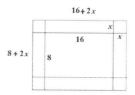

xample 8. Each spring a 12 meter × 12 meter rectangular garden has its length increased by 2 meters but its width decreased by 50 centimeters. What will be the maximum attainable area of the garden?
A. 144 m^2 B. 176 m^2 C. 189 m^2 D. 200 m^2 E. 225 m^2

Solution: E.

The area will be $(12 + 2y)(12 - 0.5y) = 144 + 24y - 6y - y^2 = 144 + 18y - y^2$, where y is the number of years. This quadratic expression has a maximum at its vertex, which occurs when $y = -b/(2a) = -18/(-2) = 9$. The area when $y = 9$ is $144 + 18(9) - 9^2 = 225$.

Example 9. A rectangular park with dimensions 125 feet by 230 feet is surrounded by a sidewalk of uniform width of x feet along the sides of the park. If the area of the sidewalk is between 2900 square feet and 7900 square feet and if x is an integer, find the sum of all distinct possible values for x. Assume all edges of the sidewalk are composed of straight line segments.

Solution: 49.
$2900 \leq (125 + 2x)(230 + 2x) - 125 \times 230 \leq 7900$.
$2900 \leq 4x^2 + 710x \leq 7900 \implies 1450 \leq x(2x + 355) \leq 3950$.
Since $1450 \div 355 \approx 4.1$, we test $x = 4$: $x(2x + 355) = 1452$. So $x = 4$ is good.
Since $3950 \div 355 \approx 11.13$, we test $x = 11$: $x(2x + 355) = 4147$. So $x = 11$ is not good.
We also see that when $x = 10$, $x(2x + 355) = 3750$. So $x = 10$ is not good.
Thus the sum of all values for x is $4 + 5 + \ldots + 10 = 49$.

4. AREAS AND VOLUMES

4.1. Given a triangle with side a, b, and c and height h on the side a, the area of this triangle is A and

$$A = \frac{1}{2}ah \qquad A = \frac{1}{2}a \times b \times \sin C$$

$$A = \sqrt{s(s-a)(s-b)(s-c)}$$

$$s = \frac{1}{2}(a+b+c)$$

$$R = \frac{a}{2\sin A} = \frac{b}{2\sin B} = \frac{c}{2\sin C} = \frac{abc}{4A}$$

$$r = \frac{A}{s} = \frac{2A}{perimeter}$$

4.2. Surface area of a sphere: $A = 4\pi r^2$. Volume of a sphere: $V = \frac{4}{3}\pi r^3 = \frac{1}{6}\pi d^3$.

Example 10. The area of a triangle is 600 square feet. If the height is three times the base, what is the height of the triangle?
A. $10\sqrt{2}$ feet   B. 20 feet   C. 60 feet   D. $30\sqrt{2}$ feet   E. $10\sqrt{6}/3$ feet.

Solution: C.

$A = \frac{1}{2} b \times h \quad \Rightarrow \quad 600 = \frac{1}{2} \times \frac{1}{3} h \times h \quad \Rightarrow \quad h^2 = 600 \times 2 \times 3 = 6^2 \times 10^2 \quad \Rightarrow \quad h = 60$.

Example 11. In square $ABCD$, X lies on DC such that $DX : XC = 5 : 2$ and Y lies on BC such that $BY : YC = 3 : 4$. The ratio of the area of $\triangle AXC$ to the area of $\triangle ABY$ is
A. 2:7.   B. 2:3.   C. 3:4.   D. 4:9.   E. 9:16.

Solution: B.
Since 5 + 2 = 7, and 3 + 4 = 7, each side length is a multiple of 7, so 5 : 2 can be represented as $5a : 2a$ and 3 : 4 can be represented at $3a : 4a$.
Then the area $\triangle AXC$: area$\triangle ABY = \frac{1}{2}(2a) \times 7 : \frac{1}{2}(3a) \times 7 = 2 : 3$.

Example 12. The number of cubic centimeters in the volume of a sphere is the same as the number of square millimeters in its surface area. The radius of the sphere in millimeters is:
A. 300   B. 30   C. 3   D. 0.03   E. none of these

Solution: C.
Surface area of a sphere: $A = 4\pi r^2$. Volume of a sphere: $V = \frac{4}{3}\pi r^3$.
So $\frac{4}{3}\pi r^3 = 4\pi r^2 \quad \Rightarrow \quad \frac{1}{3} r = 1 \quad \Rightarrow \quad r = 3$.

Example 13. The height of a triangle is 6 units less than the length of its base. If the area of this triangle is 42 square units, how many units is the length of its base?
A. $\sqrt{90} + 3$ B. $\sqrt{90} - 3$ C. $\sqrt{60}$ D. 10   E. $\sqrt{42} - \sqrt{5}$.

Solution: A.
Let b be the length of the base. $42 = \frac{1}{2}b(b-1) \Rightarrow b^2 - 6b - 84 = 0$.
$b = \frac{6 + \sqrt{36 + 4 \times 84}}{2} = 3 + \sqrt{93}$, and $b = 3 - \sqrt{93}$ (ignored).

PROBLEMS

Problem 1. If the length and width of a rectangle were increased by 1, the area would be 84. The area would be 48 if the length and width were diminished by 1. Find the perimeter P of the original rectangle.
A. $10 < P < 20$ B. $20 < P < 30$ C. $30 < P < 40$ D. $40 < P < 50$ E. none of the above

Problem 2. The altitudes of a triangle are three distinct integers, the larger two of which are 12 and 66. Find the length of the shortest altitude. Which of the following intervals contains the integer solution?
A. $(0, 3)$ B. $[3, 6)$ C. $[6, 9)$ D. $[9, 12)$ E. none of the above

Problem 3. Suppose the lengths of three of the four lateral edges of a pyramid with a rectangular base are 8, 5, and 1 in that order. Find the length of the fourth lateral edge.
A. $\sqrt{79}$ B. $\sqrt{47}$ C. $2\sqrt{10}$ D. $2\sqrt{11}$ E. $2\sqrt{3}$.

Problem 4. An isosceles right triangle region of area 36 is cut from a corner of a rectangular region with sides of length $6\sqrt{2}$ and $6(\sqrt{2}+1)$. What is the perimeter of the resulting trapezoid?
A. 36   B. $18\sqrt{2}+18$ C. 30   D. $12\sqrt{2}+24$ E. $24\sqrt{2}+12$.

Problem 5. Pictured are two semicircles. AB is tangent to the smaller semicircle and parallel to CD. If AB is 16, find the area of the shaded region.
a. 16π b. 32π c. 8π d. π e. 18π

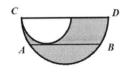

Problem 6. A triangle has one side of length a and two sides of length b. The area of the triangle is

A. $\dfrac{a}{4}\sqrt{4b^2-a^2}$ B. $\dfrac{a}{2}\sqrt{4b^2-a^2}$ C. $ab\sqrt{\dfrac{3}{4}}$ D. $\dfrac{b}{4}\sqrt{4a^2-b^2}$ E. $\dfrac{b}{2}\sqrt{4a^2-b^2}$

Problem 7. A 25 foot tall ladder is placed along the vertical wall of a house. The foot of the ladder is 20 feet from the bottom of the house. If the top of the ladder slips 8 feet, then the foot of the ladder will slide how many feet?
A. 3 ft.   B. 5 ft.   C. 8 ft.   D. 4 ft.   E. 7 ft.

Algebra II Through Competitions Chapter 24 Geometry

Problem 8. The perimeter of a rectangle is 100 cm, and its diagonal has length x. Express the area of the rectangle as a function of x.

A. $\dfrac{(50-x)^2}{2}$ cm^2 B. $\dfrac{(50+x)^2}{2}$ cm^2 C. $\dfrac{2500-x^2}{2}$ cm^2 D. $2x^2 - 2500$ cm^2

E. $x^2 + 2500$ cm^2.

Problem 9. The volume of wood V in a tree varies jointly as the height h and the square of the girth g (girth is the distance around). If the volume of a redwood tree is 216m³ when the height is 30 m and the girth is 1.5 m, what is the height of a tree whose volume is 960m³ and girth is 2m?
A. 14.8 m.   B. 45 m.   C. 50 m.   D. 75 m.   E. 95 m.

Problem 10. A street sign 20 feet from a lamp post casts a shadow 5 feet long. If the lamp post were 1 foot taller the shadow would only be 4 feet 8 inches long. How high is the lamp on the post?
A. 15 ft.   B. 17 1/2 ft.   C. 25 1/3 ft.   D. 35 ft.   E. 70 ft.

Problem 11. A rectangular computer image in enclosed with a two pixel wide frame composed of 504 pixels. If the width of the image is increased by 10% then the frame needs 524 pixels. How many pixels does the image have?
A. 3,600 pixels   B. 3,721 pixels   C. 3,800 pixels   D. 15,200 pixels   E. None of these.

Problem 12. If the sum of the squares of the lengths of all the sides of a rectangle is 100, then the length of a diagonal of the rectangle is
A. $2\sqrt{5}$ B. $2\sqrt{13}$ C. $4\sqrt{3}$ D. $5\sqrt{2}$ E. 10.

Problem 13. A circle is centered at the vertex of the right angle of an isosceles right triangle. The circle passes through the trisection points of the hypotenuse of the triangle. If the length of the radius of the circle is 10, find the area of the triangle.
A. 45.   B. 60.   C. 75.   D. 90.   E. 105.

Problem 14. An open box is to be created from a nine-inch by twelve-inch piece of posterboard by cutting congruent squares from each corner and folding up the sides. The goal is to cut squares of the size that will produce the box with maximum volume. Determine that volume to the nearest cubic inch.
A. 82.   B. 98.   C. 118.   D. 150.   E. 164.

Problem 15. The perimeter of a right triangle is six times longer than its shortest side. What is the value of the ratio of the longer leg to the shorter one?

A. 1.083   B. 1.2   C. 2.4   D. 2.6   E. $2.\overline{6}$.

Problem 16. The lengths of the three sides of a right triangle are consecutive multiples of three. What is the area of the triangle?
A. 108   B. 54   C. 36   D. 90   E. 45

Problem 17: Alicia is 5 feet tall and walks on a perfectly horizontal flat surface at the rate of 5 feet per second away from a streetlight with its lamp 16 feet above horizontal flat surface. Alicia is now directly under the lamp and walk away from the lamp for 22 seconds. Find the number of feet in the length of Alicia's shadow directly after those 22 seconds.

Problem 18. A square is inscribed in another square, such that each vertex divides a side of the outside square into intervals of length x and y, where $x > y$. What is x/y, if the area of the inscribed square is 4/5 of the area of the outside square?
A. $4-\sqrt{15}$ B. $2+\sqrt{6}$ C. $\dfrac{\sqrt{5}}{2}$ D. $4+\sqrt{15}$ E. $2-\sqrt{6}$.

Problem 19. When a right triangle of area 3 square units is rotated 360° about its shortest leg, the solid that results has a volume of 30 cubic units. What is the volume, in cubic units, of the solid that results when the same right triangle is rotated about its longer leg?

Problem 20. A rectangular garden measuring 80 by 60 meters has its area doubled by adding a border of uniform width along both shorter sides and one longer side. Find the width of the border.
A. 10   B. 15   C. 20   D. 25   E. 30.

Problem 21. In a regular pentagon, each diagonal has a length of 1. The perimeter of the pentagon can be written in reduced simplest radical form $\dfrac{k\sqrt{w}-p}{f}$, where k, w, p, and f represent positive integers. Find the smallest possible value of $(k + w + p + f)$.

SOLUTIONS

Problem 1. Solution: C.
$(L+1)(W+1) = 84$ and $(L-1)(W-1) = 48$ so we get
$LW + L + W + 1 = 84$
$LW - L - W + 1 = 48$
Subtracting these last two equations yields $2L + 2W = 36$, so we could go on and solve for L and W, however, we are looking for the perimeter, so we are done.

Problem 2. Solution: D.
One extreme for the shortest altitude will occur when the two shortest altitudes are both 12, in an isosceles triangle. The other extreme occurs in a figure similar to the one below. From similar triangles
$\triangle BMA \sim \triangle COA$ that $\dfrac{12}{x} = \dfrac{66}{AO} \Rightarrow AO = \dfrac{11x}{2}$.

Also from similar triangles $\triangle BMC \sim \triangle ANC$ that $\dfrac{12}{x+u} = \dfrac{h}{w} \Rightarrow h = \dfrac{12w}{x+u}$.

But we know that $u > w$ and $\dfrac{11x}{2} < u$, so

$h = \dfrac{12w}{x+u} < \dfrac{12u}{x+u} < \dfrac{12u}{\dfrac{2}{11}u+u}$.

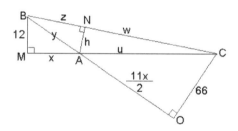

Simplifying this we have so $h < \dfrac{12u}{\dfrac{13}{11}u} = \dfrac{132u}{13u} = 10\dfrac{2}{3}$.

Thus $12 \le x < 10\dfrac{2}{3}$ and the only distinct integer length for the shortest altitude is 11.

Problem 3. Solution: C.

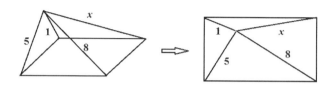

By Pythagorean Theorem, $1^2 + 8^2 = 5^2 + x^2 \Rightarrow x = 2\sqrt{19}$.

Problem 4. Solution: D.

$x = 6(\sqrt{2}+1) - 6\sqrt{2} = 6$

$y = 6\sqrt{2} \cdot \sqrt{2} = 12$.

Perimeter $= 6 + 6\sqrt{2} + 6(\sqrt{2}+1) + 12 = 24\sqrt{2} + 12$.

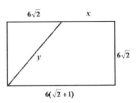

Problem 5. Solution:

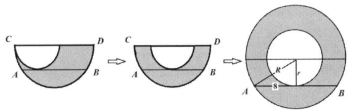

The shaded area is $\frac{1}{2}(\pi R^2 - \pi r^2) = \frac{\pi}{2}(R^2 - r^2) = \frac{\pi}{2}(8^2) = 32\pi$.

Problem 6. Solution: A.

The area is $\frac{1}{2} a \times h = \frac{a}{2}\sqrt{b^2 - (\frac{a}{2})^2} = \frac{a}{2}\sqrt{\frac{4b^2 - a^2}{4}} = \frac{a}{4}\sqrt{4b^2 - a^2}$

Problem 7. Solution: D.
The foot of the 25-foot ladder is 20 feet from the wall and the top of the ladder is 15 feet from the ground. (This is a 3-4-5 right triangle). When the top slips down 8 feet. The base will be $\sqrt{25^2 - 7^2} = \sqrt{625 - 49} = \sqrt{576} = 24$ feet from the wall, so it has moved 4 feet.

Problem 8. Solution: D.
The area is $A = l \times w$, and the perimeter $2(l + w) = 100$, so $l + w = 50$ and $(l + w)^2 = 50^2$. $\Rightarrow l^2 + w^2 + 2lw = 2500$. This last expression can be rewritten as $l^2 + w^2 + 2lw = d^2 + 2A$
$= 2500$ or $x^2 + 2A = 2500$ $\Rightarrow A = \frac{2500 - x^2}{2}$.

Problem 9. Solution: D.
We have $V = k \cdot h \cdot g^2$ and know that $216 = k \cdot 30 \cdot 1.5^2 = 67.5k$ $\Rightarrow$ $k = 3.2$. Thus $960 = 3.2 \cdot h \cdot 2^2$ $\Rightarrow$ $h = 75$.

Problem 10. Solution: B.
In the figure, we have $\frac{5}{h} = \frac{25}{x}$ $\Rightarrow$ $x = 5h$ $\quad$ (1)

$$\frac{4'8"}{h} = \frac{24'8"}{x+1} \Rightarrow \frac{4\frac{2}{3}}{h} = \frac{24\frac{2}{3}}{x+1} \Rightarrow \frac{14}{h} = \frac{74}{x+1} \quad (2)$$

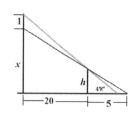

Substituting (1) into (2): $\frac{14}{h} = \frac{74}{5h+1} \Rightarrow 14(5h+1) = 74h \Rightarrow$

$h = \frac{7}{2}$. So $x = 5h = 5 \times \frac{7}{2} = 17\frac{1}{2}$.

Problem 11. Solution: A.
Since the frame is only 2 pixels wide, it is composed of $2(2L) + 2(2W) - 16$.

Note that the 16 pixels form the corners of the rectangle. So we have $4L + 4W - 16 = 504 \Leftrightarrow 4L + 4W = 520$. If we increase the width by 10%, we have $2(2L \cdot 1.1) + 2(2W) - 16 = 524 \Leftrightarrow 4.4L + 4W = 540$.
Now subtract to find $0.4L = 16 \Rightarrow L = 40$. From either of the two original equations, we see that $W = 90$, and the image has $40(90) = 3600$ pixels.

Problem 12. Solution: D.
Let x and y be the dimensions of the rectangle. Then $2x^2 + 2y^2 = 100 \Rightarrow x^2 + y^2 = 50$.
The length of the diagonal is $\sqrt{x^2 + y^2} = \sqrt{50} = 5\sqrt{2}$.

Problem 13. Solution: D.
In the diagram, let M be the midpoint of AB. Points P and Q trisect AB as shown. (see Figure) Let $AB = 6x$. Then $AM = 3x$. Since $AP = 2x$, $PM = x$. But $OM = AM$, so $OM = 3x$. Since $OP^2 = PM^2 + OM^2$, $10^2 = x^2 + (3x)^2$ and $x = \sqrt{10}$.
The area of $\triangle AOB$ is $\frac{1}{2}(AB)(OM) = \frac{1}{2}(6\sqrt{10})(3\sqrt{10}) = 90$.

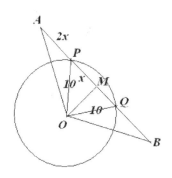

Problem 14. Solution: A.
Consider the figure as shown. The open box has length $12 - 2x$, width $9 - 2x$ and height x. So the volume $V = x(12 - 2x)(9 - 2x)$ is maximized when $x \approx 1.697$ and $V \approx 81.872$.
This approximate solution was found using a graphing calculator. Once you have had some Calculus, you will be able to find the exact solution.

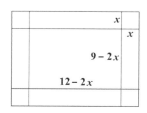

Problem 15. Solution: C.
Let a, b, and c be the lengths of the sides of a right triangle in increasing order of size.
$a + b + c = 6a \Rightarrow (b/a) + (c/a) = 5$ and $1 + (b/a)^2 = (c/a)^2 \Rightarrow 1 + (b/a)^2 = (5 - b/a)^2$
Thus $1 = 25 - 10(b/a) \Rightarrow (b/a) = 2.4$

Problem 16. Solution: B.
Let the lengths of three sides be a, b, and c. and $a = 3(x - 1)$, $b = 3x$, and $c = 3(x + 1)$.
$c^2 = a^2 + b^2 \Rightarrow [3(x+1)]^2 = [3(x-1)]^2 + (3x)^2 \Rightarrow$
$9x^2 + 18x + 9 = 9x^2 - 18x + 9 + 9x^2 \Rightarrow 9x^2 + 18x + 9 = 9x^2 - 18x + 9 + 9x^2$
$9x^2 - 36x = 0 \Rightarrow x = 4^2$.
So the lengths of three sides are 9, 12, and 15. The area is $9 \times 12/2 = 54$.

Problem 17: Solution: 50 feet.
Alicia walks a distance of $5 \times 22 = 110$ feet in 22
seconds. $\triangle ABC \sim \triangle DEC$, then $\dfrac{AB}{BC} = \dfrac{DE}{EC} \Rightarrow$

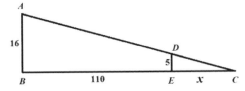

$\dfrac{16}{110 + x} = \dfrac{5}{x} \Rightarrow 16x = 5(110 + x) \Rightarrow 11x = 5 \times 110 \Rightarrow x = 50$.

Problem 18. Solution: D.
The area of the inside square is $EF^2 = x^2 + y^2$
The area of the outside square is $(x + y)^2$.
Since their ratio is 4/5, we have $\dfrac{x^2 + y^2}{(x+y)^2} = \dfrac{4}{5} \Rightarrow$

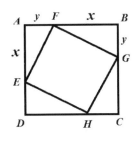

$\dfrac{(\frac{x}{y})^2 + 1}{(\frac{x}{y})^2 + 2\frac{x}{y} + 1} = \dfrac{4}{5}$.

Let $z = \dfrac{x}{y}$. $\dfrac{z^2 + 1}{z^2 + 2z + 1} = \dfrac{4}{5} \Rightarrow 5(z^2 + 1) = 4(z^2 + 2z + 1) \Rightarrow z^2 - 8z + 1 = 0$

$\Rightarrow z = \dfrac{-(-8) \pm \sqrt{(-8)^2 - 4 \times 1 \times 1}}{2} = \dfrac{8 \pm \sqrt{60}}{2} = 4 \pm \sqrt{15}$.

Since $x > y$, $x/y = 4 + \sqrt{15}$.

Problem 19. Solution: $\frac{4}{5}\pi^2$.

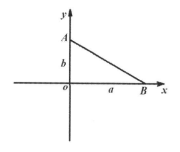 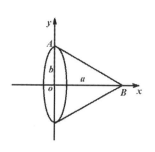 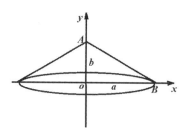

Let the volumen be V.
We have $a \times b = 6$ \hfill (1)

$\frac{1}{3}\pi b^2 \times a = 30$ \hfill (2)

$\frac{1}{3}\pi a^2 \times b = V$ \hfill (3)

We square both sides of (1): $a^2 b^2 = 36$ \hfill (4)

(4) ÷ (2): $a = \frac{2}{5}\pi$ \hfill (5)

From (3), we get $V = \frac{1}{3}\pi a^2 \times b = \frac{1}{3}\pi(a \times b) \times a = \frac{1}{3}\pi(6) \times a = 2\pi a = 2\pi \times \frac{2}{5}\pi = \frac{4}{5}\pi^2$.

Problem 20. Solution: C.
The area of the original rectangular garden: $60 \times 80 = 4800$.
The area of the new rectangular garden:
$4800 \times 2 = (80 + 2x)(60 + x) \Rightarrow 4800 \times 2 = 4800 + 200x + 2x^2$

$\Rightarrow 2x^2 + 200x = 4800 \Rightarrow x^2 + 100x - 2400 = 0.$
$(x - 20)(x + 120) = 0 \Rightarrow x = 20.$

Problem 21. Solution: 17.
Since $ABCDE$ is a regular pentagon, we know that AF and AG trisect the angle BAE.
We label each line segment as follows:

Applying the angle bisector theorem to $\triangle ABG$:
$\frac{AB}{BF} = \frac{AG}{GF} \Rightarrow \frac{x}{y} = \frac{y}{1-2y}$ \hfill (1)

Applying the angle bisector theorem to $\triangle ABD$:

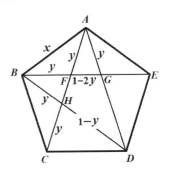

$\dfrac{AB}{BH} = \dfrac{AD}{DH} \Rightarrow \dfrac{x}{y} = \dfrac{1}{1-y}$ (2)

Solving (1) and (2): $\dfrac{y}{1-2y} = \dfrac{1}{1-y} \Rightarrow y^3 - 3y + 1 = 0$.

$y = \dfrac{3-\sqrt{5}}{2}$ and $y = \dfrac{3+\sqrt{5}}{2}$ (ignored since it is greater than 1).

So $x = \dfrac{y}{1-y} = \dfrac{\dfrac{3-\sqrt{5}}{2}}{1-\dfrac{3-\sqrt{5}}{2}} = \dfrac{\sqrt{5}-1}{2}$.

The perimeter of the pentagon is $5x = 5 \times \dfrac{\sqrt{5}-1}{2} = \dfrac{5\sqrt{5}-5}{2}$.

Thus $k + w + p + f = 5 + 5 + 5 + 2 = 17$.

A

absolute value, 46, 48, 54, 57, 58, 162, 210, 347
acute angle, 41, 295, 296, 315
angle, 36, 37, 38, 41, 43, 44, 275, 293, 295, 296, 297, 300, 301, 304, 305, 306, 307, 308, 313, 315, 318, 323, 376, 378, 389
area, 34, 37, 40, 41, 43, 44, 49, 50, 51, 55, 56, 57, 64, 65, 81, 90, 137, 138, 146, 275, 284, 302, 355, 359, 379, 380, 381, 382, 383, 384, 386, 387, 388, 389
arithmetic sequence, 217, 225, 226, 227, 228, 229, 234, 237, 240, 242, 245, 255
average, 1, 2, 3, 9, 12, 55, 58, 80, 239, 246
axis of symmetry, 68

B

base, 37, 138, 150, 154, 167, 174, 190, 246, 379, 380, 381, 382, 386
base 10, 167

C

Cauchy inequality, 352
center, 326, 327, 328, 329, 330, 331, 332, 334, 336, 337, 338, 339, 340, 342, 343, 344, 346, 347, 348, 349, 350, 351, 355
chord, 331
circle, 264, 270, 275, 292, 326, 327, 328, 329, 330, 331, 332, 333, 334, 335, 336, 337, 338, 339, 340, 341, 346, 350, 351, 354, 355, 356, 377, 378, 383
coefficient, 64, 68, 114, 115, 119, 121, 125, 197, 198, 199, 200, 201, 202, 203, 204, 205, 206
collinear, 38
combination, 134, 259, 366
common divisor, 251, 365, 371
common factor, 176, 180, 182, 366, 374
common fraction, 19, 26, 41, 225, 246, 284, 337
complex number, 68, 207, 208, 209, 210, 214, 216, 217, 314
concentric, 275
congruent, 339, 377, 383
Conjugate, 347

constant, 1, 5, 7, 48, 64, 67, 79, 81, 102, 116, 119, 125, 179, 188, 231, 236, 250, 342, 346, 347
counting, 154, 367
cube, 148

D

decimal, 26, 154, 172, 174, 250, 275
degree, 88, 114, 116, 119, 126, 127, 184, 187, 209, 218, 224
denominator, 18, 19, 26, 30, 74, 91, 110, 137, 143, 145, 182, 184, 185, 186, 187, 312, 347
diagonal, 146, 150, 293, 383, 384, 387
diameter, 275, 326, 327, 333, 334, 339
difference, 19, 26, 30, 65, 79, 83, 94, 97, 120, 124, 128, 132, 162, 222, 225, 228, 229, 234, 240, 265, 269, 304, 309, 347, 361, 365, 369, 373
digit, 18, 20, 174, 235, 251, 255, 259, 261, 265, 266, 267, 269, 270, 273, 281, 361, 362, 364, 369, 371, 372, 375
Divisibility, 361
divisible, 18, 119, 120, 122, 123, 129, 132, 134, 247, 265, 267, 269, 271, 361, 362, 363, 365, 369, 370, 372, 373
divisor, 364

E

edge, 284, 288, 382
ellipse, 342, 343, 344, 345, 346, 350, 351, 353, 354, 355, 356, 358
Ellipse, 342, 353
equation, 22, 24, 35, 36, 37, 39, 41, 42, 44, 48, 49, 50, 51, 52, 53, 54, 55, 56, 57, 58, 59, 60, 61, 62, 63, 65, 67, 68, 69, 70, 71, 73, 74, 76, 77, 82, 84, 85, 88, 94, 104, 105, 114, 118, 125, 126, 129, 130, 135, 136, 143, 144, 147, 148, 151, 152, 153, 155, 156,炅157, 158, 159, 160, 161, 162, 163, 168, 174, 176, 178, 181, 182, 193, 194, 195, 208, 210, 213, 214, 217, 218, 226, 228, 232, 237, 240, 270, 287, 290, 301, 302, 310, 311, 312, 316, 324, 325, 326, 327, 328, 329, 330, 331, 332, 333, 334, 335, 336, 337, 338, 339, 340, 341, 342, 343, 344, 346, 348, 349, 350,

351, 353, 354, 355, 356, 357, 358, 359, 360, 370, 373
equilateral, 40, 43, 298
equilateral triangle, 298
evaluate, 107
even number, 257, 364
event, 261, 273, 277, 278, 279, 280, 286
exponent, 104, 136, 154, 167, 233
expression, 27, 28, 29, 32, 64, 65, 96, 97, 102, 105, 128, 131, 136, 141, 142, 146, 153, 164, 170, 171, 173, 179, 182, 186, 187, 194, 200, 201, 204, 233, 281, 296, 380, 386

F

face, 68, 273, 283, 284, 286
factor, 18, 53, 62, 96, 119, 120, 121, 122, 123, 124, 125, 128, 132, 133, 186, 194, 238, 364, 366, 370, 375
Factor Theorem, 123, 125, 134
factorial, 259
Fibonacci sequence, 251, 257
finite, 241, 250
foci, 342, 343, 344, 345, 347, 349, 350, 356
formula, 14, 15, 33, 34, 35, 37, 45, 50, 51, 52, 60, 65, 102, 132, 133, 140, 141, 226, 227, 229, 230, 233, 237, 238, 239, 240, 250, 255, 262, 274, 281, 296, 302, 304, 306, 309, 310, 312, 318, 319, 320, 321, 322, 323, 343, 346, 355, 378
fraction, 18, 20, 21, 23, 26, 60, 137, 154, 186, 187, 238
function, 27, 43, 63, 64, 65, 99, 100, 101, 102, 103, 104, 105, 106, 107, 108, 109, 110, 111, 113, 114, 115, 119, 124, 125, 130, 132, 136, 145, 176, 179, 182, 183, 185, 190, 192, 218, 224, 241, 242, 249, 296, 297, 309, 310, 314, 315, 316, 317, 325, 383

G

geometric mean, 61
geometric sequence, 225, 231, 232, 243, 245, 255
graph, 41, 47, 55, 56, 57, 67, 69, 100, 102, 103, 104, 119, 124, 184, 185, 192, 196, 342, 345, 347, 348, 351, 352, 354, 355, 357

H

Hyperbola, 353
hypotenuse, 378, 383

I

improper fraction, 370
independent events, 278
inequality, 55, 56, 86, 87, 88, 89, 90, 91, 92, 93, 94, 95, 96, 97, 98, 194, 284
infinite series, 25, 233, 236, 241
infinity, 140, 146
integer, 21, 41, 45, 48, 49, 52, 67, 92, 94, 95, 108, 111, 114, 125, 128, 130, 135, 136, 147, 148, 154, 172, 174, 177, 195, 197, 207, 209, 217, 218, 223, 226, 234, 236, 237, 244, 249, 250, 253, 263, 269, 281, 283, 295, 296, 364, 366, 367, 369, 370, 374, 375, 380, famous 382, 385
integers, 21, 26, 78, 79, 82, 87, 92, 108, 109, 125, 129, 130, 134, 138, 142, 146, 158, 176, 182, 190, 199, 203, 210, 218, 226, 234, 235, 238, 242, 247, 249, 251, 259, 260, 261, 265, 267, 270, 281, 283, 285, 362, 365, 366, 369, 370, 371, 372, 373, 374, 375, 377, 382, 384
intercept, 41, 42, 196
intersecting lines, 350, 354, 357
intersection, 40, 41, 42, 43, 53, 58, 61, 72, 79, 103, 160, 161, 330, 333, 337, 340, 347, 354, 357
inverse, 104, 105, 108, 109, 110, 176, 179
irrational number, 191
isosceles, 146, 299, 305, 377, 382, 383, 385
isosceles triangle, 299, 305, 377, 385

L

lateral edge, 382
least common multiple, 20
line, 33, 34, 35, 36, 37, 38, 39, 40, 41, 42, 43, 44, 45, 61, 67, 69, 70, 88, 89, 100, 104, 110, 121, 183, 184, 185, 196, 262, 270, 329, 330, 331, 332, 333, 334, 337, 338, 339, 340, 341, 342, 344, 347, 351, 352, 355, 358, 380, 389
line segment, 33, 35, 44, 333, 342, 347, 380, 389
linear equation, 79

locus, 353
lowest terms, 18, 26, 182, 187, 225, 284, 298, 314, 337, 370

M

mean, 197
median, 37
midpoint, 33, 37, 342, 347, 387
multiple, 126, 132, 135, 143, 222, 279, 281, 361, 373, 381

N

natural number, 26, 138, 155, 171, 183, 234, 241, 250, 366, 369
natural numbers, 155, 171, 183, 234, 241
negative number, 61, 86, 167
number line, 46, 88, 89, 90, 91, 97
numerator, 19, 26, 30, 97, 182, 184, 185, 187

O

obtuse angle, 294
odd number, 235, 239, 364
operation, 24, 32, 103
ordered pair, 81, 371, 373
origin, 35, 40, 328, 331, 336, 340, 342, 347, 348
outcome, 278

P

Parabola, 65, 353
parallel, 38, 354, 382
parallelogram, 33
pentagon, 384, 389, 390
percent, 191, 195
perimeter, 4, 81, 84, 146, 382, 383, 384, 385, 386, 390
permutation, 259
perpendicular, 38, 39, 40, 42, 342, 347
plane, 5, 41, 50, 81, 188, 266, 342, 347
point, 7, 34, 35, 36, 37, 39, 40, 41, 42, 43, 44, 45, 46, 51, 61, 70, 100, 103, 104, 270, 275, 326, 328, 329, 330, 331, 332, 333, 334, 335, 336, 337, 339, 340, 341, 344, 348, 350, 351, 353, 354, 356, 357, 377, 378
polynomial, 88, 114, 115, 116, 117, 118, 119, 120, 122, 123, 124, 125, 126, 127, 129, 130, 132, 133, 134, 135, 182, 199, 200, 202, 209, 210, 217, 218, 224
positive number, 73, 94, 95, 109, 236
power, 137, 364
prime factorization, 180, 366, 367, 372
prime number, 28, 41, 132, 265, 283, 361, 363, 366, 370
principal square root, 136
probability, 68, 71, 188, 273, 274, 275, 276, 277, 278, 279, 280, 281, 282, 283, 284, 285, 286, 287, 288, 289, 290, 291
product, 7, 8, 25, 67, 77, 79, 80, 82, 101, 102, 116, 118, 121, 128, 131, 142, 207, 210, 252, 255, 258, 260, 261, 269, 270, 279, 283, 287, 308, 318, 366, 367, 369, 371, 373, 374, 375
proper fraction, 234
pyramid, 382
Pythagorean Theorem, 33, 318, 385

Q

quadrant, 37, 295, 296, 297, 300, 301, 315, 323, 330
quadrilateral, 34, 41, 44
quotient, 116, 117, 119, 122, 123, 129, 133, 195, 364, 365

R

radius, 270, 275, 326, 327, 328, 329, 332, 333, 334, 336, 338, 339, 340, 381, 383
random, 273, 275, 276, 281, 283, 284, 285
range, 100, 145, 183, 312
rate, 1, 2, 6, 7, 67, 68, 70, 189, 193, 195, 384
ratio, 75, 125, 140, 168, 169, 182, 225, 229, 230, 231, 238, 239, 240, 293, 305, 342, 346, 348, 356, 357, 359, 376, 378, 381, 384, 388
rational number, 18, 88, 298, 314, 374
real number, 24, 28, 46, 49, 51, 55, 60, 61, 63, 67, 68, 86, 92, 93, 103, 109, 114, 128, 129, 130, 136, 139, 145, 147, 162, 167, 168, 173, 182, 183, 191, 207, 209, 210, 217, 315, 351

real numbers, 24, 28, 46, 49, 51, 60, 63, 86, 92, 93, 103, 109, 114, 130, 145, 167, 168, 173, 182, 183, 191, 207, 209, 217, 351
reciprocal, 332
rectangle, 81, 146, 150, 347, 348, 382, 383, 387
relatively prime, 129, 257, 285, 375
remainder, 9, 116, 117, 118, 119, 121, 122, 129, 133, 134, 184, 195, 364, 365, 366, 369, 372, 373
Remainder Theorem, 117, 118, 125, 133, 134
repeating decimal, 18, 19, 20, 26
rhombus, 49, 57, 293
right angle, 44, 293, 383
right triangle, 33, 146, 150, 313, 318, 378, 379, 382, 383, 384, 386, 388
root, 72, 77, 89, 115, 119, 122, 123, 125, 126, 127, 128, 132, 135, 136, 144, 150, 152, 156, 210, 217, 223, 224
rotation, 264

S

semicircle, 275, 382
sequence, 49, 225, 228, 232, 234, 235, 236, 241, 242, 245, 246, 248, 250, 251, 255
set, 19, 22, 89, 90, 91, 93, 103, 117, 154, 196, 241, 265, 267, 269, 270, 271, 283, 287, 342, 347, 370, 373, 374
similar, 385
simplifying, 144, 151, 230, 343, 346, 352
slope, 35, 36, 37, 38, 40, 41, 42, 43, 44, 70, 105, 106, 111, 331, 332, 337, 339, 340, 341
solution set, 22, 81, 92, 93, 94, 136, 173
sphere, 380, 381
square, 50, 53, 54, 57, 58, 60, 61, 65, 95, 136, 141, 144, 151, 153, 190, 234, 237, 266, 270, 275, 288, 326, 327, 351, 356, 363, 379, 380, 381, 383, 384, 388, 389
square root, 60, 136, 141
sum, 11, 16, 17, 20, 25, 26, 30, 40, 53, 54, 55, 56, 58, 59, 61, 62, 66, 67, 69, 77, 80, 83, 87, 94, 97, 101, 105, 108, 109, 112, 120, 122, 123, 124, 125, 127, 128, 130, 131, 133, 134, 146, 147, 148, 151, 152, 153, 161, 162, 165, 174, 176, 187, 195, 197, 200, 录201, 202, 215, 225, 226, 227, 228, 229, 230, 231, 232, 233, 234, 235, 236, 237, 238, 239, 241, 242, 243, 244, 245, 246, 249, 250, 251, 253, 255, 257, 261, 266, 271, 279, 283, 284, 285, 287, 290, 301, 302, 303, 309, 316, 318, 324, 337, 340, 342, 344, 361, 362, 366, 367, 368, 369, 370, 372, 373, 374, 375, 380, 383
surface area, 190, 381

T

tangent line, 330, 331, 332, 333, 337, 340, 341
trapezoid, 44, 146, 382
triangle, 34, 37, 40, 41, 43, 44, 137, 313, 376, 378, 380, 381, 382, 383, 384
trisect, 387, 389

V

variable, 73, 76, 86, 99, 153, 179, 182, 329
vertex, 43, 64, 65, 67, 69, 70, 95, 295, 342, 347, 349, 355, 380, 383, 384
volume, 5, 381, 383, 384, 387

W

whole number, 136, 363

X

x-axis, 38, 49, 51, 68, 101, 102, 184, 196, 295, 296, 327, 330
x-coordinate, 67, 96, 161, 329, 337, 344
x-intercept, 36, 41, 45, 68, 71, 114, 124, 125, 167, 192, 336

Y

y-axis, 37, 49, 101, 351
y-coordinate, 67, 329
y-intercept, 36, 37, 41, 45, 105, 124, 192, 331, 332

Z

zero, 38, 46, 61, 77, 88, 91, 97, 111, 121, 122, 123, 125, 126, 129, 132, 142, 153, 156, 182, 184, 185, 187, 189, 191, 192, 232, 240, 296, 352, 358, 361

Made in the USA
Lexington, KY
23 June 2016